Movement Equations 1

Non-deformable Solid Mechanics Set

coordinated by
Abdelkhalak El Hami

Volume 1

Movement Equations 1

Location, Kinematics and Kinetics

Michel Borel
Georges Vénizélos

iSTE

WILEY

First published 2016 in Great Britain and the United States by ISTE Ltd and John Wiley & Sons, Inc.

ISTE Ltd
27-37 St George's Road
London SW19 4EU
UK

www.iste.co.uk

John Wiley & Sons, Inc.
111 River Street
Hoboken, NJ 07030
USA

www.wiley.com

Library of Congress Control Number: 2016946047

British Library Cataloguing-in-Publication Data
A CIP record for this book is available from the British Library
ISBN 978-1-78630-032-4

Contents

Introduction

The idea of mechanics is probably one of the first scientific thoughts that captivated the human mind when, for example, man wanted to use a lever to lift the stone that he needed for building a shelter. Then, with the passing of time, as the mechanics developed and enabled the description of the motion of bodies, man used it to conceive the functioning of machines, which at first were an aid in fulfilling his tasks, and then even replaced him.

Moreover, man entertained this idea, though unconsciously, when he took an interest in the cycle of the seasons, as it had occurred to him that he could provide for his subsistence by cultivating the land. Subsequently, astral phenomena started occupying his thoughts more and more, since he had always sensed they were influencing his life; and little by little, throughout the centuries he learnt how to analyze, comprehend and formalize these phenomena.

Then, Newton came along and formulated his laws, which led to the equations of motion! Their unquestionable power – though the resulting path equations are far more complex – was once more illustrated recently by the Rosetta spacecraft, which traveled for several years covering phenomenal distances through space before entering orbit around the Chury comet, on which it delivered the Philae lander, whose mission was to observe and analyze the comet. And, the challenge was met! A further recent proof of this power of the equations is the “theoretical discovery” of a planet on the frontier of the solar system, a result of the observation of the paths of neighboring astral bodies that were obviously influenced by its presence. It was in this way that Neptune was discovered in 1845.

There is something fascinating, even mysterious, about the equations of motion of a solid body, though they have nothing to do with magic, but simply with man-made science.

At the French National Conservatory of Arts and Crafts (*Conservatoire National des Arts et Métiers* – CNAM), this science is taught to very diverse audiences. Professor Michel Cazin, who led the chair of mechanics for over two decades, was very keen to provide the audience with clear and mathematically well-structured teaching, which was always well received, whether in Paris or in the centers throughout the country. Those of us who have worked by his side and have drawn inspiration from his ideas and approach, which he continuously sought to improve, have wished to preserve this manner of conceiving the teaching of mechanics, to the benefit of generations to come. This is the ambition that gives birth to this set of five volumes on *Non-deformable Solid Mechanics*.

In order to clearly distinguish the essential stages in the development of the formalism leading to the establishment of the equations of motion, which is in a way the core of the discipline, with its major types of application, Volumes 1 to 3 out of the Set will focus specifically on movement equations, the first of which is the present book.

This Volume 1 relates to the elements required for the establishment of the equations of motion, namely how to position in space the solid body whose motion is to be described, how to formulate its kinematics, which expresses this motion in terms of velocity and acceleration fields, and how to characterize its mass, inertia and energy properties.

Volume 2 is special. Throughout the development of the formalism, the mechanics scientist uses various mathematical tools – vectors, torsors, for example – that he is required to handle proficiently. In many real-life instances, having to a certain extent forgotten these tools, the practitioner needs information on how to properly handle some of these tools and achieve the desired result; it may happen that the information he finds is not always in a form that is readily usable, requiring time to adapt and properly use it. Volume 2 gathers a certain number of useful mathematical tools in a form that makes them ready to use in the application of the present work.

Volume 3 is dedicated first to the positioning of the solid body in its environment, while taking into account the forces acting on it, then to the introduction of the fundamental law of dynamics, and to the equations of motion that result from it either as a direct expression or as a consequence of the energy dimension of motion. This presentation is preceded by a methodological form that resumes the main formulae developed in Volume 1, aiming to guide the mechanics scientist in preparing the material required for setting up the equations of motion.

Throughout this third volume and the next ones, these equations will be used in the study of the small motions and vibrations of a solid body or of the conditions of stationary motion, as well as to express the motion of systems of solid bodies and introduce robotics.

Now, regarding the present volume, it consists of three main chapters, illustrated by a certain number of exercises that are presented either within each chapter, allowing for the content developed to be readily applied, or at the end of the chapter, if going further into the use of its content may be useful.

With a focus on locating solid bodies, the first chapter identifies the main types of systems of reference that can be used and shows how to identify the solid body in relation to them, how to describe its position, orientation and also evolution during the motion. The method for determining local systems of reference connected to curves or surfaces will be developed in the corresponding chapters of Volume 2. This chapter also presents the case where a rotation about an axis should be taken into account in order to define the location of a solid body; however, vector rotation will be more extensively developed in Volume 2.

The second chapter introduces motion descriptors such as velocity and acceleration, and then provides an extensive description of how the fields of velocity and acceleration of a solid body in motion can be expressed in torsor form, given that the torsor notation proves to be a particularly beneficial tool in the mechanics of the solid body. It examines the formulation of these fields in the specific cases of motion with fixed plane, with the very physical notions of fixed centroid and mobile centroid, the combination of motions when the systems of reference are moving relative to each other, and when the solid bodies are in contact with one another during their motion, by characterizing the nature of this contact in terms of displacements and velocities.

The third chapter introduces the kinetic properties of a solid that play an essential role in how its motion unfolds, namely mass and inertia. It introduces the two essential notions of kinetic torsor and dynamic torsor that amalgamate the two kinematic and kinetic aspects in the same concept; the expression of these two torsors is one of the last stages leading to the fundamental law of dynamics, the last being the energetic dimension of motion. This chapter approaches in its last part the energetic aspect by defining and developing the concept of kinetic energy of a moving solid body, by insisting on the torsor form of its expression, which will facilitate its subsequent use.

To conclude the introduction to this volume, the *Non-deformable Solid Mechanics* set such as described above is intended to be an essential tool for readers who want to gain quite rigorous knowledge of the discipline, such as students or those who, in the exercise of their profession, feel the need to develop their own approach to the problem they are dealing with. Digital technologies can certainly help engineers to solve the problems they are faced with, but among a practitioner's preoccupations there may also be a need to go beyond the framework defined by software, hence the importance of having a tool to guide his efforts.

This is the spirit in which we have conceived this set and the feedback given by the CNAM audience throughout the years of teaching has been a great encouragement for us.

Table of Notations

M	material point
t	time
$[t_i, t_f]$	time interval
$m(\mathscr{D})$	mass of a continuous mechanical set $(\mathscr{D})$
$\rho(\mathrm{M})$	density of the material point M
$\mu(\mathscr{D})$	measure on a continuous mechanical set $(\mathscr{D})$
δ_{ij}	Kronecker symbol
ε_{ijk}	three-index permutation symbol
$\underline{\phi}^{(\mathscr{D})}{}_{/\mu}$	average of the scalar or vector function ϕ on the continuous mechanical set $(\mathscr{D})$
$\vec{V}$	vector
$(\lambda) \equiv (\overrightarrow{x_\lambda}\ \overrightarrow{y_\lambda}\ \overrightarrow{z_\lambda})$	basis
$\langle\lambda\rangle \equiv \langle \mathrm{O}_\lambda \lvert \overrightarrow{x_\lambda}\ \overrightarrow{y_\lambda}\ \overrightarrow{z_\lambda} \rangle$	system of reference

$\begin{array}{c\|ccc} (e) \backslash (E) & \vec{X} & \vec{Y} & \vec{Z} \\ \hline \vec{x} & p_{11} & p_{12} & p_{13} \\ \vec{y} & p_{21} & p_{22} & p_{23} \\ \vec{z} & p_{31} & p_{32} & p_{33} \end{array}$	table of passage from basis $(e) \equiv (\vec{x}\,\vec{y}\,\vec{z})$ to basis $(E) \equiv (\vec{X}\,\vec{Y}\,\vec{Z})$
$p(e,E) = \begin{bmatrix} p_{11} & p_{12} & p_{13} \\ p_{21} & p_{22} & p_{23} \\ p_{31} & p_{32} & p_{33} \end{bmatrix}$	matrix of passage from basis $(e) \equiv (\vec{x}\,\vec{y}\,\vec{z})$ to basis $(E) \equiv (\vec{X}\,\vec{Y}\,\vec{Z})$
x, y, z	Cartesian coordinates of a point
r, α, z	cylindrical-polar coordinates of a point
R, α, β	spherical coordinates of a point
ψ, θ, φ	Euler angles: precession, nutation and spin, respectively
$\Pi(\vec{V},\vec{W})$	plane of the two vectors $\vec{V}$ and $\vec{W}$
$\Pi(\mathrm{O}\|\vec{V},\vec{W})$	plane of the two vectors $\vec{V}$ and $\vec{W}$ passing through the point O
$\overrightarrow{\mathrm{OM}}$	vector from point O (tail) to point M (head)
$\overrightarrow{\mathrm{O}_\lambda \mathrm{O}_S}$	position vector of the point O_S relative to the origin O_λ of the chosen system of reference $\langle \lambda \rangle$
$\widehat{\vec{V},\vec{W}}$	angle between two vectors, oriented from $\vec{V}$ to $\vec{W}$
$\|\vec{V}\|$	norm of vector $\vec{V}$
$\vec{V} \cdot \vec{W}$	scalar product of two vectors $\vec{V}$ and $\vec{W}$
$\vec{V} \wedge \vec{W}$	vector product of two vectors $\vec{V}$ and $\vec{W}$
$\vec{u}(\alpha)$	polar vector in cylindrical-polar coordinates

$\vec{k}(\alpha,\beta)$	polar vector in spherical coordinates
$\mathscr{R}_{\vec{u},\alpha} \equiv \left[\vec{u}\middle\|\alpha\right]$	vector rotation by angle α about the axis defined by vector $\vec{u}$
$\Gamma_{\lambda}^{(t_i,t_f)}(\mathrm{M})$	trajectory, in the system of reference $\langle\lambda\rangle$,of the material point M, during the time interval $[t_i,t_f]$
$\overrightarrow{D_{\lambda}}^{(t_i,t_f)}(\mathrm{M})$	displacement, in the system of reference $\langle\lambda\rangle$, of the material point M during the time interval $[t_i,t_f]$
$\overrightarrow{v^{(\lambda)}}(\mathrm{M}\|t)$ ou $\overrightarrow{v^{(\lambda)}}(\mathrm{M})$	velocity at time t of physical point M as it moves through reference system $\langle\lambda\rangle$
$\overrightarrow{J^{(\lambda)}}(\mathrm{M}\|t)$ ou $\overrightarrow{J^{(\lambda)}}(\mathrm{M})$	acceleration at time t of physical point M as it moves through reference system $\langle\lambda\rangle$
$\overrightarrow{\omega_S^{\lambda}}$	rotation vector of solid (S) as it moves in relation to reference system $\langle\lambda\rangle$
$\vec{v_e}\left(\mathrm{M}\|_{\mu}^{\lambda}\right)$	drive velocity of physical point M in relation to the movement of reference system $\langle\mu\rangle$ relative to reference system $\langle\lambda\rangle$
$\vec{J_e}\left(\mathrm{M}\|_{\mu}^{\lambda}\right)$	drive acceleration of physical point M in relation to the movement of reference system $\langle\mu\rangle$ relative to reference system $\langle\lambda\rangle$
$\vec{J_c}\left(\mathrm{M}\|_{\mu}^{\lambda}\right)$	Coriolis acceleration as applied to physical point M in relation to the movement of reference system $\langle\mu\rangle$ relative to reference system $\langle\lambda\rangle$

$\dfrac{d^{(\lambda)}}{dt}\vec{V}$	derivative with respect to time of vector $\vec{V}$ in the system of reference $\langle\lambda\rangle$
$\{\mathscr{T}\}_{\mathrm{P}} = \left[\vec{s}\{\mathscr{T}\} \middle\vert \overrightarrow{\mathscr{M}_{\mathrm{P}}}\{\mathscr{T}\}\right]$	torsor characterized by its two reduction elements in point P
$\vec{s}\{\mathscr{T}\}$	sum of torsor $\{\mathscr{T}\}$: 1st reduction element of the torsor
$\overrightarrow{\mathscr{M}_{\mathrm{P}}}\{\mathscr{T}\}$	moment in P of the torsor $\{\mathscr{T}\}$: 2nd element of reduction of the torsor for which the relation is verified $\forall$Q $\overrightarrow{\mathscr{M}_{\mathrm{Q}}}\{\mathscr{T}\} = \overrightarrow{\mathscr{M}_{\mathrm{P}}}\{\mathscr{T}\} + \overrightarrow{\mathrm{QP}} \wedge \vec{s}\{\mathscr{T}$
$\mathscr{I} = \vec{s}\{\mathscr{T}\} \cdot \overrightarrow{\mathscr{M}_{\mathrm{P}}}\{\mathscr{T}\}$	scalar invariant of torsor $\{\mathscr{T}\}$, independent of point P
$\{\mathscr{T}_1\} \otimes \{\mathscr{T}_2\}$	product of two torsors
$\left\{{}^{\lambda}_{S}\right\}_{\mathrm{P}_S} = \left[\overrightarrow{\omega_S^{\lambda}} \middle\vert \overrightarrow{v^{(\lambda)}}(\mathrm{P}_S)\right]$	torsor distributor of velocities or kinematic torsor associated to the motion of the material point P_s of the solid (*S*)
$\{p_S^{\lambda}\}$	kinetic torsor associated with the motion of the solid (S) in the system of reference $\langle\lambda\rangle$
$\{\mathscr{A}_S^{\lambda}\}$	dynamic torsor associated with the motion of solid (S) in the system of reference $\langle\lambda\rangle$
$I_{\mathrm{O}_S}(S\vert m)$	inertia operator of the solid (S) having the measure of mass m
$\left[I_{\mathrm{O}_S}(S\vert m)\right]_{/(S)} = \begin{bmatrix} A & -F & -E \\ -F & B & -D \\ -E & -D & C \end{bmatrix}$	matrix representation in the basis (S) of the inertia operator of the solid (S) in one point of the solid, taken as reference point, here O_S

A, B, C	moments of inertia of the solid (S) relative to the three axes of the system of reference (S) passing through the reference point of the operator,
D, E, F	products of inertia of the solid (S) relative to the three planes formed by the axes of the system of reference (S) and passing through the reference point of the operator
$T^{(\lambda)}(S)$	kinetic energy of the solid (S) in its motion relative to the system of reference $\langle\lambda\rangle$
$2T^{(\lambda)}(S)$	*vis viva* of the solid (S) in its motion relative to the system of reference $\langle\lambda\rangle$

If the location of the solid (S) relative to the system of reference $\langle\lambda\rangle$ is represented by the parameters q_i with $i \leq 6$, we write: $\left\{ {}^{\lambda}_{S} \right\} = \left\{ {}^{\lambda}_{S,i} \right\} q_i'$ where:

$\left\{ {}^{\lambda}_{S,i} \right\}_{O_s} = \left[\overrightarrow{{}^{i}\delta(\lambda, S)} \middle\vert \overrightarrow{{}^{i}d_{O_S}} \right]$	partial torsor distributor relative to the variable q_i
$\overrightarrow{{}^{i}\delta(\lambda, S)}$	partial rotation quantity relative to variable q_i, component of the variable q_i' of rotation quantity such as $\overrightarrow{\omega_S^{\lambda}} = \overrightarrow{{}^{i}\delta(\lambda, S)} q_i'$
$\overrightarrow{{}^{i}d_{O_S}}$	component of the variable q_i' of the velocity of the point O_S, such that $\overrightarrow{v^{(\lambda)}(O_S)} = \overrightarrow{{}^{i}d_{O_S}} q_i'$

1

Location of Solid Bodies

Mechanics is the science of relations between:

– the motion of an arbitrary physical system $(\mathscr{D})$ in the system of reference $\langle \lambda \rangle$ from which it is observed;

– and the forces that act upon it.

But the motion of a physical system in space is a very relative notion, which depends in an essential manner on the observer. Before any development concerning the physical system, it is, therefore, fundamental to locate it relative to its observer. This is the focus of the present chapter, in which two considerations will be examined, namely how to represent it in order to locate it, and then how to determine its location.

For all that follows, the physical system considered is the solid body, which is a continuous and rigid (this latter notion will be subsequently specified) material set.

1.1. The notion of system of reference

The system of reference is used in mechanics to set a reference, based on which it is possible to locate a mechanical element (solid body or part of an elastic medium), then to observe the evolution of its location and study its motion.

At the terrestrial scale, the favored domain of mechanics is the physical space seen from a geometrical perspective, as an affine space of a three-dimensional vector space, represented by a *basis* consisting of three independent vectors; this representation is important because any vector in the geometrical space of terrestrial mechanics can be expressed as a linear combination of these three vectors of the basis.

The principle of the reference system of an element located in this space thus relies on two simultaneous measurements:

– that of the distance between this element and a point in space considered as fixed reference for the motion examined;

– that of the orientation of the segment "origin-element" relative to the three independent directions that form the reference trihedron.

The set comprising the point of reference and the system of three independent axes forms a system of reference.

1.2. Frame of reference

1.2.1. *Setting up a frame of reference*

When a system of reference is used for the study of a motion, it is also called a frame of reference. This notion is of particular importance in the study of motion, since its choice depends on:

– the observer, meaning the context in which the motion evolves;

– the forces that act on the mechanical system under study, as the set of these forces should be inside the environment of reference, in order to be able to apply to this motion the principles that govern it.

1.2.1.1. *Choosing the elements of a frame of reference*

In principle, the frame of reference used to locate a solid body and follow its evolution should consist of elements that are independant of the body and of the motion to be described. When the latter takes place in the terrestrial environment, the ideal frame of reference meets the following conditions:

– the point of reference chosen as origin of the system of reference is the center of inertia of the solar system.

– its independent directions are symbolized by three axes that are defined by three stars considered to be fixed relative to the motion of the Earth.

It is the solar frame of reference, noted $\langle g \rangle$, which ultimately illustrates the fundamental notion of the Galilean system of reference, which will be specified later on in this chapter.

This frame of reference is, however, poorly suited for the vast majority of motions taking place in the terrestrial environment, if only because of the measurement of the distance "origin-element", which fixes a scale that is completely disproportionate relative to the dimensions of the body whose motion is studied or to its field of evolution.

Moreover, since the Earth is moving relative to this system of reference, its axes, which would provide three independent directions, cannot be considered fixed for the motion studied. If we resorted to such a frame of reference, we would constantly be in *relative motion*, which would further complicate the formulation of the problem even though, for most of the motions studied in the terrestrial space, the *drive* terms are negligible compared with the *relative* ones; but this is not the case for slow motions such as those of icebergs or the continental drift.

Other systems of reference should, therefore, be considered. An example could be a system of reference linked to the Earth, having the origin in its center and the axes alongside three fixed directions of the terrestrial globe. But, this would pose the same scale difficulty mentioned above. Consequently, the study of the motion of a body requires a frame of reference adapted to its field of evolution, meaning a point in this field whose fixed position does not depend on the motion, so that it can serve as the origin, and three axes that are in their turn considered independent.

We shall subsequently see that these conditions for locating a moving body are not sufficient for the application without precautions of what constitutes the basis of the mechanical expression of solids, the *fundamental law of dynamics*, which is essentially expressed in the Galilean frame of reference.

1.2.1.2. *Units of measurement*

The study of the motion of a body begins by locating it in the frame of reference chosen as best adapted to its evolution and, among other things, by expressing the "origin-element" distance according to the three directions of reference. It is, therefore, necessary to define the origin of the system of reference and a unit of measurement in each direction.

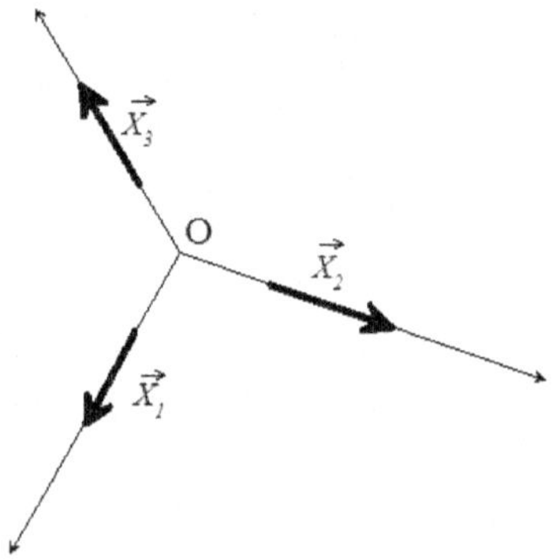

Figure 1.1. *Reference trihedron*

The three directions, defined in general by three independent vectors $\left(\overrightarrow{X_1}, \overrightarrow{X_2}, \overrightarrow{X_3}\right)$ – which are neither all three of them coplanar, nor at least two of them collinear – are then represented as oriented axes with a common origin, thus forming a trihedron.

1.2.1.3. *Normed system of reference*

A vector is associated with each of these axes, and its norm can serve as a unit for measuring the projection of the "origin-element" vector on the axis. It is desirable, and even advisable, that measurements performed along each axis should be comparable, in the sense that the same unit should be taken on each axis unless otherwise required; in this case, the axes are characterized by the following unit vectors, respectively:

$$\vec{x}_1 = \frac{\overrightarrow{X_1}}{\left\|\overrightarrow{X_1}\right\|}; \quad \vec{x}_2 = \frac{\overrightarrow{X_2}}{\left\|\overrightarrow{X_2}\right\|}; \quad \vec{x}_3 = \frac{\overrightarrow{X_3}}{\left\|\overrightarrow{X_3}\right\|},$$

where $\vec{x}_1^{\,2} = \vec{x}_1 \cdot \vec{x}_1 = 1$; $\quad \vec{x}_2^{\,2} = \vec{x}_2 \cdot \vec{x}_2 = 1$; $\quad \vec{x}_3^{\,2} = \vec{x}_3 \cdot \vec{x}_3 = 1$.

In this case, the system of reference or the frame of reference, denoted as $\left\langle O \mid \vec{x}_1\ \vec{x}_2\ \vec{x}_3 \right\rangle$, is said to be *normed*.

1.2.1.4. *Orthogonal system of reference*

The measure of the "origin-element" segment is expressed in units defined below by projecting this segment on the three axes of the system of reference; the direction along which the projection is made should be chosen for each axis. This choice should preferably lead to setting simple and direct relations between these projections and the projected quantity.

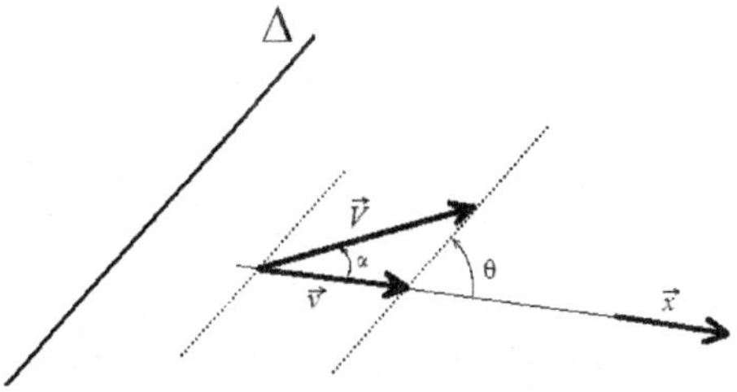

Figure 1.2. *Projection parallel to an axis*

Let us consider the plane defined by a vector $\vec{V}$ and an axis of unit vector $\vec{x}$, with an angle α between these two vectors; let us further consider a direction Δ at an angle θ with the axis. According to the above diagram, the projection of vector $\vec{V}$ on this axis, parallel to Δ, leads to a quadratic relation between the modules of vector $\vec{V}$ and its projection $\vec{v}$. Based on relations in general and right triangles, the following formula can be established

$$\left\|\vec{v}\right\|^2 - 2\left\|\vec{V}\right\| \times \left\|\vec{v}\right\| \cos\alpha + \left\|\vec{V}\right\|^2 \left(1 - \frac{\sin^2\alpha}{\sin^2\theta}\right) = 0 \text{, where}$$

$$\left\|\vec{v}\right\| = \left\|\vec{V}\right\| \cos\alpha,$$

which is in line with the result provided by trigonometry for orthogonal projection ($\theta = \pi / 2$), though not quite convenient to use.

Since, the conditions in which vectors are projected on the three axes of a system of reference should preferably be simplified and standardized, the choice will go as a rule (except for very specific cases that are not the object of this work) for a frame of reference with three mutually orthogonal directions that form a trirectangular trihedron; the orthogonal projections of the "origin-element" vector are easily expressed using the basic trigonometric functions of a right triangle.

Such a system of reference is called *orthogonal*; it is characterized by three vectors $\vec{x_1}$, $\vec{x_2}$, $\vec{x_3}$ that are connected by the relations

$$\vec{x_1} \cdot \vec{x_2} = 0, \quad \vec{x_2} \cdot \vec{x_3} = 0, \quad \vec{x_3} \cdot \vec{x_1} = 0$$

1.2.1.5. *Direct orthonormal system of reference – properties*

In conclusion to the above analyses, a system of reference that is adequate for studying the motion of bodies is at the same time normed and orthogonal, i.e. *orthonormal.*

But, in order to accurately account for the motion of bodies, their location should be perfected by taking into consideration various facets of their motion. As a rule, we decompose it into two essential elementary phases whose combination gives the motion of bodies: translation and rotation. While translation can be readily represented by a vector that defines its direction and orientation, the axis about which rotation takes place and its sense should be specified.

It is, therefore, important to specify how we define orientation in the previously introduced system of three axes, and it does not take an oenologist to memorize, and all the more so to use, Maxwell's corkscrew rule, which is presented below.

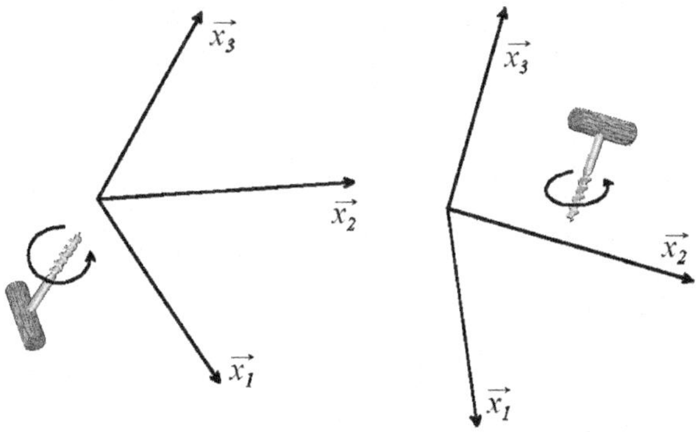

Figure 1.3. *Illustration of how Maxwell's corkscrew rule is applied*

Being placed perpendicularly to the plane $\Pi\left(\vec{x_1},\vec{x_2}\right)$, the corkscrew is rotated in the sense defined by the oriented right angle $\widehat{\vec{x_1},\vec{x_2}}$; the orientation of its advance then defines the orientation of the third axis $\vec{x_3}$ of the system of reference.

Such a system of reference is called *direct*. Under these conditions, the normed and mutually orthogonal basis vectors satisfy the three vector relations

$$\vec{x_1} \wedge \vec{x_2} = \vec{x_3}\,, \quad \vec{x_2} \wedge \vec{x_3} = \vec{x_1}\,, \quad \vec{x_3} \wedge \vec{x_1} = \vec{x_2}$$

The direct orthonormal system of reference is commonly used in solid mechanics, with the exception of some particular cases that would require the use of other types of systems of reference. In particular, this choice makes it easier to use vector calculus because of the standardization of the interpretation of its results. It is also advisable to have a clear summary of the properties that characterize the direct orthonormal system of reference, namely the three relations previously introduced

$$\begin{cases} \overrightarrow{x_1}^2 = 1, & \overrightarrow{x_2}^2 = 1, & \overrightarrow{x_3}^2 = 1 \\ \overrightarrow{x_1} \cdot \overrightarrow{x_2} = 0, & \overrightarrow{x_2} \cdot \overrightarrow{x_3} = 0, & \overrightarrow{x_3} \cdot \overrightarrow{x_1} = 0 \\ \overrightarrow{x_1} \wedge \overrightarrow{x_2} = \overrightarrow{x_3}, & \overrightarrow{x_2} \wedge \overrightarrow{x_3} = \overrightarrow{x_1}, & \overrightarrow{x_3} \wedge \overrightarrow{x_1} = \overrightarrow{x_2} \end{cases}$$

1.2.1.6. *Symbolic formulation*

The first two sets of the above relations can be symbolically written as a single expression

$$\overrightarrow{x_i} \cdot \overrightarrow{x_j} = \delta_{ij},$$

and the last one as

$$\overrightarrow{x_i} \wedge \overrightarrow{x_j} = \varepsilon_{ijk} \overrightarrow{x_k},$$

where the indices i, j, k take the three values $1, 2$ or 3 and the terms δ_{ij} and ε_{ijk} are defined as follows:

– the Kronecker symbol δ_{ij} defined by

$$\begin{matrix} \delta_{ij} = 1 & \text{if } i = j \\ \delta_{ij} = 0 & \text{if } i \neq j \end{matrix};$$

– ε_{ijk}, the three-index permutation symbol defined by:

- $\varepsilon_{ijk} = 1$ if the values of three indices i, j, k are not a circular permutation of the ordered values of the triplet 1,2,3,

- $\varepsilon_{ijk} = -1$ if the values of three indices i, j, k are a circular permutation of the ordered values of the triplet 1,3,2,

- $\varepsilon_{ijk} = 0$ if the values of at least two of the indices i, j, k are equal.

1.2.2. *Various types of frames of reference*

The expression of the equations that reflect the fundamental law of dynamics depends on the frame of reference relative to which the motion is considered; corrective terms can be introduced to account for the real environment in which the observer is located and for the environment in which the observed motion unfolds, and for the influence that these environments can have on the observed motion.

Up until the 18th Century, mechanics was the domain of astronomers and mathematicians who were essentially interested in the study of celestial bodies. Newton's second law or the *fundamental law of dynamics* aimed at expressing the motion of planets in the solar system under the action of forces of attraction exerted by the other celestial bodies. Akin to celestial mechanics, which relied heavily on time measurement, the field of watchmaking was also rapidly expanding, owing to the study of pendular motions.

The industrial development during the 19th Century and the invention of the machine as the main production tool led the mathematicians and physicists into studying the mechanics of systems. The fundamental law had already been formulated, but it had to be applied to a different environment, as the top priority was the local observation of motion, and this went beyond the field of watchmaking, which had nevertheless opened the path.

The study of motion had to take into consideration other external forces in addition to those exerted by the gravitational field; they were either acting at a distance, such as the electromagnetic forces, or by contact, when the physical system in motion was subject to stresses, or still when there were physical links between the system and its external environment or among the components of the system.

Foucault's works, most notably those related to the gyroscope and pendulum motion, have revealed the important role played by the choice of the system of reference. In particular, they have shown that when a system was isolated from its close environment and, more specifically, freed from the actions that the environment could exert on it, the observed motion of the gyroscope or pendulum could not be explained in a system of reference linked to the place of observation.

In the case of the pendulum, the locally observed rotation of its oscillation plane could only be explained when the description of its motion took into account the motion of the Earth.

The pendulum devised by Foucault, and used in the demonstration he conducted in the Pantheon in 1857, consisted of a heavy sphere hanging from a high ceiling on a 67 m long silver wire in such a way that the disruption of the pendulum oscillation because of the stress exerted by its anchor would be minimal. In addition to the fact that the suspension wire was chosen to minimize the deformation stresses, the weight of the sphere was such that it served the same purpose. The pendulum was, therefore, isolated from where its motion apparently took place.

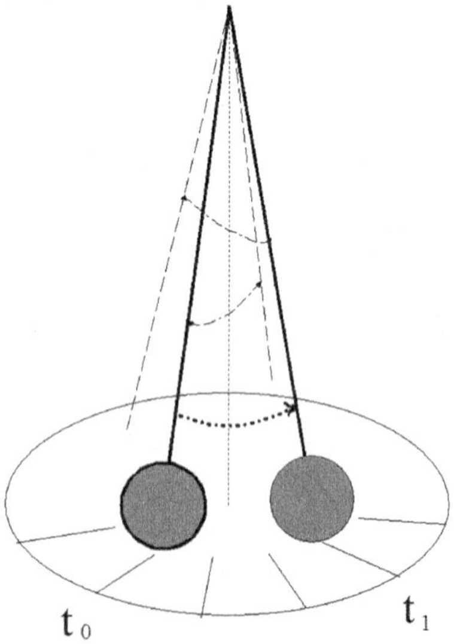

Figure 1.4. *The principle of Foucault's pendulum*

The pendulum was put into oscillation in a vertical plane located at the moment t_0 and during its pendular motion the behavior of its oscillation plane was observed, which indicated that it rotated about the vertical axis.

This observation can only be explained if the motion of the pendulum takes place in a system of reference that differs from the one of the observer linked to a local system of reference, and relative to which it rotates. But, the only rotation undergone by the local system of reference is the rotation of the Earth, to which it is connected; the observation indicates that the motion of the pendulum takes place in a system of reference with axes oriented independently of the local rotation, being, for example, collinear with the fixed axes of the solar frame of reference. This explanation could be corroborated by an adequate interpretation of the measurement of the angle of rotation of the pendulum oscillation plane during the time interval $[t_0, t_1]$.

In the mechanics it is, therefore, important to have a relevant classification of the systems of reference in order to conveniently express the consequences of the application of the fundamental law.

1.2.2.1. *Galilean frame of reference*

The fundamental law of dynamics states that there is at least one geometric system of reference (called *Galilean*) and a system of time reference (*privileged time scale*) for which:

– *a material point* (that is a body whose dimensions are negligible at the scale of observations and whose motion can at each moment be assimilated to a translation) *which is not subjected to any force is moving with acceleration zero, therefore has a rectilinear and uniform motion;*

– a material point subjected to one or several forces has, at each instant, an acceleration that is proportional to the resultant of these forces;

– the coefficients of proportionality between acceleration and the resultant of forces are constants characteristic to various materials that can be considered. By definition, these constants are the masses.

This notion of the Galilean system of reference is of paramount importance in the formulation of the mechanics, so its meaning must be clearly determined, as it is intimately connected to the nature of the motion under consideration.

The choice of the Galilean system of reference, indispensable for the application of the fundamental laws of dynamics, depends on the motion to be observed. All the actuators susceptible to influence the motion of the mechanical system ($\mathcal{D}$) to be described and studied should be included in the system of reference, in the sense that it should be possible to define their location and describe their motion as well; and in the absence of these actions, the mechanical system considered should have a rectilinear and uniform motion.

The system is then said to be mechanically isolated for the motions described.

For the study of motions in the terrestrial environment, the ideal Galilean system of reference has its origin in the center G of the solar system and the directions of its axes given by the three stars most distant from this system and whose angular position relative to an observer of the solar system would be considered invariable. It is worth noting here that the closest star to the Sun is already distant enough to accurately define an axis direction.

1.2.2.2. *Pseudo-Galilean frame of reference*

The approach of most problems in mechanics requires that they should be expressed in such systems of reference that the orders of magnitude of the variables involved can be compared with the other orders of magnitude of the problem; furthermore, all the elements involved in the study of a motion should be part of it, and clearly identified and located. This point is particularly important, since it heavily influences the accuracy of results and their interpretation, in the case of very high or very low values, which are not related to the physical environment of the study, and in disproportion with the others.

The need for a frame of reference that is better adapted to solving a problem of mechanics has led to the introduction of a concept of the pseudo-Galilean system of reference, in order to apply the fundamental law of dynamics under more convenient conditions; this should, therefore, have properties comparable to those of the Galilean system of reference, and more specifically, a material point that is not subjected to any force in this system should also have a rectilinear uniform motion.

Ideally, this system of reference should, therefore, have fixed orientations of axes and undergo uniform translation relative to the Galilean system of reference; in reality, few practical systems of reference meet such criteria. This means that:

– to the extent that forces are inevitably acting, to account for the relative motion of the pseudo-Galilean system of reference relative to the Galilean system of reference, which is not strictly translation nor a rectilinear one,

– and because rotations inevitably generate accelerations, corrective terms are added to the expressions of the fundamental law.

But, if the orders of magnitude of these corrective terms generated by the motion of the working system of reference are negligible relative to those of the problem, and do not significantly influence the study of a motion, then the system of reference can be considered pseudo-Galilean and can serve as basic frame of reference for the application of the fundamental law.

1.2.2.3. *Other local systems of reference*

The study of the motion of a body starts necessarily by determining its location in a properly chosen system of reference that allows for a correct description, and which in practice is almost never an ideal Galilean system of reference, and does not even meet the criteria of the pseudo-Galilean system of reference.

Such a system of reference is nevertheless critically important, as it allows for the practical location of the solid body whose motion is to be studied. But, it is obviously not adapted to the direct application of the fundamental law; it is, therefore, important to introduce in the formulation of this law the corrective terms that are essential for describing the motion proper to this system of reference relative to the frame of reference chosen for studying the motion.

Chapter 2 of this volume provides the information required for expressing these complementary terms.

1.3. Location of a solid body

In order to illustrate this statement, we shall consider the relative location of two rectangular parallelepipeds considered solid, and therefore impenetrable, (S_1) and (S_2).

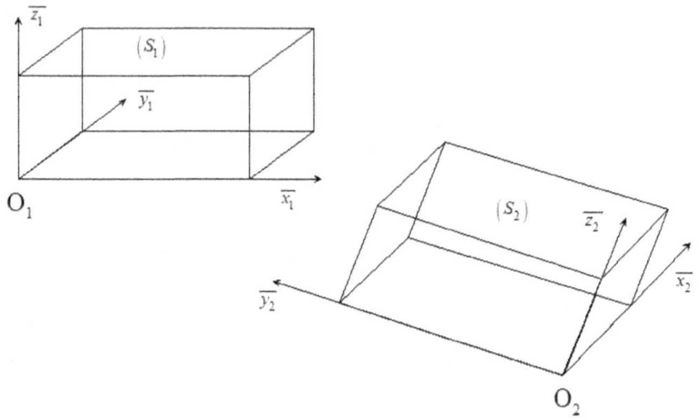

Figure 1.5. *The principle of locating a solid body*

To the vertex O_1 of (S_1), we connect a direct orthonormal system of axes with the basis $\left(\overrightarrow{x_1}\ \overrightarrow{y_1}\ \overrightarrow{z_1}\right)$; to the vertex O_2 of (S_2), we similarly connect a direct orthonormal system of axes $\left(\overrightarrow{x_2}, \overrightarrow{y_2}, \overrightarrow{z_2}\right)$ and its associated basis $\left(\overrightarrow{x_2}\ \overrightarrow{y_2}\ \overrightarrow{z_2}\right)$.

In order to locate (S_2) relative to (S_1), it is sufficient:

– to locate the position of O_2 relative to O_1 by the *position vector* $\overrightarrow{O_1O_2}$;

– to express the orientation of the $(\vec{x_2}\ \vec{y_2}\ \vec{z_2})$ basis relative to $(\vec{x_1}\ \vec{y_1}\ \vec{z_1})$.

NOTE 1.1.– While representing series of vectors inside brackets, in order to differentiate the notation of a sequence of vectors, for example the system $(\vec{x_1}, \vec{y_1}, \vec{z_1})$ composed of these three vectors separated by comma, from the reference entity that forms a basis, in the rest of this work we shall use for the latter the notation $(\vec{x_1}\ \vec{y_1}\ \vec{z_1})$ without separators between the vectors to express the basis associated with the solid (S_1), for example, which will be written $(S_1) \equiv (\vec{x_1}\ \vec{y_1}\ \vec{z_1})$.

Similarly, the system of reference associated with the solid (S_1), consisting of the origin O_1 and the basis (S_1), is denoted $\langle S_1 \rangle \equiv \langle O_1 | \vec{x_1}\ \vec{y_1}\ \vec{z_1} \rangle$.

1.3.1. *The principle of locating a solid*

In order to locate a solid (S) in a system of reference $\langle \lambda \rangle$ – an essential condition for studying the motion in this system of reference – the solid should, therefore, be connected to a system of reference $\langle S \rangle$ that characterizes it. This implies choosing a fixed origin O_S in the solid and associating with it a direct orthonormal system of axes with the function of basis $(S) \equiv (\vec{x_S}\ \vec{y_S}\ \vec{z_S})$.

1.3.2. *Location parameters of a solid*

The location parameters of a solid result from the above considerations; they are of two types, linear or angular when the point taken as origin of the system of reference $\langle S \rangle$ is located, which means that the components of its position vector are determined, angular when the system of reference needs to be oriented relative to

$\langle \lambda \rangle$, which serves as frame of reference for determining and expressing the location of the solid.

1.3.3. *Coordinates of the position vector*

Let us, therefore, consider a system of reference $\langle \lambda \rangle \equiv \left\langle O_\lambda \,|\, \overrightarrow{x_\lambda}\ \overrightarrow{y_\lambda}\ \overrightarrow{z_\lambda} \right\rangle$ chosen for the study of motion and another system $\langle S \rangle \equiv \left\langle O_S \,|\, \overrightarrow{x_S}\ \overrightarrow{y_S}\ \overrightarrow{z_S} \right\rangle$ connected to the solid. The components of the position vector $\overrightarrow{O_\lambda O_S}$ depend on the chosen representation.

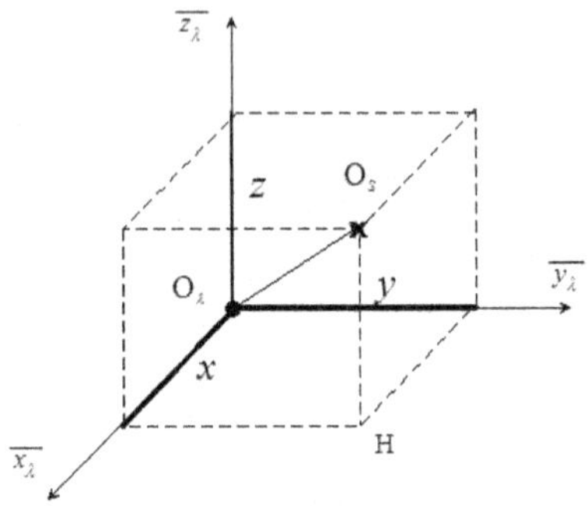

Figure 1.6. *The Cartesian coordinates*

1.3.3.1. *Cartesian coordinates x,y,z*

$$\overrightarrow{O_\lambda O_S} = x\overrightarrow{x_\lambda} + y\overrightarrow{y_\lambda} + z\overrightarrow{z_\lambda}$$

1.3.3.2. *Cylindrical or cylindrical-polar coordinates r, α, z*

$$\overrightarrow{O_\lambda O_S} = r\overrightarrow{u}(\alpha) + z\overrightarrow{z_\lambda}$$

H being the orthogonal projection of $\mathbf{O_S}$ on the plane $\Pi\left(O_\lambda \,|\, \overrightarrow{x_\lambda}, \overrightarrow{y_\lambda}\right)$, that is

$$\overrightarrow{O_\lambda O_S} = \overrightarrow{O_\lambda H} + \overrightarrow{HO_S} \quad \text{with} \quad \overrightarrow{O_\lambda H} = x\overrightarrow{x_\lambda} + y\overrightarrow{y_\lambda} \quad \text{and} \quad \overrightarrow{HO_S} = z\overrightarrow{z_\lambda},$$

and $\vec{u}(\alpha)$ is the unit vector collinear with $\overrightarrow{O_\lambda H}$, oriented from O_λ to H, or

$$\vec{u}(\alpha)=\frac{\overrightarrow{O_\lambda H}}{\left\|\overrightarrow{O_\lambda H}\right\|} \quad \text{with} \quad \left\|\overrightarrow{O_\lambda H}\right\|=r=\sqrt{x^2+y^2}>0,$$

then,

$$\vec{u}(\alpha)=\frac{x\overrightarrow{x_\lambda}+y\overrightarrow{y_\lambda}}{\sqrt{x^2+y^2}}=\cos\alpha\,\overrightarrow{x_\lambda}+\sin\alpha\,\overrightarrow{y_\lambda}.$$

The angle α is modulo 2π defined by the following two relations

$$\cos\alpha=\frac{x}{\sqrt{x^2+y^2}} \quad , \quad \sin\alpha=\frac{y}{\sqrt{x^2+y^2}}.$$

NOTE 1.2.– The notation $\vec{u}(\alpha)$ is conventional; it expresses the vector in the plane $\Pi\left(\overrightarrow{x_\lambda},\overrightarrow{y_\lambda}\right)$ which makes an oriented angle α with $\overrightarrow{x_\lambda}$, namely $\alpha\equiv\widehat{\overrightarrow{x_\lambda},\vec{u}(\alpha)}$.

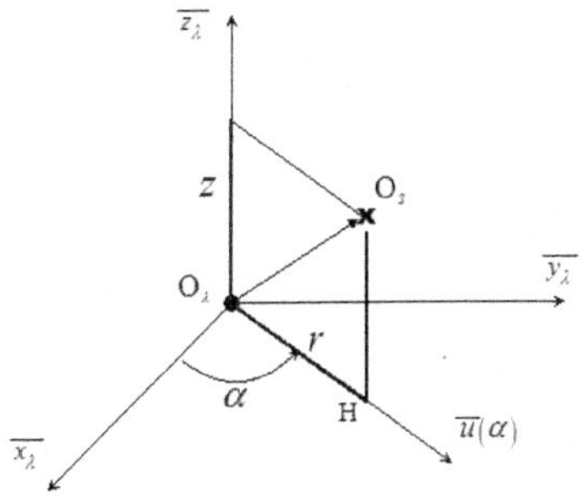

Figure 1.7. *Cylindrical-polar coordinates*

Likewise, we define the following vectors

$$\vec{u}\left(\alpha+\pi/2\right)\ ,\ \vec{u}(\alpha+\pi)=-\vec{u}(\alpha)\ ,\ \vec{u}\left(\alpha+3\pi/2\right)=-\vec{u}\left(\alpha+\pi/2\right),$$

according to Figure 1.8.

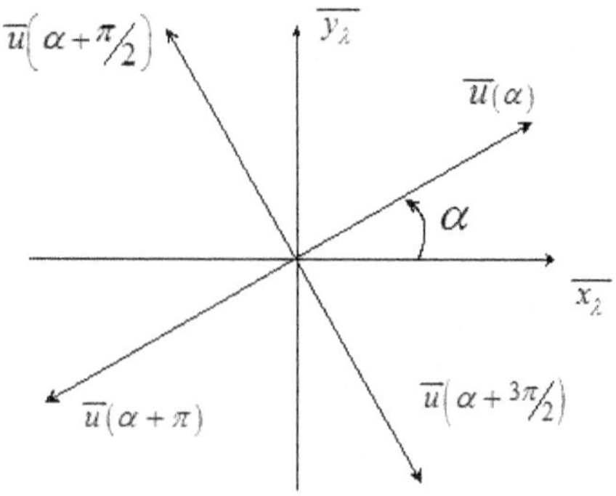

Figure 1.8. *Principle of the notation* $\vec{u}(\alpha)$

1.3.3.3. *Spherical coordinates* R, α, β

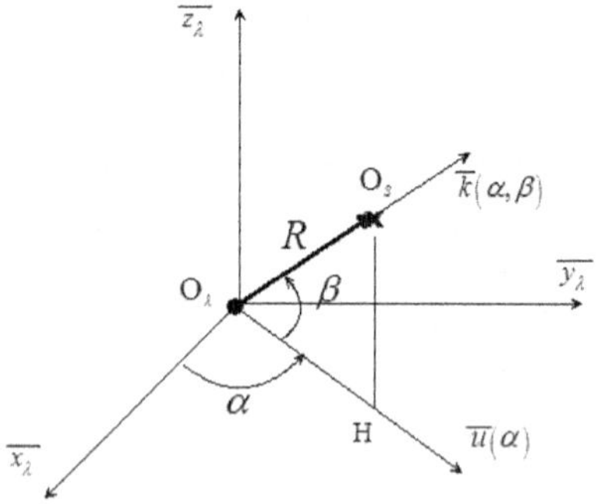

Figure 1.9. *Spherical coordinates*

$$\overrightarrow{O_\lambda O_S} = R\,\vec{k}(\alpha,\beta)$$

where, $\vec{k}(\alpha,\beta)$ is the unit vector collinear with $\overrightarrow{O_\lambda O_S}$ oriented from O_λ to O_S, which is

$$\vec{k}(\alpha,\beta) = \frac{\overrightarrow{O_\lambda O_S}}{\left\|\overrightarrow{O_\lambda O_S}\right\|} \quad \text{with} \quad \left\|\overrightarrow{O_\lambda O_S}\right\| = R = \sqrt{x^2+y^2+z^2} > 0,$$

$$\Rightarrow \vec{k}(\alpha,\beta)=\frac{x\overrightarrow{x_\lambda}+y\overrightarrow{y_\lambda}+z\overrightarrow{z_\lambda}}{\sqrt{x^2+y^2+z^2}}=\frac{x\overrightarrow{x_\lambda}+y\overrightarrow{y_\lambda}}{\sqrt{x^2+y^2+z^2}}+\frac{z\overrightarrow{z_\lambda}}{\sqrt{x^2+y^2+z^2}}.$$

The angle $\beta\left(-\frac{\pi}{2}<\beta<\frac{\pi}{2}\right)$ is chosen as follows

$$\sin\beta=\frac{z}{\sqrt{x^2+y^2+z^2}} \quad \text{and} \quad \cos\beta=\sqrt{1-\sin^2\beta}=\frac{\sqrt{x^2+y^2}}{\sqrt{x^2+y^2+z^2}}>0.$$

Hence, we deduce

$$\frac{1}{\sqrt{x^2+y^2+z^2}}=\frac{\cos\beta}{\sqrt{x^2+y^2}},$$

and the expression of vector $\vec{k}(\alpha,\beta)$

$$\vec{k}(\alpha,\beta)=\cos\beta\,\frac{x\overrightarrow{x_\lambda}+y\overrightarrow{y_\lambda}}{\sqrt{x^2+y^2}}+\sin\beta\,\overrightarrow{z_\lambda}.$$

Using the relations established for α in cylindrical-polar coordinates, we obtain

$$\vec{k}(\alpha,\beta)=\cos\beta\left(\cos\alpha\,\overrightarrow{x_\lambda}+\sin\alpha\,\overrightarrow{y_\lambda}\right)+\sin\beta\,\overrightarrow{z_\lambda}=\cos\beta\,\vec{u}(\alpha)+\sin\beta\,\overrightarrow{z_\lambda}.$$

The relations that express the Cartesian coordinates (x,y,z) as a function of spherical coordinates (R,α,β) are:

$$\begin{cases} x=R\cos\beta\cos\alpha \\ y=R\cos\beta\sin\alpha \\ z=R\sin\beta \end{cases}.$$

NOTE 1.3.– It is considered by convention that $\vec{K}(\alpha,\beta)=\vec{k}(\alpha,\beta+\pi/2)$.

1.3.3.4. *General coordinates*

It is also possible to locate the position in the system of reference $\langle\lambda\rangle$ of the point O_S relative to O_λ with an arbitrary system of three parameters Q_1, Q_2, Q_3, which can be expressed as

$$\overrightarrow{O_\lambda O_S} = \vec{F}(Q_1, Q_2, Q_3).$$

The use of this system of coordinates is of particular interest in software applications, when the formalism developed in this manner is designed to apply to a high number of problems and to offer the possibility of choosing the system of coordinates to be used.

An illustration of this case is provided by the example below, in which O_λ is located in the middle of the segment PN. Two groups of three parameters are thus obtained, (α, r_1, r_2) and $(\alpha, \alpha_1, \alpha_2)$, and they both allow for locating the point O_S.

This type of coordinate, called bifocal, is used in particular in the space domain. The angle α_1 represents the 1st azimuth of the O_S to be located.

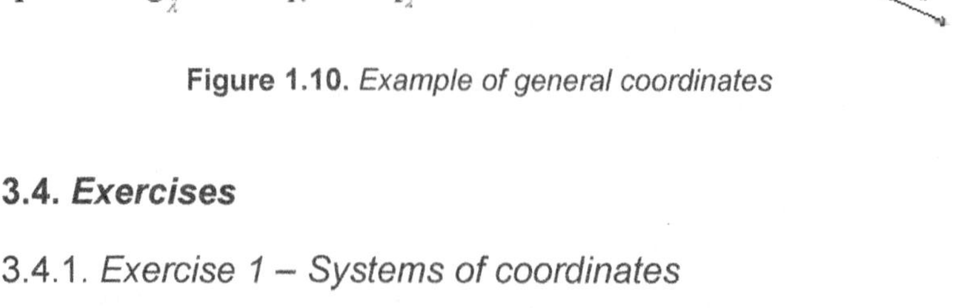

Figure 1.10. *Example of general coordinates*

1.3.4. *Exercises*

1.3.4.1. *Exercise 1 – Systems of coordinates*

In the system of reference $\langle 0 \rangle \equiv \langle O | \vec{x}\,\vec{y}\,\vec{z} \rangle$, let M be a point identified by its spherical coordinates (R, α, β) which take the values

$$R = 10m,\ \alpha = \beta = \frac{\pi}{4} rad$$

Question: Determine the cylindrical-polar coordinates (R, α, z) and the Cartesian coordinates (x, y, z) of point M.

$$\vec{k}(\alpha, \beta) = \cos\beta\,\vec{u}(\alpha) + \sin\beta\,\vec{z} \text{ with } \vec{u}(\alpha) = \cos\alpha\,\vec{x} + \sin\alpha\,\vec{y},$$

$$\overrightarrow{OM} = R\vec{k}(\alpha, \beta) = r\vec{u}(\alpha) + z\vec{z} = x\vec{x} + y\vec{y} + z\vec{z}.$$

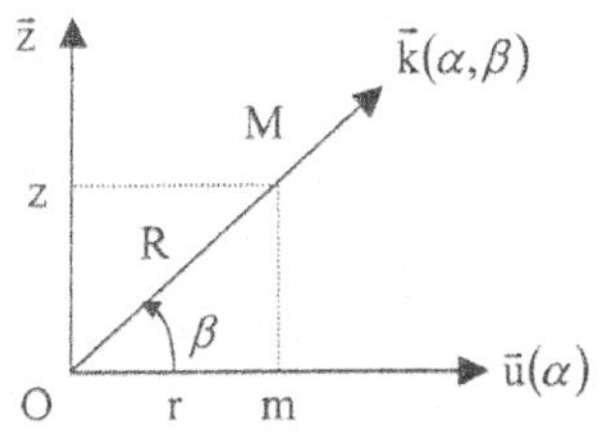

Meridian Plane

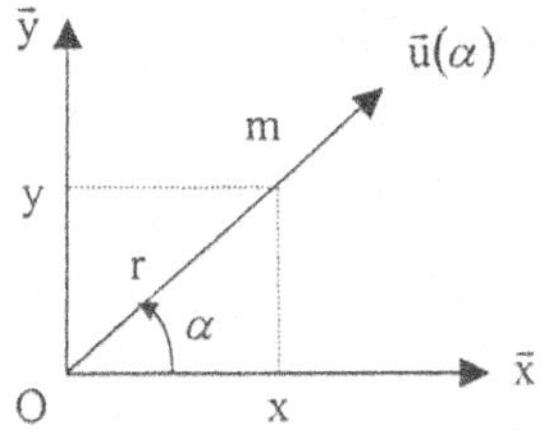

Equatorial plane

Hence, we deduce that

$$r = R\cos\beta = 10m \times \frac{\sqrt{2}}{2} = 5\sqrt{2}\,m$$

$$z = R\sin\beta = 5\sqrt{2}\,m$$

$$x = r\cos\alpha = 5\sqrt{2}\,m \times \frac{\sqrt{2}}{2} = 5m$$

$$y = r\sin\alpha = 5m$$

1.3.4.2. *Exercise 2 – System of coordinates*

In the system of reference $\langle 0 \rangle \equiv \langle O | \vec{x}\,\vec{y}\,\vec{z} \rangle$, let M be a point identified by its Cartesian coordinates (x, y, z) that take the values

$$x = 3m,\ y = -4m,\ z = 5m.$$

Question: Determine its cylindrical-polar coordinates (R, α, z) and then its spherical coordinates (R, α, β).

$$\overrightarrow{\mathrm{OM}} = x\vec{x} + y\vec{y} + z\vec{z} = r\cos\alpha\,\vec{x} + r\sin\alpha\,\vec{y} + z\vec{z}$$

$$\Rightarrow \quad (r\cos\alpha)^2 + (r\sin\alpha)^2 = r^2 = x^2 + y^2; \qquad \tan\alpha = \frac{y}{x}.$$

According to the above figure, we also have

$$R = \sqrt{r^2 + z^2}$$

$$\cos\beta = \frac{r}{R}, \qquad \sin\beta = \frac{z}{R}.$$

Therefore, we obtain

$$r = \sqrt{9+16} = 5m, \tan\alpha = -\frac{4}{3} \quad \Rightarrow \quad \alpha = -53{,}13° = -0{,}92\,rad$$

$$R = \sqrt{25+25} = 5\sqrt{2}, \sin\beta = \frac{5}{5\sqrt{2}} = \frac{\sqrt{2}}{2} = \cos\beta \quad \Rightarrow \quad \beta = \frac{\pi}{4}.$$

1.4. Positioning of a system of reference connected to a solid

1.4.1. *Several examples of location systems of reference*

Before we approach this aspect of the question of locating a solid before any study of its motion, it may help to take a look at the systems of reference that could be used in the locating stage. In the section above we have sketched the configuration of three of them, directly connected to the common systems of coordinates described above.

In certain configurations of motion on a given curve or surface, one may tend to favor the description starting from local data that characterize it. This leads to using the attached local trihedrons. This is the case of the Frenet trihedron, which is locally defined in a point of a curve, or that of Darboux–Ribaucour trihedron, in the case of a surface.

1.4.1.1. *The Cartesian system of reference*

This is the system of reference $\left\langle O_\lambda \middle| \overrightarrow{x_\lambda}\ \overrightarrow{y_\lambda}\ \overrightarrow{z_\lambda} \right\rangle$.

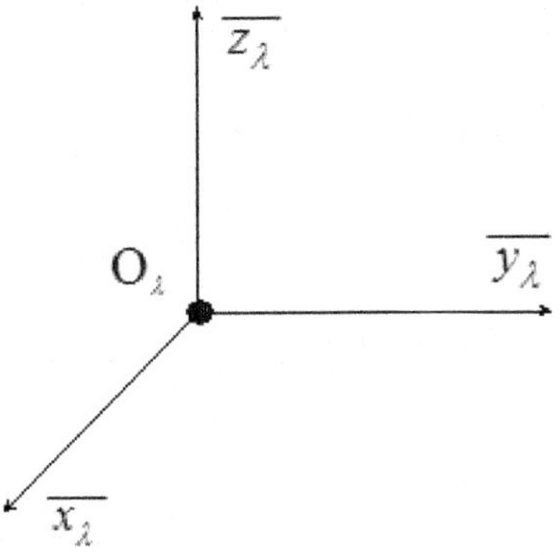

Figure 1.11. *The Cartesian system of reference*

1.4.1.2. *Cylindrical-polar system of reference*

This is the system of reference $\left\langle O_\lambda \middle| \vec{u}(\alpha)\ \vec{u}\left(\alpha + \pi/2\right)\ \overrightarrow{z_\lambda} \right\rangle$.

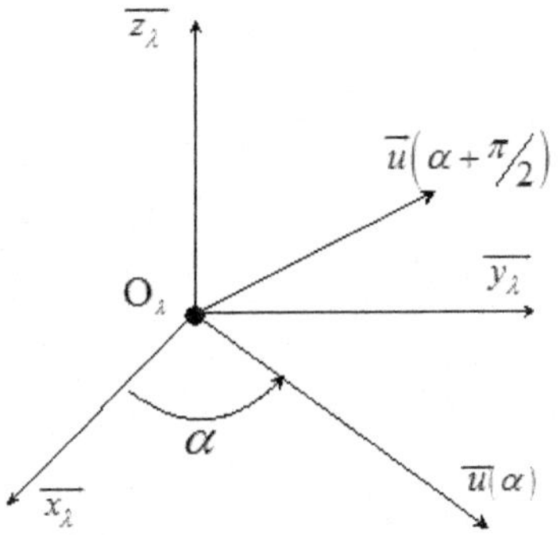

Figure 1.12. *Cylindrical-polar system of reference*

1.4.1.3. *Spherical system of reference*

This is the system of reference $\left\langle O_\lambda \middle| \vec{k}(\alpha,\beta)\ \vec{u}\left(\alpha+\pi/2\right)\ \vec{K}(\alpha,\beta) \right\rangle$.

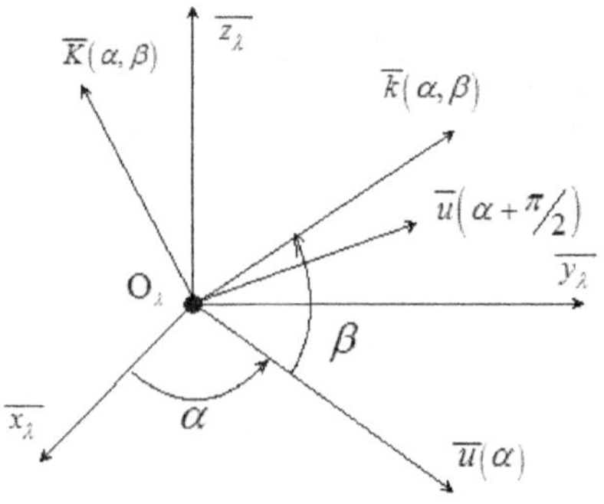

Figure 1.13. *Spherical system of reference*

1.4.1.4. *Frenet trihedron*

Defined in a point M of a curve Γ, the Frenet trihedron consists of three unit vectors forming a basis, with which two scalar quantities are associated, which define the curve in that given point. These elements are defined as follows: their presentation will be completed in the chapter dedicated to left curved lines in Volume 2 [BOR 16a].

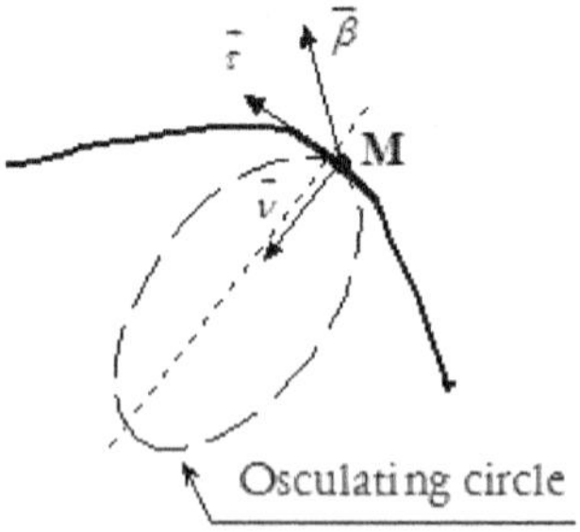

Figure 1.14. *Diagram of the Frenet trihedron*

– $\vec{\tau}$ unit vector tangent in M to the curve Γ;

– in a first approximation, if in the vicinity of the point M the curve is equivalent to an arc of circle of radius R (called radius of curvature), when passing to the limit, the length of the arc tending to 0, the limit plane that contains this arc is the osculating plane;

– $\vec{\nu}$ unit vector orthogonal to $\vec{\tau}$ in M, contained in the osculating plane and oriented from point M to the curvature center of the arc;

– $\vec{\beta}$ unit vector orthogonal in M to the osculating plane $\Pi\left(M|\vec{\tau},\vec{\nu}\right)$, such that $\vec{\beta}=\vec{\tau}\wedge\vec{\nu}$;

– $\left\langle M|\vec{\tau}\ \vec{\nu}\ \vec{\beta}\right\rangle$ is the local Frenet system of reference.

1.4.1.5. *Darboux–Ribaucour trihedron*

A surface is represented by a vector function of two variables $\vec{f}(q_1,q_2)$. If one set of values is fixed for the parameter q_1, then to each one corresponds a curve Γ_{q_2} along which evolves the parameter q_2; similarly, if we fix a set of values for q_2, then to each one corresponds a curve C_{q_1}.

To each point M on the surface S studied, there corresponds a pair of values (q_1,q_2), and consequently a curve C_{q_1} and a curve Γ_{q_2}, a vector $\vec{\tau_1}$ tangent in M to the former and a vector $\vec{\tau_2}$ tangent in M to the latter. We then consider the vector $\vec{n}$, perpendicular to the surface S, such that $\vec{n}=\vec{\tau_1}\wedge\vec{\tau_2}$.

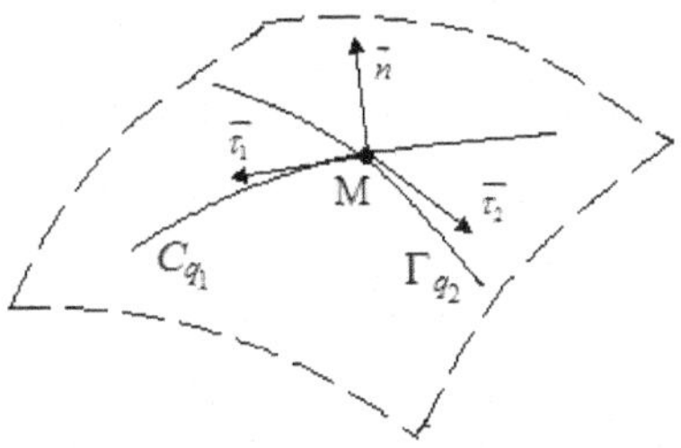

Figure 1.15. *Diagram of the Darboux–Ribaucour trihedron*

$\left\langle M|\vec{\tau_1}\ \vec{\tau_2}\ \vec{n}\right\rangle$ is the local Darboux–Ribaucour system of reference.

The establishment of Frenet and Darboux–Ribaucour trihedrons will be further elaborated in the respective chapters of Volume 2 [BOR 16a] in relation to vector functions of 1 or 2 variables.

1.4.2. *General location parameters*

In order to define a principle for establishing a set of parameters representative for their relative positioning, let us consider two direct orthonormal bases $(x) \equiv \left(\overrightarrow{x_1}\ \overrightarrow{x_2}\ \overrightarrow{x_3}\right)$ and $(X) \equiv \left(\overrightarrow{X_1}\ \overrightarrow{X_2}\ \overrightarrow{X_3}\right)$. By assigning them to the same origin O, we shall attempt to locate one relative to the other, by means of three angular parameters Q_4, Q_5, Q_6, the two systems of reference $\left\langle \mathrm{O} \middle| \overrightarrow{x_1}\ \overrightarrow{x_2}\ \overrightarrow{x_3} \right\rangle$ and $\left\langle \mathrm{O} \middle| \overrightarrow{X_1}\ \overrightarrow{X_2}\ \overrightarrow{X_3} \right\rangle$ in the general case when the two systems of axes are completely distinct $\left(\overrightarrow{X_i} \neq \overrightarrow{x_j}, \forall i, j = 1,2 \text{ or } 3\right)$.

Let $\Pi\left(\overrightarrow{X_i}, \overrightarrow{x_j}\right)$ be a plane defined by the two vectors $\overrightarrow{X_i}$ and $\overrightarrow{x_j}$, and the line Δ_{ij} orthogonal to this plane, of unit vector $\overrightarrow{n_{ij}}$. The angle Q_5 formed by the two vectors $\overrightarrow{X_i}$ and $\overrightarrow{x_j}$ is defined by the relation

$$\overrightarrow{X_i} \wedge \overrightarrow{x_j} = \overrightarrow{n_{ij}} \sin Q_5 .$$

The following diagram indicates the choice of angles.

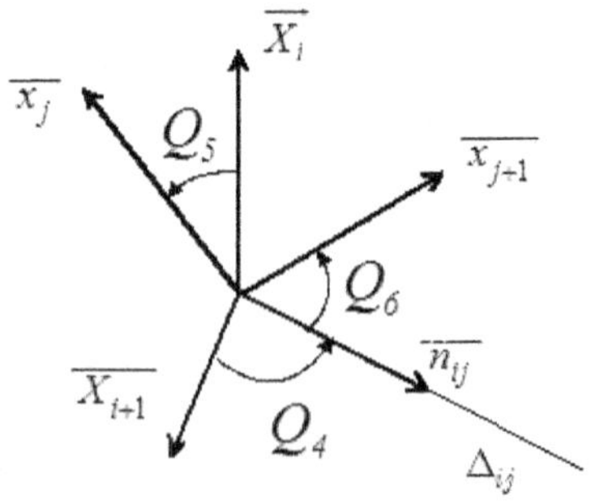

Figure 1.16. *Orientation of one basis relative to the other*

NOTE 1.4.– The indices i and j of various unit vectors of the two considered bases vary strictly according to a cyclic permutation of the numbers 1, 2 and 3 (for example, if j = 3, j + 1=1).

The transformation that leads from the basis (X) to the basis (x) results from three successive rotations:

– rotation by angle $Q_4 = \widehat{\overrightarrow{X_{i+1}}, \overrightarrow{n_{ij}}}$ about the axis $\overrightarrow{X_i}$. It is said to be of axis $\overrightarrow{X_i}$;

– rotation by angle $Q_5 = \widehat{\overrightarrow{X_i}, \overrightarrow{x_j}}$ of axis $\overrightarrow{n_{ij}}$;

– rotation by angle $Q_6 = \widehat{\overrightarrow{n_{ij}}, \overrightarrow{x_{j+1}}}$ of axis $\overrightarrow{x_j}$.

It can be noted that the transformation introduces two intermediary bases (1) and (2) to get from (X) to (x). By setting $\overrightarrow{v_{ij}} = \overrightarrow{X_i} \wedge \overrightarrow{n_{ij}}$ and $\overrightarrow{\mu_{ij}} = \overrightarrow{x_j} \wedge \overrightarrow{n_{ij}}$, the two intermediary bases are

$$\begin{cases} (1) \equiv \left(\overrightarrow{n_{ij}}\ \overrightarrow{v_{ij}}\ \overrightarrow{X_i}\right) \\ (2) \equiv \left(\overrightarrow{n_{ij}}\ \overrightarrow{\mu_{ij}}\ \overrightarrow{x_j}\right) \end{cases},$$

according to the three successive passage schemes given in Figure 1.17.

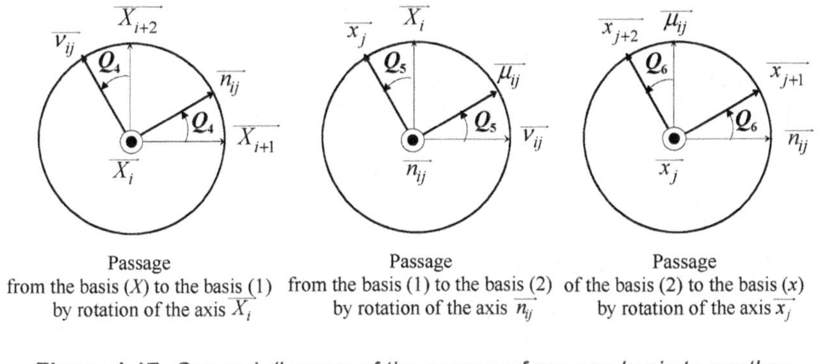

Figure 1.17. *General diagram of the passage from one basis to another*

By successive projections in the basis (1), then (2), then (x), the formulation of the projection of this latter basis on the basis (X) is obtained.

$$\begin{cases} \overrightarrow{x_j} = \sin Q_4 \sin Q_5 \overrightarrow{X_{i+1}} - \cos Q_4 \sin Q_5 \overrightarrow{X_{i+2}} + \cos Q_5 \overrightarrow{X_i} \\ \overrightarrow{x_{j+1}} = (\cos Q_4 \cos Q_6 - \sin Q_4 \cos Q_5 \sin Q_6)\overrightarrow{X_{i+1}} \ldots \\ \quad \ldots + (\sin Q_4 \cos Q_6 + \cos Q_4 \cos Q_5 \sin Q_6)\overrightarrow{X_{i+2}} + \sin Q_5 \sin Q_6 \overrightarrow{X_i}. \\ \overrightarrow{x_{j+2}} = (-\cos Q_4 \sin Q_6 - \sin Q_4 \cos Q_5 \cos Q_6)\overrightarrow{X_{i+1}} \ldots \\ \quad \ldots + (-\sin Q_4 \sin Q_6 + \cos Q_4 \cos Q_5 \cos Q_6)\overrightarrow{X_{i+2}} + \sin Q_5 \cos Q_6 \overrightarrow{X_i} \end{cases}$$

1.4.3. *Euler angles*

The question is now to apply the above described principle to define the location of a solid (S) to which the basis $\left(\overrightarrow{x_S}\ \overrightarrow{y_S}\ \overrightarrow{z_S}\right)$ is associated, which has to be located relative to a reference basis $\left(\overrightarrow{x_\lambda}\ \overrightarrow{y_\lambda}\ \overrightarrow{z_\lambda}\right)$.

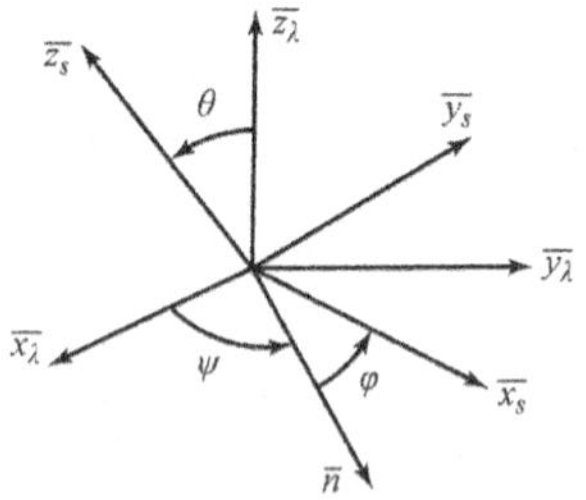

Figure 1.18. *Euler angles diagram*

Commonly used Euler angles have specific names; they are defined as follows:

$$\overrightarrow{X_i} \equiv \overrightarrow{z_\lambda} \Rightarrow \overrightarrow{X_{i+1}} \equiv \overrightarrow{x_\lambda}, \quad \overrightarrow{X_{i+2}} \equiv \overrightarrow{y_\lambda},$$
$$\overrightarrow{x_j} \equiv \overrightarrow{z_S} \Rightarrow \overrightarrow{x_{j+1}} \equiv \overrightarrow{x_S}, \quad \overrightarrow{x_{j+2}} \equiv \overrightarrow{y_S}.$$

This leads to identifying the angles as

$Q_4 \equiv \psi$ *precession angle*

$Q_5 \equiv \theta$ *nutation angle,*

$Q_6 \equiv \varphi$ *spin angle,*

according to Figure 1.18, and defining them by:

$$\begin{cases} \overrightarrow{z_\lambda} \wedge \overrightarrow{z_S} = \vec{n}\sin\theta \\ \overrightarrow{z_\lambda}\cdot\overrightarrow{z_S} = \cos\theta \end{cases} \qquad \begin{cases} \overrightarrow{x_\lambda}\cdot\vec{n} = \cos\psi \\ \overrightarrow{y_\lambda}\cdot\vec{n} = \sin\psi \end{cases} \qquad \begin{cases} \vec{n}\cdot\overrightarrow{x_S} = \cos\varphi \\ \vec{n}\cdot\overrightarrow{y_S} = -\sin\varphi \end{cases},$$

where $\vec{n}$ is the *nodal vector*.

NOTE 1.5.– The previous angular representation has a limitation in the processes used in numerical calculation. Indeed, when the angle θ (or Q_5) is very small, which happens when the planes $\Pi\left(\overrightarrow{x_\lambda}, \overrightarrow{y_\lambda}\right)$ and $\Pi\left(\overrightarrow{x_S}, \overrightarrow{y_S}\right)$ merge or become parallel, it becomes impossible to distinguish precession ψ from spin φ. This is particularly the case of numerical calculation where, should such a situation occur, it may lead to distorted results. It may be required to choose another angular representation, provided that the three chosen angles are independent.

1.4.4. *Changes of basis in the Euler representation*

The requirement to locate a solid before expressing its motion, and then to study it, leads to choosing systems of reference that are best suited for conducting each of these stages. It is, therefore, important to be able to make the changes of basis that help getting to the essential stage, which is the application of the fundamental law of dynamics in the adequate Galilean or pseudo-Galilean system of reference.

To make ideas clear, let us consider two bases

$$(\lambda) \equiv \left(\overrightarrow{x_\lambda}\ \overrightarrow{y_\lambda}\ \overrightarrow{z_\lambda}\right) \text{ and } (S) \equiv \left(\overrightarrow{x_S}\ \overrightarrow{y_S}\ \overrightarrow{z_S}\right),$$

with a common origin O_λ and a point M belonging to the physical system being considered $(\mathscr{D})$.

The position vector $\overrightarrow{O_\lambda M}$ can be projected in the two bases

$$\overrightarrow{O_\lambda M} = x_\lambda \overrightarrow{x_\lambda} + y_\lambda \overrightarrow{y_\lambda} + z_\lambda \overrightarrow{z_\lambda} = x_S \overrightarrow{x_S} + y_S \overrightarrow{y_S} + z_S \overrightarrow{z_S}.$$

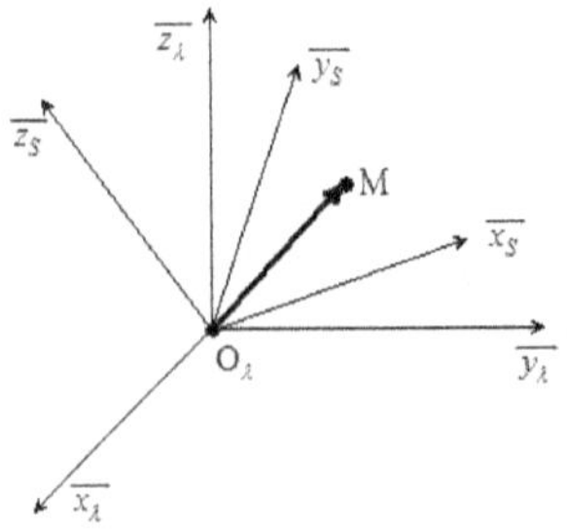

Figure 1.19. *Euler angles, location of the systems of reference*

The task is then to express the relations that connect the components of the position vector in the two bases as a function of the three angles that define the orientation of the basis (S) relative to the basis (λ).

The change of basis $(\lambda) \to (S)$ results from three successive rotations, by the corresponding angles ψ, θ and φ.

1.4.4.1. *The first rotation by angle ψ, or precession*

This rotation about the axis $\overrightarrow{z_\lambda}$, called precession, makes the passage from the basis (λ) to the basis $(1) \equiv \left(\vec{n}\ \vec{n_1}\ \overrightarrow{z_\lambda}\right)$ according to Figure 1.20.

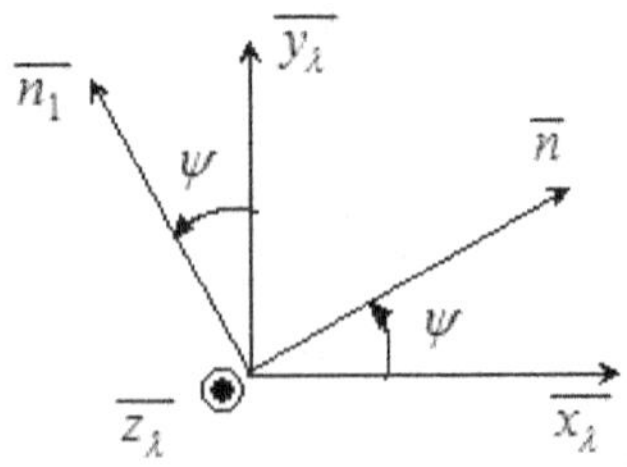

Figure 1.20. *Precession*

The projection of basis (1) on the basis (λ) yields:

$$\begin{cases} \vec{n} = \cos\psi\, \vec{x_\lambda} + \sin\psi\, \vec{y_\lambda} \\ \vec{n_1} = -\sin\psi\, \vec{x_\lambda} + \cos\psi\, \vec{y_\lambda}\,, \\ \vec{z_\lambda} \end{cases}$$

a result that can be represented by the following table of the change of basis:

$(\lambda) \backslash (1)$	$\vec{n}$	$\vec{n_1}$	$\vec{z_\lambda}$
$\vec{x_\lambda}$	$\cos\psi$	$-\sin\psi$	0
$\vec{y_\lambda}$	$\sin\psi$	$\cos\psi$	0
$\vec{z_\lambda}$	0	0	1

,

and by the associated matrix

$$M(\lambda,1) = \begin{bmatrix} \cos\psi & -\sin\psi & 0 \\ \sin\psi & \cos\psi & 0 \\ 0 & 0 & 1 \end{bmatrix}.$$

The relation between the two bases is then written in matrix form:

$$\begin{bmatrix} \vec{x_\lambda} \\ \vec{y_\lambda} \\ \vec{z_\lambda} \end{bmatrix} = M(\lambda,1) \begin{bmatrix} \vec{n} \\ \vec{n_1} \\ \vec{z_\lambda} \end{bmatrix}.$$

If the vector $\overrightarrow{O_\lambda M}$ is expressed in both bases

$$\overrightarrow{O_\lambda M} = x_1 \vec{n} + y_1 \vec{n_1} + z_1 \vec{z_\lambda} = x_\lambda \vec{x_\lambda} + y_\lambda \vec{y_\lambda} + z_\lambda \vec{z_\lambda}\,,$$

$$\overrightarrow{O_\lambda M} = x_1\left(\cos\psi \vec{x_\lambda} + \sin\psi \vec{y_\lambda}\right) + y_1\left(-\sin\psi \vec{x_\lambda} + \cos\psi \vec{y_\lambda}\right) + z_1 \vec{z_\lambda}\,,$$

$$\overrightarrow{O_\lambda M} = (x_1 \cos\psi - y_1 \sin\psi)\overrightarrow{x_\lambda} + (x_1 \sin\psi + y_1 \cos\psi)\overrightarrow{y_\lambda} + z_1 \overrightarrow{z_\lambda}.$$

$\overrightarrow{O_\lambda M}$ can be written in matrix form

$$\begin{bmatrix} x_\lambda \\ y_\lambda \\ z_\lambda \end{bmatrix} = \begin{bmatrix} \cos\psi & -\sin\psi & 0 \\ \sin\psi & \cos\psi & 0 \\ 0 & 0 & 1 \end{bmatrix} \times \begin{bmatrix} x_1 \\ y_1 \\ z_1 \end{bmatrix} = M(\lambda,1) \begin{bmatrix} x_1 \\ y_1 \\ z_1 \end{bmatrix}.$$

It can be noted that since it is possible to write

$$\begin{bmatrix} x_1 \\ y_1 \\ z_1 \end{bmatrix} = \begin{bmatrix} \cos\psi & \sin\psi & 0 \\ -\sin\psi & \cos\psi & 0 \\ 0 & 0 & 1 \end{bmatrix} \times \begin{bmatrix} x_\lambda \\ y_\lambda \\ z_\lambda \end{bmatrix} = M^{-1}(\lambda,1) \begin{bmatrix} x_\lambda \\ y_\lambda \\ z_\lambda \end{bmatrix},$$

this yields

$$M^{-1}(\lambda,1) = M(1,\lambda) = \overline{M(\lambda,1)},$$

where $\overline{M(\lambda,1)}$ is the transpose matrix of $M(\lambda,1)$.

1.4.4.2. *The second rotation by angle θ, or nutation*

This rotation about the axis $\vec{n}$, called nutation, makes the passage from the basis (1) to the basis $(2) \equiv (\vec{n}\ \vec{w}\ \vec{z_S})$ according to Figure 1.21.

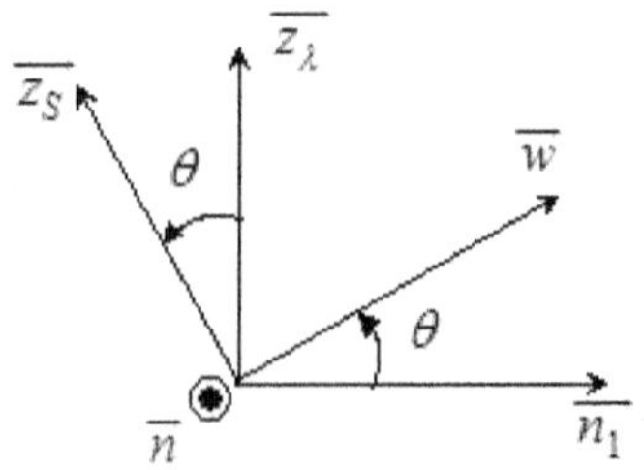

Figure 1.21. *Nutation*

The table of the change of basis and the associated matrix are

(1) | (2)	$\vec{n}$	$\vec{w}$	$\vec{z_S}$
$\vec{n}$	1	0	0
$\vec{n_1}$	0	$\cos\theta$	$-\sin\theta$
$\vec{z_\lambda}$	0	$\sin\theta$	$\cos\theta$

,

$$M(1,2)=\begin{bmatrix}1 & 0 & 0\\ 0 & \cos\theta & -\sin\theta\\ 0 & \sin\theta & \cos\theta\end{bmatrix},$$

hence the relation between the two bases is

$$\begin{bmatrix}\vec{n}\\ \vec{n_1}\\ \vec{z_\lambda}\end{bmatrix}=M(1,2)\begin{bmatrix}\vec{n}\\ \vec{w}\\ \vec{z_S}\end{bmatrix}.$$

If the vector $\overrightarrow{O_\lambda M}$ is expressed in the basis (2) by $\overrightarrow{O_\lambda M}=x_2\vec{n}+y_2\vec{w}+z_2\vec{z_S}$, then the matrix form is obtained:

$$\begin{bmatrix}x_1\\ y_1\\ z_1\end{bmatrix}=\begin{bmatrix}1 & 0 & 0\\ 0 & \cos\theta & -\sin\theta\\ 0 & \sin\theta & \cos\theta\end{bmatrix}\times\begin{bmatrix}x_2\\ y_2\\ z_2\end{bmatrix}=M(1,2)\begin{bmatrix}x_2\\ y_2\\ z_2\end{bmatrix},$$

and moreover

$$M^{-1}(1,2)=M(2,1)=\overline{M(1,2)}.$$

1.4.4.3. *The third rotation by angle φ, or spin*

This rotation about the axis $\vec{z_S}$, called spin, makes the passage from the basis (2) to the basis (S) according to Figure 1.22.

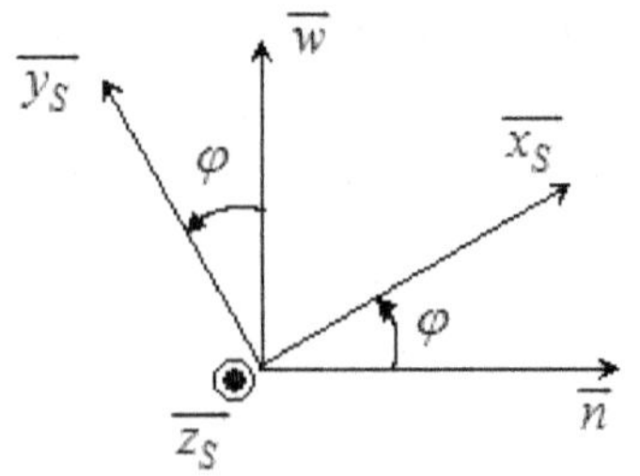

Figure 1.22. *Spin*

hence the table of the change of basis and the associated matrix are

$(2)\|(S)$	$\overrightarrow{x_S}$	$\overrightarrow{y_S}$	$\overrightarrow{z_S}$
$\vec{n}$	$\cos\varphi$	$-\sin\varphi$	0
$\vec{w}$	$\sin\varphi$	$\cos\varphi$	0
$\overrightarrow{z_S}$	0	0	1

,

$$M(2,S)=\begin{bmatrix}\cos\varphi & -\sin\varphi & 0\\ \sin\varphi & \cos\varphi & 0\\ 0 & 0 & 1\end{bmatrix},$$

and the two matrix relations are

$$\begin{bmatrix}\vec{n}\\ \vec{w}\\ \overrightarrow{z_S}\end{bmatrix}=M(2,S)\begin{bmatrix}\overrightarrow{x_S}\\ \overrightarrow{y_S}\\ \overrightarrow{z_S}\end{bmatrix}$$

$$\begin{bmatrix}x_2\\ y_2\\ z_2\end{bmatrix}=\begin{bmatrix}\cos\varphi & -\sin\varphi & 0\\ \sin\varphi & \cos\varphi & 0\\ 0 & 0 & 1\end{bmatrix}\times\begin{bmatrix}x_S\\ y_S\\ z_S\end{bmatrix}=M(2,S)\begin{bmatrix}x_S\\ y_S\\ z_S\end{bmatrix},$$

and moreover

$$M^{-1}(2,S)=M(S,2)=\overline{M(2,S)}.$$

1.4.4.4. *Relation between (λ) and (S) bases*

Based on these results, the following relations between the vectors of the two bases are obtained:

$$\begin{bmatrix}\overrightarrow{x_\lambda}\\ \overrightarrow{y_\lambda}\\ \overrightarrow{z_\lambda}\end{bmatrix}=M(\lambda,1)\times M(1,2)\times M(2,S)\begin{bmatrix}\overrightarrow{x_S}\\ \overrightarrow{y_S}\\ \overrightarrow{z_S}\end{bmatrix}=M(\lambda,S)\begin{bmatrix}\overrightarrow{x_S}\\ \overrightarrow{y_S}\\ \overrightarrow{z_S}\end{bmatrix}$$

or

$$\begin{bmatrix}\overrightarrow{x_\lambda}\\ \overrightarrow{y_\lambda}\\ \overrightarrow{z_\lambda}\end{bmatrix}=\begin{bmatrix}\cos\psi\cos\varphi-\sin\psi\cos\theta\sin\varphi & -\cos\psi\sin\varphi-\sin\psi\cos\theta\cos\varphi & \sin\psi\sin\theta\\ \sin\psi\cos\varphi+\cos\psi\cos\theta\sin\varphi & -\sin\psi\sin\varphi+\cos\psi\cos\theta\cos\varphi & -\cos\psi\sin\theta\\ \sin\theta\sin\varphi & \sin\theta\cos\varphi & \cos\theta\end{bmatrix}\begin{bmatrix}\overrightarrow{x_S}\\ \overrightarrow{y_S}\\ \overrightarrow{z_S}\end{bmatrix},$$

Therefore, the relation between the components of vector $\overrightarrow{O_\lambda M}$ in these two bases is:

$$\begin{bmatrix}x_\lambda\\ y_\lambda\\ z_\lambda\end{bmatrix}=M(\lambda,1)\times M(1,2)\times M(2,S)\begin{bmatrix}x_S\\ y_S\\ z_S\end{bmatrix},$$

which leads to the relation between the passage matrices

$$M(\lambda,S)=M(\lambda,1)\times M(1,2)\times M(2,S).$$

NOTE 1.6.– To ensure coherence of the terms of a matrix of change of basis, and considering that this is a relation between unit vectors, the sum of squared terms of the same row or the same column of this matrix should always be equal to 1. For example, it can be verified that $\overrightarrow{x_\lambda}^2=1$ or

$$(\cos\psi\cos\varphi - \sin\psi\cos\theta\sin\varphi)^2 + (-\cos\psi\sin\varphi - \sin\psi\cos\theta\cos\varphi)^2 \ldots$$
$$\ldots + \sin^2\psi\sin^2\theta = 1$$

and the sum of the member-by-member products of the terms of two rows or of two columns should be zero in order to express that the basis is orthogonal, for example $\overrightarrow{y_\lambda}\cdot\overrightarrow{z_\lambda} = 0$ or

$$(\sin\psi\cos\varphi + \cos\psi\cos\theta\sin\varphi)\sin\theta\sin\varphi \ldots$$
$$\ldots + (-\sin\psi\sin\varphi + \cos\psi\cos\theta\cos\varphi)\sin\theta\cos\varphi - \cos\psi\sin\theta\cos\theta = 0 \, .$$

1.4.5. *Exercises*

1.4.5.1. *Exercise 3 – Euler angles*

Let (S) be a solid and two vectors $\overrightarrow{x_S}$ and $\overrightarrow{y_S}$ associated with it in the direct orthonormal basis $(\lambda) \equiv \left(\overrightarrow{x_\lambda}\,\overrightarrow{y_\lambda}\,\overrightarrow{z_\lambda}\right)$.

Question 1(a): Given the vectors

$$\overrightarrow{x_S} = \frac{\sqrt{6}}{4}\overrightarrow{x_\lambda} + \frac{\sqrt{2}}{4}\overrightarrow{y_\lambda} + \frac{\sqrt{2}}{2}\overrightarrow{z_\lambda}; \quad \overrightarrow{y_S} = -\frac{\sqrt{6}}{4}\overrightarrow{x_\lambda} - \frac{\sqrt{2}}{4}\overrightarrow{y_\lambda} + \frac{\sqrt{2}}{2}\overrightarrow{z_\lambda}$$

verify that they are unit and orthogonal vectors. Calculate $\overrightarrow{z_S} = \overrightarrow{x_S} \wedge \overrightarrow{y_S}$.

$\overrightarrow{x_S}$ and $\overrightarrow{y_S}$ are unit vectors $\Rightarrow \overrightarrow{x_S}^2 = 1$, $\overrightarrow{y_S}^2 = 1$,

$\overrightarrow{x_S}$ and $\overrightarrow{y_S}$ are orthogonal $\overrightarrow{x_S}\cdot\overrightarrow{y_S} = 0$

$$\frac{6}{16} + \frac{2}{16} + \frac{2}{4} = \frac{16}{16} = 1 \quad ; \quad -\frac{\sqrt{6}}{4}\times\frac{\sqrt{6}}{4} - \frac{\sqrt{2}}{4}\times\frac{\sqrt{2}}{4} + \frac{\sqrt{2}}{2}\times\frac{\sqrt{2}}{2} = 0$$

Therefore, the two vectors are unit and orthogonal vectors

$$\vec{z_S} = \vec{x_S} \wedge \vec{y_S} = \left(\frac{\sqrt{6}}{4}\vec{x_\lambda} + \frac{\sqrt{2}}{4}\vec{y_\lambda} + \frac{\sqrt{2}}{2}\vec{z_\lambda} \right) \wedge \left(-\frac{\sqrt{6}}{4}\vec{x_\lambda} - \frac{\sqrt{2}}{4}\vec{y_\lambda} + \frac{\sqrt{2}}{2}\vec{z_\lambda} \right).$$

$$'' \quad '' \qquad = \frac{1}{2}\vec{x_\lambda} - \frac{\sqrt{3}}{2}\vec{y_\lambda}$$

It can be verified that this vector is also a unit vector orthogonal to $\vec{x_S}$ and $\vec{y_S}$.

Question 1(b): Write the matrix $[p(\lambda, S)]$ of passage from (λ) to the basis $(S) \equiv \left(\vec{x_S}\ \vec{y_S}\ \vec{z_S}\right)$ connected to the solid (S).

The table of passage from the basis (λ) to the basis (S) can be written:

$p(\lambda, S) \quad \Leftrightarrow$

	$\vec{x_S}$	$\vec{y_S}$	$\vec{z_S}$
$\vec{x_\lambda}$	$\frac{\sqrt{6}}{4}$	$-\frac{\sqrt{6}}{4}$	$\frac{1}{2}$
$\vec{y_\lambda}$	$\frac{\sqrt{2}}{4}$	$-\frac{\sqrt{2}}{4}$	$-\frac{\sqrt{3}}{2}$
$\vec{z_\lambda}$	$\frac{\sqrt{2}}{2}$	$\frac{\sqrt{2}}{2}$	0

,

Therefore, the matrix of passage is

$$[p(\lambda, S)] = \begin{bmatrix} \frac{\sqrt{6}}{4} & -\frac{\sqrt{6}}{4} & \frac{1}{2} \\ \frac{\sqrt{2}}{4} & \frac{\sqrt{2}}{4} & -\frac{\sqrt{3}}{2} \\ \frac{\sqrt{2}}{2} & \frac{\sqrt{2}}{2} & 0 \end{bmatrix}.$$

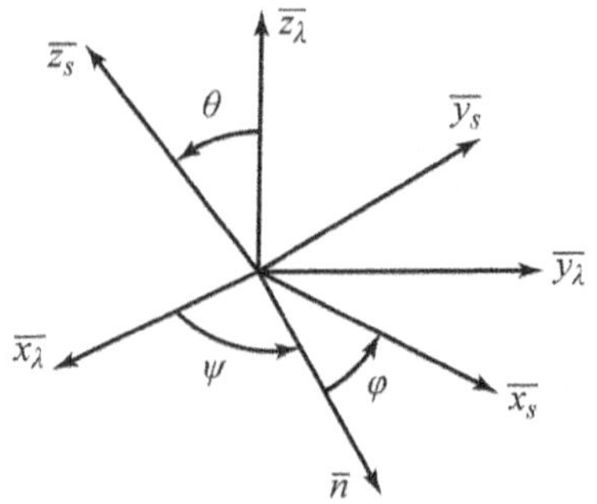

Question 1(c): Define the orientation of the basis (S) relative to (λ) by means of the Euler angles. Calculate the value of these angles, and specify the nodal vector $\vec{n}$:

– $\overrightarrow{z_\lambda} \cdot \overrightarrow{z_S} = \cos\theta = 0 \quad \Rightarrow \quad \theta = \frac{\pi}{2} + k\pi \quad$ with $\quad \theta \in [-\pi, \pi]$,

$$\Rightarrow \begin{cases} \theta = \dfrac{\pi}{2} & \Rightarrow \quad \sin\theta = 1 \\ \theta = -\dfrac{\pi}{2} & \Rightarrow \quad \sin\theta = -1 \end{cases},$$

– $\overrightarrow{z_\lambda} \wedge \overrightarrow{z_S} = \vec{n}\sin\theta$

$$\overrightarrow{z_\lambda} \wedge \left(\frac{1}{2}\overrightarrow{x_\lambda} - \frac{\sqrt{3}}{2}\overrightarrow{y_\lambda}\right) = \frac{\sqrt{3}}{2}\overrightarrow{x_\lambda} + \frac{1}{2}\overrightarrow{y_\lambda} = \cos\frac{\pi}{6}\overrightarrow{x_\lambda} + \sin\frac{\pi}{6}\overrightarrow{y_\lambda} = \vec{u}\left(\frac{\pi}{6}\right) = \vec{n}\sin\theta$$

$$\Rightarrow \begin{cases} \sin\theta = 1 & \Rightarrow \quad \vec{n} = \cos\dfrac{\pi}{6}\overrightarrow{x_\lambda} + \sin\dfrac{\pi}{6}\overrightarrow{y_\lambda} = \vec{u}\left(\dfrac{\pi}{6}\right) \\ \sin\theta = -1 & \Rightarrow \quad \vec{n} = \cos\left(\pi + \dfrac{\pi}{6}\right)\overrightarrow{x_\lambda} + \sin\left(\pi + \dfrac{\pi}{6}\right)\overrightarrow{y_\lambda} = \vec{u}\left(\pi + \dfrac{\pi}{6}\right) \end{cases},$$

or

$$\frac{\sqrt{2}}{2}\left(\overrightarrow{x_S} + \overrightarrow{y_S}\right) \wedge \overrightarrow{z_S} = \frac{\sqrt{2}}{2}\overrightarrow{x_S} - \frac{\sqrt{2}}{2}\overrightarrow{y_S} = \cos\frac{\pi}{4}\overrightarrow{x_S} - \sin\frac{\pi}{4}\overrightarrow{y_S} = \vec{n}\sin\theta$$

$$\Rightarrow \begin{cases} \sin\theta = 1 & \Rightarrow \quad \vec{n} = \cos\dfrac{\pi}{4}\overrightarrow{x_S} - \sin\dfrac{\pi}{4}\overrightarrow{y_S} \\ \sin\theta = -1 & \Rightarrow \quad \vec{n} = \cos\left(\pi + \dfrac{\pi}{4}\right)\overrightarrow{x_S} - \sin\left(\pi + \dfrac{\pi}{4}\right)\overrightarrow{y_S} \end{cases},$$

$$-\begin{cases} \vec{n}\cdot\vec{x_\lambda} = \cos\psi = \begin{cases} \cos\dfrac{\pi}{6} \\ \cos\left(\pi+\dfrac{\pi}{6}\right) \end{cases} \\ \vec{n}\cdot\vec{y_\lambda} = \sin\psi = \begin{cases} \sin\dfrac{\pi}{6} \\ \sin\left(\pi+\dfrac{\pi}{6}\right) \end{cases} \end{cases} \Rightarrow \begin{cases} \sin\theta = 1 \ , \ \psi = \dfrac{\pi}{6} + 2k\pi \\ \sin\theta = -1 \ , \ \psi = \pi + \dfrac{\pi}{6} + 2k\pi \end{cases},$$

$$-\begin{cases} \vec{n}\cdot\vec{x_S} = \cos\varphi = \begin{cases} \cos\dfrac{\pi}{4} \\ \cos\left(\pi+\dfrac{\pi}{4}\right) \end{cases} \\ \vec{n}\cdot\vec{y_S} = \sin\varphi = \begin{cases} \sin\dfrac{\pi}{4} \\ \sin\left(\pi+\dfrac{\pi}{4}\right) \end{cases} \end{cases} \Rightarrow \begin{cases} \sin\theta = 1 \ , \ \varphi = \dfrac{\pi}{4} + 2k\pi \\ \sin\theta = -1 \ , \ \phi = \pi + \dfrac{\pi}{4} + 2k\pi \end{cases}.$$

Question 2: Answer the same questions for the vectors

$$\vec{x_S} = \frac{1}{\sqrt{3}}\vec{x_\lambda} - \frac{1}{\sqrt{3}}\vec{y_\lambda} - \frac{1}{\sqrt{3}}\vec{z_\lambda}; \quad \vec{y_S} = \frac{1}{\sqrt{2}}\vec{x_\lambda} + \frac{1}{\sqrt{2}}\vec{z_\lambda}.$$

Using the same approach employed above, it can be verified that

$$\vec{x_S}^2 = \frac{3}{3} = 1 \ , \ \vec{y_S}^2 = \frac{2}{2} = 1 \ , \ \vec{x_S}\cdot\vec{y_S} = \frac{1}{\sqrt{6}} - \frac{1}{\sqrt{6}} = 0,$$

$$\vec{z_S} = \vec{x_S} \wedge \vec{y_S} = -\frac{1}{\sqrt{6}}\vec{x_\lambda} - \frac{2}{\sqrt{6}}\vec{y_\lambda} + \frac{1}{\sqrt{6}}\vec{z_\lambda}$$

$$[p(\lambda,S)] = \begin{bmatrix} \dfrac{1}{\sqrt{3}} & \dfrac{1}{\sqrt{2}} & -\dfrac{1}{\sqrt{6}} \\ -\dfrac{1}{\sqrt{3}} & 0 & -\dfrac{2}{\sqrt{6}} \\ -\dfrac{1}{\sqrt{3}} & \dfrac{1}{\sqrt{2}} & \dfrac{1}{\sqrt{6}} \end{bmatrix},$$

$$\vec{z_\lambda}\cdot\vec{z_S} = \cos\theta = \frac{1}{\sqrt{6}} = 0{,}408 \quad , \quad \sin\theta = \pm\sqrt{1-\cos^2\theta} = \pm\sqrt{\frac{5}{6}} = \pm 0{,}913$$

$$\vec{z_\lambda}\wedge\vec{z_S} = \vec{n}\sin\theta = \frac{2}{\sqrt{6}}\vec{x_\lambda} - \frac{1}{\sqrt{6}}\vec{y_\lambda} = \frac{1}{\sqrt{2}}\vec{x_S} + \frac{1}{\sqrt{3}}\vec{y_s} \quad ,$$

$$\Rightarrow \quad \theta = 1{,}15\, rd\, .$$

$$\begin{cases} \vec{n}\cdot\vec{x_\lambda} = \cos\psi = \dfrac{2}{\sqrt{6}\sin\theta} = 0.894 \\ \vec{n}\cdot\vec{y_\lambda} = \sin\psi = -\dfrac{1}{\sqrt{6}\sin\theta} = -0{,}447 \end{cases} \quad \begin{cases} \vec{n}\cdot\vec{x_S} = \cos\varphi = \dfrac{1}{\sqrt{2}\sin\theta} = 0{,}775 \\ \vec{n}\cdot\vec{y_S} = -\sin\varphi = \dfrac{1}{\sqrt{3}\sin\theta} = -0{,}632 \end{cases} ,$$

$$\Rightarrow \quad \psi = 0{,}465\, rd, \qquad \varphi = -0{,}685\, rd$$

1.4.5.2. *Exercise 4 – Euler angles*

The basis (e) connected to the cutting face of a tool is located relative to the basis (r) connected to the body of the tool, by means of the angles:

– k, cutting edge angle measured over $\vec{z}$;

– λ, cutting edge inclination angle measured over $\vec{u}(k)$;

– γ, *cutting angle* measured over $\vec{x_c}$,

as indicated by the diagrams below.

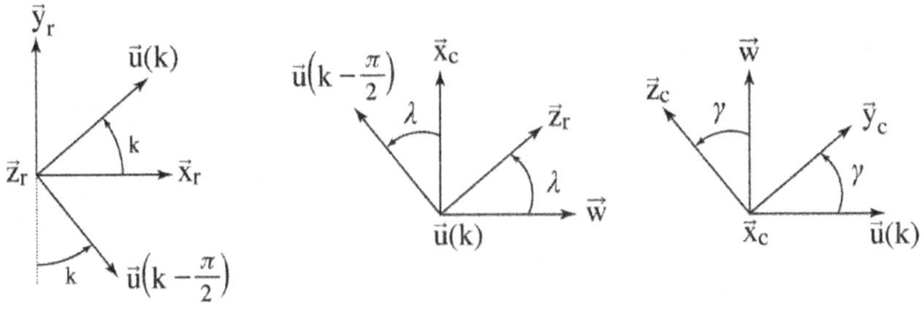

The following passage tables can be associated with these three changes of basis:

	$\overrightarrow{x_r}$	$\overrightarrow{y_r}$	$\overrightarrow{z_r}$
$\vec{u}\left(k-\frac{\pi}{2}\right)$	$\sin k$	$-\cos k$	0
$\vec{u}(k)$	$\cos k$	$\sin k$	0
$\overrightarrow{z_r}$	0	0	1

	$\vec{u}(k)$	$\overrightarrow{z_r}$	$\vec{u}\left(k-\frac{\pi}{2}\right)$
$\vec{u}(k)$	1	0	0
$\vec{w}$	0	$\cos\lambda$	$-\sin\lambda$
$\overrightarrow{x_c}$	0	$\sin\lambda$	$\cos\lambda$

,

	$\overrightarrow{x_c}$	$\vec{u}(k)$	$\vec{w}$
$\overrightarrow{x_c}$	1	0	0
$\overrightarrow{y_c}$	0	$\cos\gamma$	$\sin\gamma$
$\overrightarrow{z_c}$	0	$-\sin\gamma$	$\cos\gamma$

.

Question 1: Determine the vectors $\overrightarrow{x_r}, \overrightarrow{y_r}, \overrightarrow{z_r}$ by their components in the basis $(c) \equiv \left(\overrightarrow{x_c}\ \overrightarrow{y_c}\ \overrightarrow{z_c}\right)$ using the angle parameters k, λ, γ.

According to the previous tables of change of basis

$$\overrightarrow{x_r} = \cos k\,\vec{u}(k) + \sin k\,\vec{u}\left(k-\frac{\pi}{2}\right),$$

$$\overrightarrow{y_r} = \sin k\,\vec{u}(k) - \cos k\,\vec{u}\left(k-\frac{\pi}{2}\right)$$

$$\overrightarrow{z_r} = \cos\lambda\,\vec{w} + \sin\lambda\,\overrightarrow{x_c}$$

$$\vec{u}\left(k-\frac{\pi}{2}\right) = -\sin\lambda\,\vec{w} + \cos\lambda\,\overrightarrow{x_c},$$

$$\begin{array}{l}\vec{u}(k)=\cos\gamma\,\overrightarrow{y_c}-\sin\gamma\,\overrightarrow{z_c}\\ \vec{w}=\sin\gamma\,\overrightarrow{y_c}+\cos\gamma\,\overrightarrow{z_c}\end{array},$$

$$\Rightarrow\begin{cases}\overrightarrow{x_r}=\sin k\cos\lambda\,\overrightarrow{x_c}+(\cos k\cos\gamma-\sin k\sin\lambda\sin\gamma)\,\overrightarrow{y_c}\cdots\\ \qquad\cdots-(\cos k\sin\gamma+\sin k\sin\lambda\cos\gamma)\,\overrightarrow{z_c}\\ \overrightarrow{y_r}=-\cos k\cos\lambda\,\overrightarrow{x_c}+(\sin k\cos\gamma+\cos k\sin\lambda\sin\gamma)\,\overrightarrow{y_c}\cdots\\ \qquad\cdots-(\sin k\sin\gamma-\cos k\sin\lambda\cos\gamma)\,\overrightarrow{z_c}\\ \overrightarrow{z_r}=\sin\lambda\,\overrightarrow{x_c}+\cos\lambda\sin\gamma\,\overrightarrow{y_c}+\cos\lambda\cos\gamma\,\overrightarrow{z_c}\end{cases}.$$

Question 2(a): Next, let us consider $k=\frac{\pi}{4}$, $\lambda=0$, $\gamma=\frac{\pi}{6}$. Establish the passage table of change of basis from (r) to (c).

The change of basis from (r) to (c) is expressed by the table

	$\overrightarrow{x_c}$	$\overrightarrow{y_c}$	$\overrightarrow{z_c}$
$\overrightarrow{x_r}$	$\sin k\cos\lambda$	$\cos k\cos\gamma-\sin k\sin\lambda\sin\gamma$	$-\cos k\sin\gamma-\sin k\sin\lambda\cos\gamma$
$\overrightarrow{y_r}$	$-\cos k\cos\lambda$	$\sin k\cos\gamma+\cos k\sin\lambda\sin\gamma$	$-\sin k\sin\gamma+\cos k\sin\lambda\cos\gamma$,
$\overrightarrow{z_r}$	$\sin\lambda$	$\cos\lambda\sin\gamma$	$\cos\lambda\cos\gamma$

which for the given particular values is written,

	$\overrightarrow{x_c}$	$\overrightarrow{y_c}$	$\overrightarrow{z_c}$
$\overrightarrow{x_r}$	$\frac{\sqrt{2}}{2}$	$\frac{\sqrt{6}}{4}$	$-\frac{\sqrt{2}}{4}$
$\overrightarrow{y_r}$	$-\frac{\sqrt{2}}{2}$	$\frac{\sqrt{6}}{4}$	$-\frac{\sqrt{2}}{4}$.
$\overrightarrow{z_r}$	0	$\frac{1}{2}$	$\frac{\sqrt{3}}{2}$

Question 2(b): Deduce from the previous table the value of the angles of precession ψ, nutation θ and spin φ of the Eulerian representation of the orientation of basis (c) relative to (r). Determine the nodal vector $\vec{n}$.

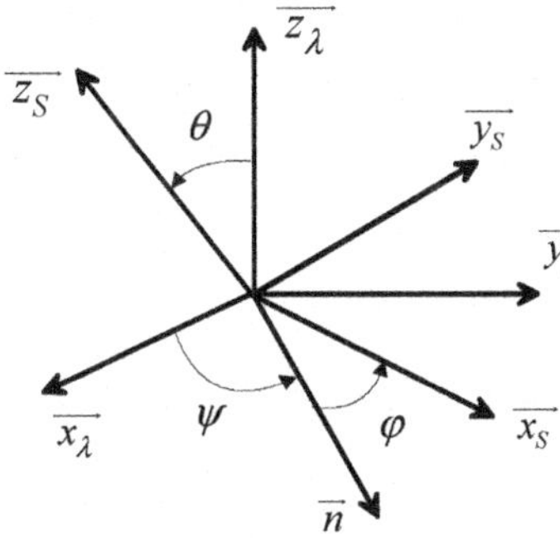

By analogy with this diagram of the principle of definition of the Euler angles, we have

$$\begin{cases} \overrightarrow{z_c} \wedge \overrightarrow{z_r} = \vec{n}\sin\theta \\ \overrightarrow{z_c} \cdot \overrightarrow{z_r} = \cos\theta \end{cases}, \quad \begin{cases} \overrightarrow{x_c} \cdot \vec{n} = \cos\psi \\ \overrightarrow{y_c} \cdot \vec{n} = \sin\psi \end{cases}, \quad \begin{cases} \vec{n} \cdot \overrightarrow{x_r} = \cos\varphi \\ \vec{n} \cdot \overrightarrow{y_r} = -\sin\varphi \end{cases}.$$

According to the previous table

$$\begin{cases} \overrightarrow{z_c} \wedge \overrightarrow{z_r} = \frac{\sqrt{2}}{4}\overrightarrow{y_r} - \frac{\sqrt{2}}{4}\overrightarrow{x_r} = -\frac{1}{2}\overrightarrow{x_c} = \vec{n}\sin\theta \Rightarrow \vec{n} = \pm\overrightarrow{x_c} \quad , \quad \sin\theta = \mp\frac{1}{2} \\ \overrightarrow{z_c} \cdot \overrightarrow{z_r} = \frac{\sqrt{3}}{2} = \cos\theta \end{cases},$$

$$\begin{cases} \overrightarrow{x_c} \cdot \vec{n} = \pm 1 = \cos\psi \\ \overrightarrow{y_c} \cdot \vec{n} = 0 = \sin\psi \end{cases},$$

$$\begin{cases} \vec{n} \cdot \overrightarrow{x_r} = \pm\frac{\sqrt{2}}{2} = \cos\varphi \\ \vec{n} \cdot \overrightarrow{y_r} = \mp\frac{\sqrt{2}}{2} = -\sin\varphi \end{cases}.$$

These results lead to two sets of values

$$-\ \vec{n} = \vec{x}_c,\ \ \theta = -\frac{\pi}{6},\ \ \psi = 0,\ \ \varphi = \frac{\pi}{4},$$

$$-\ \vec{n} = -\vec{x}_c,\ \ \theta = \frac{\pi}{6},\ \ \psi = \pi,\ \ \varphi = -\frac{3\pi}{4}.$$

1.5. Vector rotation $\mathscr{R}_{\vec{u},\alpha}$

Let $\mathscr{R}_{\vec{u},\alpha}$ be the rotation by angle α of a vector $\vec{a}$ about the axis Δ with the direction vector $\vec{u}$. Let $\vec{b}$ be the vector resulting from the rotation of $\vec{a}$.

Let Π be the plane perpendicular to vector $\vec{u}$ and passing through the point A, the origin of vector $\vec{a}$. The projection on the axis $\vec{u}$ and on plane Π can be written

$$\begin{aligned} \vec{a} &= \left(\vec{u}\cdot\vec{a}\right)\vec{u} + \vec{u}\wedge\left(\vec{a}\wedge\vec{u}\right) \\ \vec{b} &= \left(\vec{u}\cdot\vec{b}\right)\vec{u} + \vec{u}\wedge\left(\vec{b}\wedge\vec{u}\right) \end{aligned}.$$

Since $\vec{b}$ results from $\vec{a}$ by rotation about the axis $\left(A|\vec{u}\right)$, the two vectors have the same orthogonal projection on this axis, which gives

$$\left(\vec{u}\cdot\vec{a}\right)\vec{u} = \left(\vec{u}\cdot\vec{b}\right)\vec{u}.$$

The projections of the two vectors on the plane Π, orthogonal to $\vec{u}$, have the same module.

$$\left\|\vec{u}\wedge\left(\vec{a}\wedge\vec{u}\right)\right\| = \left\|\vec{u}\wedge\left(\vec{b}\wedge\vec{u}\right)\right\|.$$

The problem posed is to determine the vector $\vec{b}$ as a function of the known elements, namely $\vec{a}$, $\vec{u}$ and α. Taking into account the above, this comes down to determining the expression $\vec{u}\wedge\left(\vec{b}\wedge\vec{u}\right)$.

Let us then consider the local system of reference connected to the two vectors $\vec{u}$ and $\vec{a}$, whose unit vectors are

$$\vec{u}, \quad \vec{k} = \frac{\vec{u} \wedge \left(\vec{a} \wedge \vec{u}\right)}{\left\| \vec{u} \wedge \left(\vec{a} \wedge \vec{u}\right) \right\|}, \quad \vec{l} = \frac{\vec{u} \wedge \vec{a}}{\left\| \vec{u} \wedge \vec{a} \right\|},$$

The three vectors $\vec{u}$, $\vec{a}$ and $\vec{k}$ are indeed coplanar. Therefore, $\vec{u} \wedge \vec{a}$ defines a direction orthogonal to this plane, which leads to the vector $\vec{l}$.

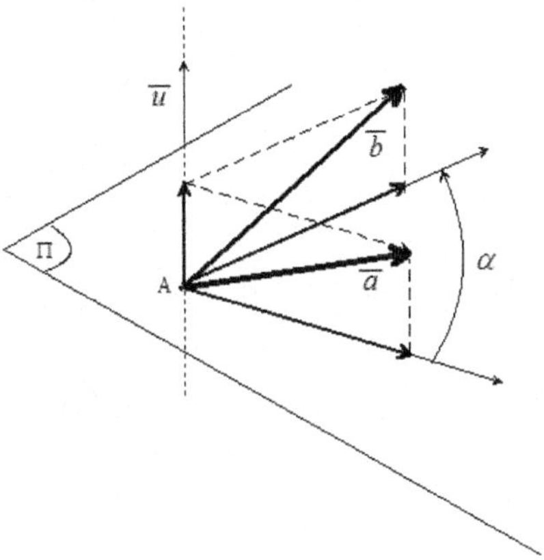

Figure 1.23. *Vector rotation* $\mathcal{R}_{\overline{u},\alpha}$

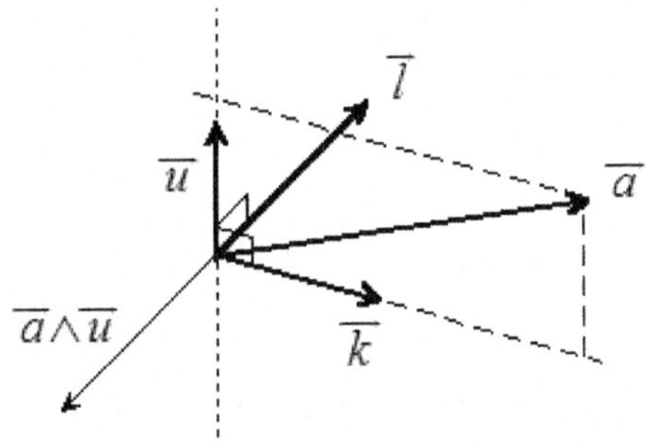

Figure 1.24. *System of reference associated with the vector rotation* $\mathcal{R}_{\overline{u},\alpha}$

It can be noted that

$$\left\|\vec{u} \wedge \left(\vec{a} \wedge \vec{u}\right)\right\| = \left\|\vec{u}\right\| \times \left\|\vec{a} \wedge \vec{u}\right\| \times \left|\sin\left(\vec{u}, \vec{a} \wedge \vec{u}\right)\right|,$$

with $\vec{u} \perp \left(\vec{a} \wedge \vec{u}\right)$ and $\left\|\vec{u}\right\| = 1$,

hence, $\left\|\vec{u} \wedge \left(\vec{a} \wedge \vec{u}\right)\right\| = \left\|\vec{a} \wedge \vec{u}\right\| = \left\|\vec{u} \wedge \vec{a}\right\|$.

Since, the vector to be determined $\vec{u} \wedge \left(\vec{b} \wedge \vec{u}\right)$ is orthogonal to $\vec{u}$, it belongs to the plane $\Pi\left(\vec{k}, \vec{l}\right)$; it can, therefore, be projected orthogonally on these two vectors, which leads to

$$\vec{u} \wedge \left(\vec{b} \wedge \vec{u}\right) = \left\{\left[\vec{u} \wedge \left(\vec{b} \wedge \vec{u}\right)\right] \cdot \vec{k}\right\}\vec{k} + \left\{\left[\vec{u} \wedge \left(\vec{b} \wedge \vec{u}\right)\right] \cdot \vec{l}\right\}\vec{l},$$

with $\left\|\vec{u} \wedge \left(\vec{b} \wedge \vec{u}\right)\right\| = \left\|\vec{u} \wedge \left(\vec{a} \wedge \vec{u}\right)\right\| = \left\|\vec{u} \wedge \vec{a}\right\|$ since $\left\|\vec{a}\right\| = \left\|\vec{b}\right\|$,

$$\Rightarrow \quad \vec{u} \wedge \left(\vec{b} \wedge \vec{u}\right) = \cos\alpha \left\|\vec{u} \wedge \left(\vec{a} \wedge \vec{u}\right)\right\|\vec{k} + \sin\alpha \left\|\vec{u} \wedge \vec{a}\right\|\vec{l}$$

$$\text{'' ''} \qquad = \cos\alpha\left[\vec{u} \wedge \left(\vec{a} \wedge \vec{u}\right)\right] + \sin\alpha\left[\vec{u} \wedge \vec{a}\right].$$

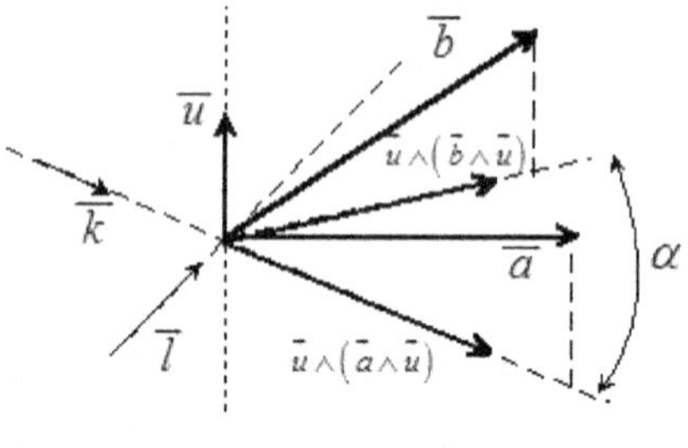

Figure 1.25. *Diagram of projection in the vector rotation* $\mathcal{R}_{\overline{u},\alpha}$

This leads us to deduce the general expression of $\vec{b}$

$$\vec{b} = \left(\vec{u} \cdot \vec{a}\right)\vec{u} + \left[\vec{u} \wedge \left(\vec{a} \wedge \vec{u}\right)\right]\cos\alpha + \left(\vec{u} \wedge \vec{a}\right)\sin\alpha.$$

But, since

$$\left(\vec{u}\cdot\vec{a}\right)\vec{u} = \vec{a} - \vec{u}\wedge\left(\vec{a}\wedge\vec{u}\right) = \vec{a} + \vec{u}\wedge\left(\vec{u}\wedge\vec{a}\right),$$

the vector $\vec{b}$, deduced from vector $\vec{a}$ by the rotation by angle α about the axis of the direction vector $\vec{u}$, has the general expression

$$\vec{b} = \mathcal{R}_{\vec{u},\alpha}\left(\vec{a}\right) = \vec{a} + \left[\vec{u}\wedge\left(\vec{u}\wedge\vec{a}\right)\right](1-\cos\alpha) + \left(\vec{u}\wedge\vec{a}\right)\sin\alpha .$$

The matrix representation $[\mathbf{F}]$ of the vector rotation by angle α about the axis of unit vector $\vec{u} = u_i\vec{x_i}$ can be written in the basis $(e) \equiv \left(\vec{x_1}\,\vec{x_2}\,\vec{x_3}\right)$

$$[\mathbf{F}]/_e = \begin{bmatrix} \cos\alpha + (1-\cos\alpha)u_1^2 & (1-\cos\alpha)u_1u_2 - \sin\alpha\, u_3 & (1-\cos\alpha)u_3u_1 + \sin\alpha\, u_2 \\ (1-\cos\alpha)u_1u_2 + \sin\alpha\, u_3 & \cos\alpha + (1-\cos\alpha)u_2^2 & (1-\cos\alpha)u_2u_3 - \sin\alpha\, u_1 \\ (1-\cos\alpha)u_3u_1 - \sin\alpha\, u_2 & (1-\cos\alpha)u_2u_3 + \sin\alpha\, u_1 & \cos\alpha + (1-\cos\alpha)u_3^2 \end{bmatrix},$$

which, by posing $\vec{a} = a_i\vec{x_i}$ and $\vec{b} = b_i\vec{x_i}$, leads us to write

$$\begin{bmatrix} b_1 \\ b_2 \\ b_3 \end{bmatrix} = \begin{bmatrix} \cos\alpha + (1-\cos\alpha)u_1^2 & (1-\cos\alpha)u_1u_2 - \sin\alpha\, u_3 & (1-\cos\alpha)u_3u_1 + \sin\alpha\, u_2 \\ (1-\cos\alpha)u_1u_2 + \sin\alpha\, u_3 & \cos\alpha + (1-\cos\alpha)u_2^2 & (1-\cos\alpha)u_2u_3 - \sin\alpha\, u_1 \\ (1-\cos\alpha)u_3u_1 - \sin\alpha\, u_2 & (1-\cos\alpha)u_2u_3 + \sin\alpha\, u_1 & \cos\alpha + (1-\cos\alpha)u_3^2 \end{bmatrix}\begin{bmatrix} a_1 \\ a_2 \\ a_3 \end{bmatrix}.$$

1.5.1. *Exercises*

1.5.1.1. *Exercise 5 – Vector rotation*

In the direct orthonormal basis $\left(\vec{X_1}\,\vec{X_2}\,\vec{X_3}\right)$, let $\mathcal{R}_{\vec{u},\alpha}$ be the vector rotation represented by the matrix

$$\left[\mathcal{R}_{\vec{u},\alpha}\right] = \frac{1}{3}\begin{bmatrix} -2 & -1 & 2 \\ 2 & -2 & 1 \\ 1 & 2 & 2 \end{bmatrix}.$$

Question: Determine the unit vector $\vec{u}$ of the axis of the vector rotation and its angle α.

We pose

$$\left[\mathcal{R}_{\vec{u},\alpha}\right]_{/(X)} = \frac{1}{3}\begin{bmatrix} -2 & -1 & 2 \\ 2 & -2 & 1 \\ 1 & 2 & 2 \end{bmatrix} = \begin{bmatrix} p_{11} & p_{12} & p_{13} \\ p_{21} & p_{22} & p_{23} \\ p_{31} & p_{32} & p_{33} \end{bmatrix}.$$

This matrix has the property

$$Trace\left[\mathcal{R}_{\vec{u},\alpha}\right] = p_{ii} = -\frac{2}{3} - \frac{2}{3} + \frac{2}{3} = -\frac{2}{3} = 1 + 2\cos\alpha,$$

$$\cos\alpha = -\frac{5}{6} \text{ with } \alpha \in [0,\pi] \quad \Rightarrow \quad \alpha = 146,44°,$$

$$\sin\alpha = \sqrt{1-\cos^2\alpha} = \frac{\sqrt{11}}{6}.$$

The vector $\vec{u} = u_i \overrightarrow{X_i}$ has the following components in the basis (*X*)

$$\begin{cases} u_1 = \dfrac{p_{32} - p_{23}}{2\sin\alpha} = \dfrac{1}{3} \times \dfrac{6}{2\sqrt{11}} \times 1 = \dfrac{\sqrt{11}}{11} \\ u_2 = \dfrac{p_{13} - p_{31}}{2\sin\alpha} = \dfrac{1}{3} \times \dfrac{6}{2\sqrt{11}} \times 1 = \dfrac{\sqrt{11}}{11} \\ u_3 = \dfrac{p_{21} - p_{12}}{2\sin\alpha} = \dfrac{1}{3} \times \dfrac{6}{2\sqrt{11}} \times 3 = \dfrac{3\sqrt{11}}{11} \end{cases}.$$

In the basis (X), the vector rotation $\mathcal{R}_{\vec{u},\alpha}$ is, therefore, defined by

$$\mathcal{R}_{\vec{u},\alpha} \equiv \left[\vec{u} = \frac{\sqrt{11}}{11}\left(\overrightarrow{X_1} + \overrightarrow{X_2} + 3\overrightarrow{X_3}\right) \middle| \alpha = 146,44°\right].$$

1.5.1.2. *Exercise 6 – Vector rotation*

In the direct orthonormal basis $(X) \equiv \left(\overrightarrow{X_1}\ \overrightarrow{X_2}\ \overrightarrow{X_3}\right)$, let $\mathcal{R}_{\vec{u}(\alpha),\alpha}$ be a vector rotation with $\vec{u}(\alpha) = \cos\alpha\overrightarrow{X_1} + \sin\alpha\overrightarrow{X_2}$, which makes the passage from the basis (X) to the basis $(x) \equiv \left(\overrightarrow{x_1}\ \overrightarrow{x_2}\ \overrightarrow{x_3}\right)$, which is also a direct orthonormal basis.

Question 1: Write as a function of α the vectors $\overrightarrow{x_1}, \overrightarrow{x_2}, \overrightarrow{x_3}$ with their components in the basis (X).

The general expression of the vector rotation that leads us to deduce the vector $\vec{b}$ from $\vec{a}$ by the vector rotation $\mathscr{R}_{\vec{u},\alpha}$ is given by the relation

$$\vec{b} = \mathscr{R}_{\vec{u},\alpha}\left(\vec{a}\right) = \vec{a} + \left[\vec{u} \wedge \left(\vec{u} \wedge \vec{a}\right)\right](1 - \cos\alpha) + \left(\vec{u} \wedge \vec{a}\right)\sin\alpha .$$

As a preliminary stage in obtaining the result, we shall consider that the vector $\overrightarrow{x_i}$, $\forall i = 1,2,3$ is the transform by rotation of vector $\overrightarrow{X_i}$; it is possible to readily repeat the calculation with another correspondence hypothesis, and this would lead in the end to a coincidence of trihedrons of the resulting bases, simply by cyclic permutation of the identities of the unit vectors of the new basis.

$$\begin{cases} \overrightarrow{x_1} = \mathscr{R}_{\vec{u},\alpha}\left(\overrightarrow{X_1}\right) = \overrightarrow{X_1} + \left[\vec{u}(\alpha) \wedge \left(\vec{u}(\alpha) \wedge \overrightarrow{X_1}\right)\right](1 - \cos\alpha) + \left(\vec{u}(\alpha) \wedge \overrightarrow{X_1}\right)\sin\alpha \\ \overrightarrow{x_2} = \mathscr{R}_{\vec{u},\alpha}\left(\overrightarrow{X_2}\right) = \overrightarrow{X_2} + \left[\vec{u}(\alpha) \wedge \left(\vec{u}(\alpha) \wedge \overrightarrow{X_2}\right)\right](1 - \cos\alpha) + \left(\vec{u}(\alpha) \wedge \overrightarrow{X_2}\right)\sin\alpha \\ \overrightarrow{x_3} = \mathscr{R}_{\vec{u},\alpha}\left(\overrightarrow{X_3}\right) = \overrightarrow{X_3} + \left[\vec{u}(\alpha) \wedge \left(\vec{u}(\alpha) \wedge \overrightarrow{X_3}\right)\right](1 - \cos\alpha) + \left(\vec{u}(\alpha) \wedge \overrightarrow{X_3}\right)\sin\alpha \end{cases},$$

$$\vec{u}(\alpha) \wedge \overrightarrow{X_i} = \cos\alpha\,\overrightarrow{X_1} \wedge \overrightarrow{X_i} + \sin\alpha\,\overrightarrow{X_2} \wedge \overrightarrow{X_i},$$

$$\begin{aligned} \vec{u}(\alpha) \wedge \left(\vec{u}(\alpha) \wedge \overrightarrow{X_i}\right) = \cos^2\alpha\,\overrightarrow{X_1} \wedge \left(\overrightarrow{X_1} \wedge \overrightarrow{X_i}\right)\ldots \\ \ldots + \cos\alpha\sin\alpha\,\overrightarrow{X_1} \wedge \left(\overrightarrow{X_2} \wedge \overrightarrow{X_i}\right)\ldots \\ \ldots + \sin\alpha\cos\alpha\,\overrightarrow{X_2} \wedge \left(\overrightarrow{X_1} \wedge \overrightarrow{X_i}\right)\ldots \\ \ldots + \sin^2\alpha\,\overrightarrow{X_2} \wedge \left(\overrightarrow{X_2} \wedge \overrightarrow{X_i}\right) \end{aligned},$$

$$\vec{u}(\alpha)\wedge\left(\vec{u}(\alpha)\wedge\overrightarrow{X_i}\right)=\cos^2\alpha\left[\left(\overrightarrow{X_1}\cdot\overrightarrow{X_i}\right)\overrightarrow{X_1}-\overrightarrow{X_i}\right]$$
$$\ldots+\sin^2\alpha\left[\left(\overrightarrow{X_2}\cdot\overrightarrow{X_i}\right)\overrightarrow{X_2}-\overrightarrow{X_i}\right]\ldots$$
$$\ldots+\sin\alpha\cos\alpha\left[\left(\overrightarrow{X_2}\cdot\overrightarrow{X_i}\right)\overrightarrow{X_1}+\left(\overrightarrow{X_1}\cdot\overrightarrow{X_i}\right)\overrightarrow{X_2}\right].$$

Therefore, the following relation is obtained

$$\vec{x_i}=\mathscr{R}_{\vec{u},\alpha}\left(\overrightarrow{X_i}\right)=\overrightarrow{X_i}+\cos^2\alpha(1-\cos\alpha)\left[\left(\overrightarrow{X_1}\cdot\overrightarrow{X_i}\right)\overrightarrow{X_1}-\overrightarrow{X_i}\right]\ldots$$
$$\ldots+\sin^2\alpha(1-\cos\alpha)\left[\left(\overrightarrow{X_2}\cdot\overrightarrow{X_i}\right)\overrightarrow{X_2}-\overrightarrow{X_i}\right]\ldots$$
$$\ldots+\sin\alpha\cos\alpha(1-\cos\alpha)\left[\left(\overrightarrow{X_2}\cdot\overrightarrow{X_i}\right)\overrightarrow{X_1}+\left(\overrightarrow{X_1}\cdot\overrightarrow{X_i}\right)\overrightarrow{X_2}\right]\ldots,$$
$$\ldots+\left(\cos\alpha\overrightarrow{X_1}\wedge\overrightarrow{X_i}+\sin\alpha\overrightarrow{X_2}\wedge\overrightarrow{X_i}\right)\sin\alpha$$

or

$$\begin{cases}\vec{x_1}=\mathscr{R}_{\vec{u},\alpha}\left(\overrightarrow{X_1}\right)=\left[1-\sin^2\alpha(1-\cos\alpha)\right]\overrightarrow{X_1}\ldots\\ \qquad\ldots+\sin\alpha\cos\alpha(1-\cos\alpha)\overrightarrow{X_2}-\sin^2\alpha\overrightarrow{X_3}\\ \vec{x_2}=\mathscr{R}_{\vec{u},\alpha}\left(\overrightarrow{X_2}\right)=\sin\alpha\cos\alpha(1-\cos\alpha)\overrightarrow{X_1}\ldots\\ \qquad\ldots+\left[1-\cos^2\alpha(1-\cos\alpha)\right]\overrightarrow{X_2}+\sin\alpha\cos\alpha\overrightarrow{X_3}\\ \vec{x_3}=\mathscr{R}_{\vec{u},\alpha}\left(\overrightarrow{X_3}\right)=\sin^2\alpha\overrightarrow{X_1}-\sin\alpha\cos\alpha\overrightarrow{X_2}+\cos\alpha\overrightarrow{X_3}\end{cases}.$$

Question 2: Establish the matrix representation $[p(X,x)]$ of this vector rotation.

The matrix representation of this vector rotation is the matrix of change of basis from (X) to (x), hence

$$\left[\mathscr{R}_{\vec{u},\alpha}\right]=[p(X,x)]=\begin{bmatrix}1-\sin^2\alpha(1-\cos\alpha) & \sin\alpha\cos\alpha(1-\cos\alpha) & \sin^2\alpha\\ \sin\alpha\cos\alpha(1-\cos\alpha) & 1-\cos^2\alpha(1-\cos\alpha) & -\sin\alpha\cos\alpha\\ -\sin^2\alpha & \sin\alpha\cos\alpha & \cos\alpha\end{bmatrix}.$$

1.6. Other exercises

1.6.1. *Exercise 7 – Location of an airplane – Euler angles*

The orientation of the airplane (S) should be determined relative to a system of reference $\langle 0 \rangle \equiv \langle O_0 | \overrightarrow{x_0}\ \overrightarrow{y_0}\ \overrightarrow{z_0} \rangle$.

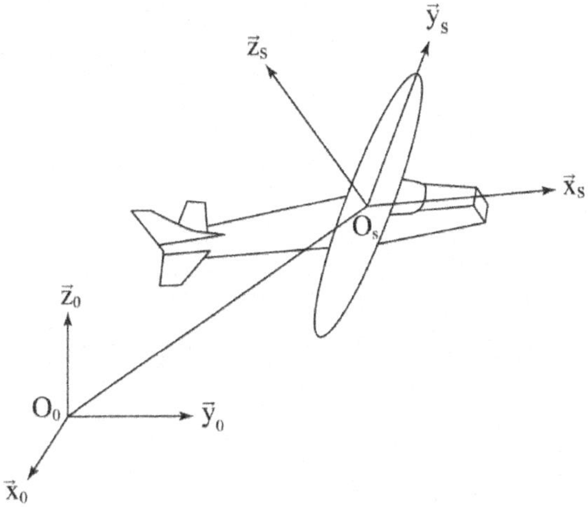

The basis connected to the airplane is defined by:

– $\overrightarrow{x_S}$, unit vector along the fuselage axis and oriented forwards;

– $\overrightarrow{y_S}$, unit vector along the longitudinal axis of the wings and oriented towards the left of the pilot so that the third axis of the trihedron directly connected to the airplane should be oriented towards the upper part of it;

– $\overrightarrow{z_S} = \overrightarrow{x_S} \wedge \overrightarrow{y_S}$, unit vector oriented "upwards" relative to the pilot.

The three angles α, β, γ that describe the orientation of the airplane are associated with three plane rotations defined below

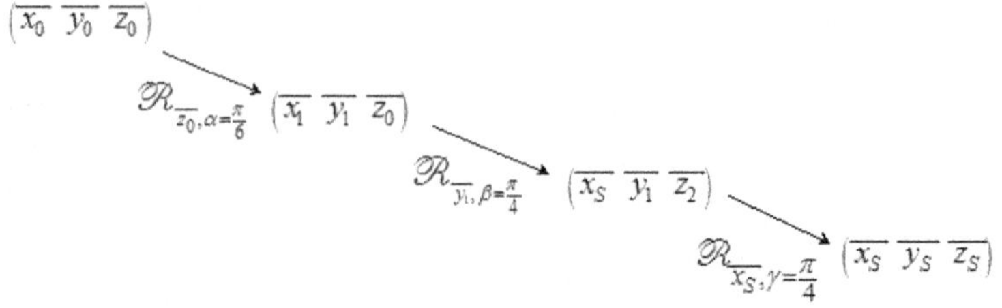

Question 1: Draw the three plane diagrams that illustrate these 3 rotations.

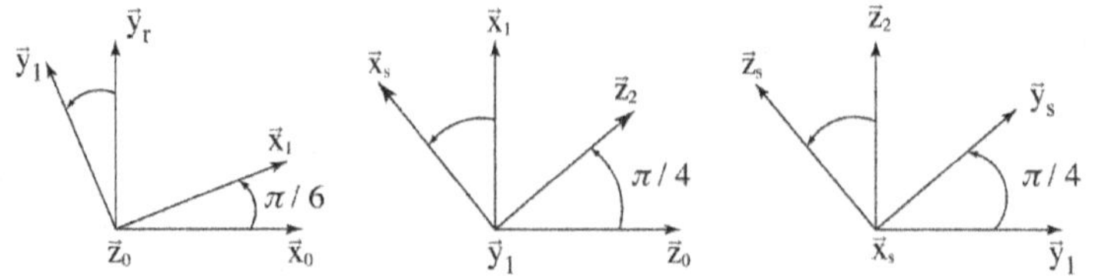

Question 2: Express $\vec{x_S}$, $\vec{y_S}$ and $\vec{z_S}$ by their components in the basis (0).

$$\vec{x_S} = \cos\frac{\pi}{4}\vec{x_1} - \sin\frac{\pi}{4}\vec{z_0} = \cos\frac{\pi}{4}\left(\cos\frac{\pi}{6}\vec{x_0} + \sin\frac{\pi}{6}\vec{y_0}\right) = \frac{\sqrt{6}}{4}\vec{x_0} + \frac{\sqrt{2}}{4}\vec{y_0} - \frac{\sqrt{2}}{2}\vec{z_0},$$

$$\vec{y_s} = \cos\frac{\pi}{4}\vec{y_1} + \sin\frac{\pi}{4}\vec{z_2} = \cos\frac{\pi}{4}\vec{y_1} + \sin\frac{\pi}{4}\left(\cos\frac{\pi}{4}\vec{z_0} + \sin\frac{\pi}{4}\vec{x_1}\right)$$

$$" = \sin^2\frac{\pi}{4}\left(\cos\frac{\pi}{6}\vec{x_0} + \sin\frac{\pi}{6}\vec{y_0}\right) + \cos\frac{\pi}{4}\left[-\sin\frac{\pi}{6}\vec{x_0} + \cos\frac{\pi}{6}\vec{y_0}\right] + \sin\frac{\pi}{4}\cos\frac{\pi}{4}\vec{z_0},$$

$$" = \left(\sin^2\frac{\pi}{4}\cos\frac{\pi}{6} - \cos\frac{\pi}{4}\sin\frac{\pi}{6}\right)\vec{x_0} + \left(\sin^2\frac{\pi}{4}\sin\frac{\pi}{6} + \cos\frac{\pi}{4}\cos\frac{\pi}{6}\right)\vec{y_0} + \sin\frac{\pi}{4}\cos\frac{\pi}{4}\vec{z_0}$$

$$\Rightarrow \vec{y_S} = \frac{\sqrt{3}-\sqrt{2}}{4}\vec{x_0} + \frac{1+\sqrt{6}}{4}\vec{y_0} + \frac{1}{2}\vec{z_0},$$

$$\vec{z}_S = -\sin\frac{\pi}{4}\vec{y}_1 + \cos\frac{\pi}{4}\vec{z}_2 = -\sin\frac{\pi}{4}\vec{y}_1 + \cos\frac{\pi}{4}\left(\cos\frac{\pi}{4}\vec{z}_0 + \sin\frac{\pi}{4}\vec{x}_1\right)$$

$$" = \sin\frac{\pi}{4}\cos\frac{\pi}{4}\left(\cos\frac{\pi}{6}\vec{x}_0 + \sin\frac{\pi}{6}\vec{y}_0\right) - \sin\frac{\pi}{4}\left[-\sin\frac{\pi}{6}\vec{x}_0 + \cos\frac{\pi}{6}\vec{y}_0\right] + \cos^2\frac{\pi}{4}\vec{z}_0,$$

$$" = \left(\sin\frac{\pi}{4}\cos\frac{\pi}{4}\cos\frac{\pi}{6} + \sin\frac{\pi}{4}\sin\frac{\pi}{6}\right)\vec{x}_0 + \left(\sin\frac{\pi}{4}\cos\frac{\pi}{4}\sin\frac{\pi}{6} - \sin\frac{\pi}{4}\cos\frac{\pi}{6}\right)\vec{y}_0 + \cos^2\frac{\pi}{4}\vec{z}_0$$

$$\Rightarrow \vec{z}_S = \frac{\sqrt{3}+\sqrt{2}}{4}\vec{x}_0 + \frac{1-\sqrt{6}}{4}\vec{y}_0 + \frac{1}{2}\vec{z}_0 .$$

Question 3: Deduce from it the matrix of passage $[p(0,S)]$.

The table of passage from the reference basis (0) to the basis (S) connected to the airplane can be written

$p(0,S) \Leftrightarrow$

	$\vec{x}_S$	$\vec{y}_S$	$\vec{z}_S$
$\vec{x}_0$	$\frac{\sqrt{6}}{4}$	$\frac{\sqrt{3}-\sqrt{2}}{4}$	$\frac{\sqrt{3}+\sqrt{2}}{4}$
$\vec{y}_0$	$\frac{\sqrt{2}}{4}$	$\frac{1+\sqrt{6}}{4}$	$\frac{1-\sqrt{6}}{4}$
$\vec{z}_0$	$-\frac{\sqrt{2}}{2}$	$\frac{1}{2}$	$\frac{1}{2}$

,

and the matrix of passage is

$$[p(0,S)] = \begin{bmatrix} \frac{\sqrt{6}}{4} & \frac{\sqrt{3}-\sqrt{2}}{4} & \frac{\sqrt{3}+\sqrt{2}}{4} \\ \frac{\sqrt{2}}{4} & \frac{1+\sqrt{6}}{4} & \frac{1-\sqrt{6}}{4} \\ -\frac{\sqrt{2}}{2} & \frac{1}{2} & \frac{1}{2} \end{bmatrix}.$$

Question 4: Determine the corresponding values of the Euler angles ψ, θ, φ for the orientation of (S) relative to (0)

$$- \ \vec{z_0} \cdot \vec{z_S} = \cos\theta = \frac{1}{2} = \cos\frac{\pi}{3} \quad \Rightarrow \quad \theta = \pm\frac{\pi}{3} + 2k\pi \quad \Rightarrow \quad \sin\theta = \pm\frac{\sqrt{3}}{2},$$

$$- \ \vec{z_0} \wedge \vec{z_S} = \vec{n}\sin\theta,$$

$$\begin{cases} \vec{z_0} \wedge \left(\dfrac{\sqrt{3}+\sqrt{2}}{4}\vec{x_0} + \dfrac{1-\sqrt{6}}{4}\vec{y_0} + \dfrac{1}{2}\vec{z_0} \right) = -\dfrac{1-\sqrt{6}}{4}\vec{x_0} + \dfrac{\sqrt{3}+\sqrt{2}}{4}\vec{y_0} = \vec{n}\sin\theta \\ \left(-\dfrac{\sqrt{2}}{2}\vec{x_S} + \dfrac{1}{2}\vec{y_S} + \dfrac{1}{2}\vec{z_S} \right) \wedge \vec{z_S} = \dfrac{1}{2}\vec{x_S} + \dfrac{\sqrt{2}}{2}\vec{y_S} = \vec{n}\sin\theta \end{cases},$$

$$\begin{cases} \text{for } \theta = \dfrac{\pi}{3}, \ \vec{n} = -\dfrac{1-\sqrt{6}}{2\sqrt{3}}\vec{x_0} + \dfrac{\sqrt{3}+\sqrt{2}}{2\sqrt{3}}\vec{y_0} = \dfrac{\sqrt{3}}{3}\vec{x_S} + \dfrac{\sqrt{6}}{3}\vec{y_S} \\ \text{for } \theta = -\dfrac{\pi}{3}, \ \vec{n} = \dfrac{1-\sqrt{6}}{2\sqrt{3}}\vec{x_0} - \dfrac{\sqrt{3}+\sqrt{2}}{2\sqrt{3}}\vec{y_0} = -\dfrac{\sqrt{3}}{3}\vec{x_S} - \dfrac{\sqrt{6}}{3}\vec{y_S} \end{cases},$$

$$- \begin{cases} \vec{n} \cdot \vec{x_0} = \cos\psi = \mp\dfrac{1-\sqrt{6}}{2\sqrt{3}} \\ \vec{n} \cdot \vec{y_0} = \sin\psi = \pm\dfrac{\sqrt{3}+\sqrt{2}}{2\sqrt{3}} \end{cases} \quad \begin{cases} \vec{n} \cdot \vec{x_S} = \cos\varphi = \pm\dfrac{\sqrt{3}}{3} \\ \vec{n} \cdot \vec{y_S} = -\sin\varphi = \pm\dfrac{\sqrt{6}}{3} \end{cases}.$$

We shall verify that $\sin^2\psi + \cos^2\psi = 1$, $\sin^2\varphi + \cos^2\varphi = 1$. In summary

$$- \ \theta = \frac{\pi}{3}, \quad \begin{cases} \cos\psi = -\dfrac{1-\sqrt{6}}{2\sqrt{3}} = 0{,}418, \ \sin\psi = \dfrac{\sqrt{3}+\sqrt{2}}{2\sqrt{3}} = 0{,}908 \\ \cos\varphi = \dfrac{\sqrt{3}}{3} = 0{,}577, \ \sin\varphi = -\dfrac{\sqrt{6}}{3} = -0{,}817 \end{cases},$$

$$\Rightarrow \ \psi = 1{,}139\, rd, \quad \varphi = -0{,}955\, rd$$

$$- \ \theta = -\frac{\pi}{3}, \quad \begin{cases} \cos\psi = \dfrac{1-\sqrt{6}}{2\sqrt{3}}, \ \sin\psi = -\dfrac{\sqrt{3}+\sqrt{2}}{2\sqrt{3}} \\ \cos\varphi = -\dfrac{\sqrt{3}}{3}, \ \sin\varphi = \dfrac{\sqrt{6}}{3} \end{cases}.$$

1.6.2. *Exercise 8 – Vector rotation*

Let $\mathcal{R}_{\vec{u}(\beta),\alpha}$ be a vector rotation with $\vec{u}(\beta) = \cos\beta\,\overrightarrow{X_1} + \sin\beta\,\overrightarrow{X_2}$, which transforms the direct orthonormal basis $(X) \equiv \left(\overrightarrow{X_1}\,\overrightarrow{X_2}\,\overrightarrow{X_3}\right)$ into $(x) \equiv \left(\overrightarrow{x_1}\,\overrightarrow{x_2}\,\overrightarrow{x_3}\right)$, which is also a direct orthonormal basis.

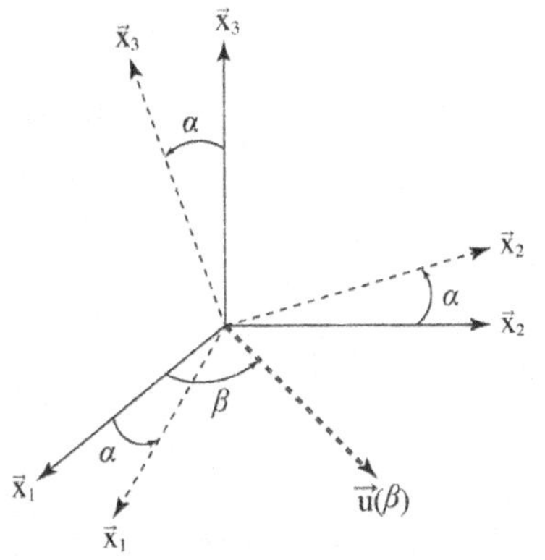

Question 1: Write the matrix representation $\left[\mathcal{R}_{\vec{u}(\beta),\alpha}\right]$ of this transformation.

The matrix representation $[\mathbf{F}]$ of the vector rotation by angle α about the axis of unit vector $\vec{u} = u_i\vec{x_i}$ can be written in the basis $(e) \equiv \left(\overrightarrow{x_1}\,\overrightarrow{x_2}\,\overrightarrow{x_3}\right)$

$$[\mathbf{F}]/_{e} = \begin{bmatrix} \cos\alpha + (1-\cos\alpha)u_1^2 & (1-\cos\alpha)u_1u_2 - \sin\alpha\, u_3 & (1-\cos\alpha)u_3u_1 + \sin\alpha\, u_2 \\ (1-\cos\alpha)u_1u_2 + \sin\alpha\, u_3 & \cos\alpha + (1-\cos\alpha)u_2^2 & (1-\cos\alpha)u_2u_3 - \sin\alpha\, u_1 \\ (1-\cos\alpha)u_3u_1 - \sin\alpha\, u_2 & (1-\cos\alpha)u_2u_3 + \sin\alpha\, u_1 & \cos\alpha + (1-\cos\alpha)u_3^2 \end{bmatrix},$$

and consequently, the vector rotation matrix can be written in the basis (X)

$$\left[\mathcal{R}_{\vec{u}(\beta),\alpha}\right]/_{(X)} = \begin{bmatrix} \cos\alpha + (1-\cos\alpha)\cos^2\beta & (1-\cos\alpha)\sin\beta\cos\beta & \sin\alpha\sin\beta \\ (1-\cos\alpha)\sin\beta\cos\beta & \cos\alpha + (1-\cos\alpha)\sin^2\beta & -\sin\alpha\cos\beta \\ -\sin\alpha\sin\beta & \sin\alpha\cos\beta & \cos\alpha \end{bmatrix}.$$

Question 2: Deduce from this, as a function of α and β, the Euler angles and the nodal vector $\vec{n}$ that define the orientation of (x) relative to (X).

The table of passage from the basis (X) to the basis (x) can be expressed as

$$p(X,x) \Leftrightarrow \begin{array}{c|ccc} & \vec{x}_1 & \vec{x}_2 & \vec{x}_3 \\ \hline \vec{X}_1 & \cos\alpha + (1-\cos\alpha)\cos^2\beta & (1-\cos\alpha)\sin\beta\cos\beta & \sin\alpha\sin\beta \\ \vec{X}_2 & (1-\cos\alpha)\sin\beta\cos\beta & \cos\alpha + (1-\cos\alpha)\sin^2\beta & -\sin\alpha\cos\beta \\ \vec{X}_3 & -\sin\alpha\sin\beta & \sin\alpha\cos\beta & \cos\alpha \end{array}.$$

In order to define the Euler representation of the relative location of these two bases, the following elements are available:

–

$$\overrightarrow{X_3}\cdot\overrightarrow{x_3} = \cos\theta = \cos\alpha \quad \Rightarrow \quad \theta = \pm\alpha + 2k\pi \quad \Rightarrow \quad \sin\theta = \pm\sin\alpha,$$

– $\overrightarrow{X_3} \wedge \overrightarrow{x_3} = \vec{n}\sin\theta$,

$$\Rightarrow \begin{cases} \vec{X}_3 \wedge \left(\sin\alpha\sin\beta\,\vec{X}_1 - \sin\alpha\cos\beta\,\vec{X}_2 + \cos\alpha\,\vec{X}_3\right) = \sin\alpha\cos\beta\,\vec{X}_1 + \sin\alpha\sin\beta\,\vec{X}_2 \\ \qquad\qquad '' \; '' \qquad\qquad = \pm\vec{n}\sin\alpha \\ \left(-\sin\alpha\sin\beta\,\vec{x}_1 + \sin\alpha\cos\beta\,\vec{x}_2 + \cos\alpha\,\vec{x}_3\right)\wedge\vec{x}_3 = \sin\alpha\cos\beta\,\vec{x}_1 + \sin\alpha\sin\beta\,\vec{x}_2 \\ \qquad\qquad '' \; '' \qquad\qquad = \pm\vec{n}\sin\alpha \end{cases},$$

$$\begin{cases} \text{for } \theta = \alpha,\ \vec{n} = \cos\beta\overrightarrow{X_1} + \sin\beta\overrightarrow{X_2} = \cos\beta\overrightarrow{x_1} + \sin\beta\overrightarrow{x_2} \\ \text{for } \theta = -\alpha,\ \vec{n} = -\left(\cos\beta\overrightarrow{X_1} + \sin\beta\overrightarrow{X_2}\right) = -\left(\cos\beta\overrightarrow{x_1} + \sin\beta\overrightarrow{x_2}\right) \end{cases},$$

$$- \begin{cases} \vec{n}\cdot\overrightarrow{X_1} = \cos\psi = \pm\cos\beta \\ \vec{n}\cdot\overrightarrow{X_2} = \sin\psi = \pm\sin\beta \end{cases} \Rightarrow \begin{cases} \text{for } \theta = \alpha,\ \psi = \beta + 2k\pi \\ \text{for } \theta = -\alpha,\ \psi = \pi + \beta + 2k\pi \end{cases},$$

$$- \begin{cases} \vec{n}\cdot\overrightarrow{x_1} = \cos\varphi = \pm\cos\beta \\ \vec{n}\cdot\overrightarrow{x_2} = -\sin\varphi = \pm\sin\beta \end{cases} \Rightarrow \begin{cases} \text{for } \theta = \alpha,\ \varphi = -\beta + 2k\pi \\ \text{for } \theta = -\alpha,\ \varphi = \pi - \beta + 2k\pi \end{cases}.$$

1.6.3. *Exercise 9 – Vector rotation*

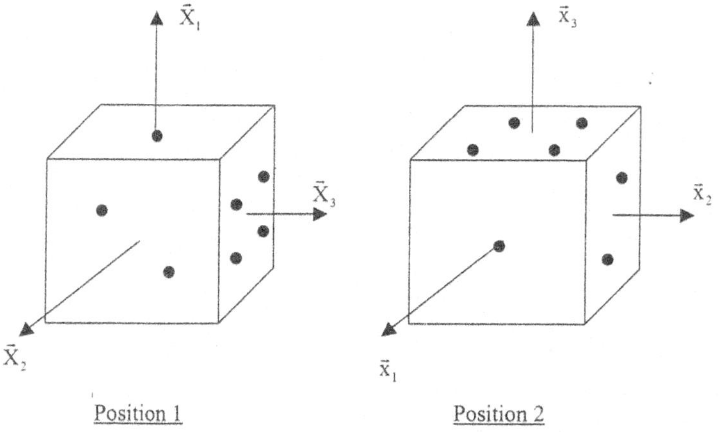

Let us consider the two orientations of the dice (S) defined by the above positions 1 and 2, with which the two direct orthonormal bases $(X) \equiv \left(\overrightarrow{X_1}\,\overrightarrow{X_2}\,\overrightarrow{X_3}\right)$ and $(x) \equiv \left(\overrightarrow{x_1}\,\overrightarrow{x_2}\,\overrightarrow{x_3}\right)$ are associated.

Question 1: Determine the matrix of change of basis $[p(X,x)]$.

In the basis (X), we have

$$\begin{aligned}\overrightarrow{x_3} &= \overrightarrow{X_1}\\ \overrightarrow{x_1} &= \overrightarrow{X_2}\\ \overrightarrow{x_2} &= \overrightarrow{X_3}\end{aligned} \quad \text{hence } [p(X,x)] = \begin{bmatrix} 0 & 0 & 1 \\ 1 & 0 & 0 \\ 0 & 1 & 0 \end{bmatrix} = \begin{bmatrix} p_{11} & p_{12} & p_{13} \\ p_{21} & p_{22} & p_{23} \\ p_{31} & p_{32} & p_{33} \end{bmatrix}.$$

Question 2: Determine the vector rotation $\mathscr{R}_{\vec{u},\alpha}$ for passing from position 1 to position 2

Since the above matrix of change of basis is the vector rotation matrix $\mathscr{R}_{\vec{u},\alpha}$, we obtain

$$Trace\left[\mathscr{R}_{\vec{u},\alpha}\right] = 1 + 2\cos\alpha = Trace[p(X,x)] = p_{ii} = 0,$$

$$\cos\alpha = -\frac{1}{2} \text{ with } \alpha \in [0,\pi] \Rightarrow \alpha = \frac{2\pi}{3} \Rightarrow \sin\alpha = \frac{\sqrt{3}}{2}.$$

The vector $\vec{u} = u_i \overrightarrow{X_i}$ is such that $\begin{cases} u_1 = \dfrac{p_{32} - p_{23}}{2\sin\alpha} = \dfrac{\sqrt{3}}{3} \\ u_2 = \dfrac{p_{13} - p_{31}}{2\sin\alpha} = \dfrac{\sqrt{3}}{3} \\ u_3 = \dfrac{p_{21} - p_{12}}{2\sin\alpha} = \dfrac{\sqrt{3}}{3} \end{cases}$,

$$\Rightarrow \mathcal{R}_{\vec{u},\alpha}(X,x) = \left[\vec{u} = \frac{\sqrt{3}}{3}\left(\overrightarrow{X_1} + \overrightarrow{X_2} + \overrightarrow{X_3}\right) \middle| \alpha = \frac{2\pi}{3} \right].$$

Question 3: Show that it is equally possible to obtain the position 2 of the dice by two successive plane rotations starting from position 1. Write the matrices of passage associated to these two rotation planes.

An intermediary position of the dice, to which the basis $(y) \equiv \left(\overrightarrow{y_1}\ \overrightarrow{y_2}\ \overrightarrow{y_3}\right)$ is assigned, should be considered. A first vector rotation is applied, such that $\vec{u} = \overrightarrow{X_1}$ and the angle is $\alpha = \frac{\pi}{2}$, then a second one such that $\vec{u} = \overrightarrow{y_2}$ and $\alpha = \frac{\pi}{2}$.

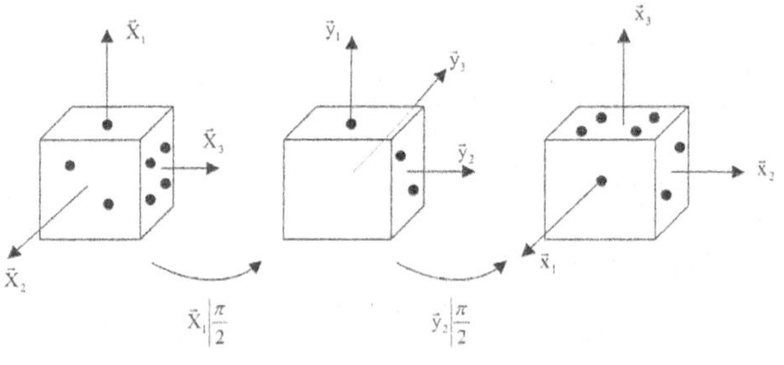

Knowing that

$$\vec{b} = \mathcal{R}_{\vec{u},\alpha}\left(\vec{a}\right) = \vec{a} + \left[\vec{u} \wedge \left(\vec{u} \wedge \vec{a}\right)\right](1 - \cos\alpha) + \left(\vec{u} \wedge \vec{a}\right)\sin\alpha,$$

– the first vector rotation $\begin{cases} \vec{y_1} = \mathcal{R}_{\vec{X_1},\frac{\pi}{2}}(\vec{X_1}) = \vec{X_1} \\ \vec{y_2} = \mathcal{R}_{\vec{X_1},\frac{\pi}{2}}(\vec{X_2}) = \vec{X_2} - \vec{X_2} + \vec{X_3} = \vec{X_3} \\ \vec{y_3} = \mathcal{R}_{\vec{X_1},\frac{\pi}{2}}(\vec{X_3}) = \vec{X_3} - \vec{X_3} - \vec{X_2} = -\vec{X_2} \end{cases}$,

– the second vector rotation $\begin{cases} \vec{x_1} = \mathcal{R}_{\vec{y_2},\frac{\pi}{2}}(\vec{y_1}) = \vec{y_1} - \vec{y_1} - \vec{y_3} = -\vec{y_3} \\ \vec{x_2} = \mathcal{R}_{\vec{y_2},\frac{\pi}{2}}(\vec{y_2}) = \vec{y_2} \\ \vec{x_3} = \mathcal{R}_{\vec{y_2},\frac{\pi}{2}}(\vec{y_3}) = \vec{y_3} - \vec{y_3} + \vec{y_1} = \vec{y_1} \end{cases}$.

We then obtain the two matrices of passage

$$[p(X,y)] = \begin{bmatrix} 1 & 0 & 0 \\ 0 & 0 & -1 \\ 0 & 1 & 0 \end{bmatrix} \quad \text{and} \quad [p(y,x)] = \begin{bmatrix} 0 & 0 & 1 \\ 0 & 1 & 0 \\ -1 & 0 & 0 \end{bmatrix},$$

and it can be verified that

$$[p(X,x)] = [p(X,y)] \times [p(y,x)].$$

1.6.4. *Exercise 10 – Vector rotation*

Let us consider two positions of the solid (S) represented by a rectangular parallelepiped with sides a, $2a$ and $3a$.

The position I of this solid is identified by the system of reference connected to it $\langle R_1 \rangle \equiv \langle \mathrm{O} | \vec{X_1}\ \vec{X_2}\ \vec{X_3} \rangle$, which is obtained starting from three of its edges and one of its vertices and defined, according to the diagram, by

$$\overrightarrow{\mathrm{OA}} = a\vec{X_1}, \quad \overrightarrow{\mathrm{OB}} = 2a\vec{X_2}, \quad \overrightarrow{\mathrm{OC}} = 3a\vec{X_3}.$$

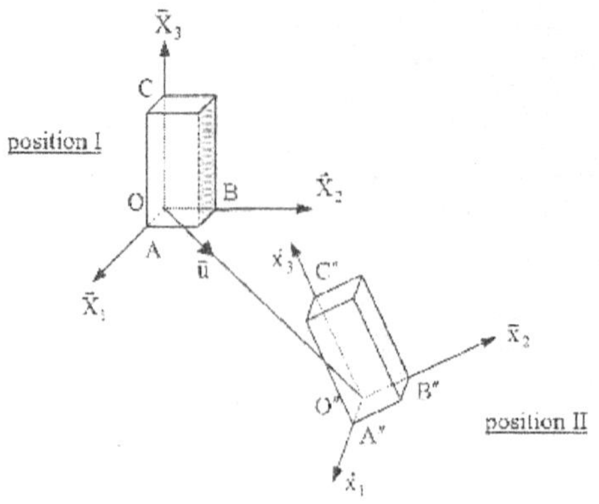

During time its location changes. Its position II is characterized by the system of reference connected to it $\langle R_2 \rangle \equiv \left\langle \mathrm{O}'' | \overrightarrow{x_1}\, \overrightarrow{x_2}\, \overrightarrow{x_3} \right\rangle$ so that

$$\begin{cases} \overrightarrow{x_1} = \dfrac{1}{18}\left[10\overrightarrow{X_1} + (2-6\sqrt{3})\overrightarrow{X_2} - (2+6\sqrt{3})\overrightarrow{X_3}\right] \\ \overrightarrow{x_2} = \dfrac{1}{18}\left[(2+6\sqrt{3})\overrightarrow{X_1} + 13\overrightarrow{X_2} + (-4+3\sqrt{3})\overrightarrow{X_3}\right]. \\ \overrightarrow{x_3} = \dfrac{1}{18}\left[(-2+6\sqrt{3})\overrightarrow{X_1} - (4+3\sqrt{3})\overrightarrow{X_2} + 13\overrightarrow{X_3}\right] \end{cases}$$

The passage from one position to the other, executed by a robot, is a combination of vector rotation $\mathscr{R}_{\vec{u},\alpha}$ and translation $\overrightarrow{\mathrm{OO}''} = 6a\vec{u}$ where $\vec{u}$ is a unit vector.

Question 1: Write the matrix of passage $[p(X,x)]$ and deduce from it $\vec{u}$ and α.

$$p(X,x) \Leftrightarrow \begin{array}{c|ccc} & \overrightarrow{x_1} & \overrightarrow{x_2} & \overrightarrow{x_3} \\ \hline \overrightarrow{X_1} & \dfrac{5}{9} & \dfrac{1+3\sqrt{3}}{9} & \dfrac{-1+3\sqrt{3}}{9} \\ \overrightarrow{X_2} & \dfrac{1-3\sqrt{3}}{9} & \dfrac{13}{18} & -\dfrac{4+3\sqrt{3}}{18} \\ \overrightarrow{X_3} & -\dfrac{1+3\sqrt{3}}{9} & \dfrac{-4+3\sqrt{3}}{18} & \dfrac{13}{18} \end{array}.$$

$$Trace[p(X,x)] = \frac{5}{9} + \frac{13}{18} + \frac{13}{18} = 2 = 1 + 2\cos\alpha \quad \Rightarrow \quad \cos\alpha = \frac{1}{2},$$

For $\alpha \in [0,\pi], \alpha = \frac{\pi}{3}$ and $\sin\alpha = \frac{\sqrt{3}}{2}$, we obtain the vector $\vec{u} = u_i \overrightarrow{X_i}$

$$\begin{cases} u_1 = \dfrac{p_{32} - p_{23}}{2\sin\alpha} = \dfrac{1}{3} \\ u_2 = \dfrac{p_{13} - p_{31}}{2\sin\alpha} = \dfrac{2}{3} \\ u_3 = \dfrac{p_{21} - p_{12}}{2\sin\alpha} = -\dfrac{2}{3} \end{cases} \quad \Rightarrow \quad \vec{u} = \frac{1}{3}\left(\overrightarrow{X_1} + 2\overrightarrow{X_2} - 2\overrightarrow{X_3}\right).$$

Question 2: Determine the coordinates in $\langle R_1 \rangle$ of the point A′, which is the position of the vertex A after vector rotation.

After the vector rotation $\mathscr{R}_{\vec{u},\alpha}$ the point A is in the position A′ so that

$$\overrightarrow{OA'} = \mathscr{R}_{\vec{u},\alpha}\left(\overrightarrow{OA}\right) = \mathscr{R}_{\vec{u},\alpha}\left(a\overrightarrow{X_1}\right) = a\overrightarrow{x_1} = a\left(\frac{5}{9}\overrightarrow{X_1} + \frac{1-3\sqrt{3}}{9}\overrightarrow{X_2} - \frac{1+3\sqrt{3}}{9}\overrightarrow{X_3}\right).$$

Question 3: Determine the coordinates in $\langle R_1 \rangle$ of the point A″, position of the vertex A as a result of the two transformations.

$$\overrightarrow{OA''} = \overrightarrow{OA'} + \overrightarrow{A'A''}$$

$$'' \; '' = a\left(\frac{5}{9}\overrightarrow{X_1} + \frac{1-3\sqrt{3}}{9}\overrightarrow{X_2} - \frac{1+3\sqrt{3}}{9}\overrightarrow{X_3}\right) + 2a\left(\overrightarrow{X_1} + 2\overrightarrow{X_2} - 2\overrightarrow{X_3}\right).$$

$$'' \; '' = a\left(\frac{23}{9}\overrightarrow{X_1} + \frac{37-3\sqrt{3}}{9}\overrightarrow{X_2} - \frac{37+3\sqrt{3}}{9}\overrightarrow{X_3}\right)$$

Question 4: Determine the displacement vector of A, $\overrightarrow{AA''}$, by its components in the basis (R_1).

$$\overrightarrow{AA''} = \overrightarrow{AO} + \overrightarrow{OA'} + \overrightarrow{A'A''}$$

$$'' \; '' = -a\overrightarrow{X_1} + a\left(\frac{5}{9}\overrightarrow{X_1} + \frac{1-3\sqrt{3}}{9}\overrightarrow{X_2} - \frac{1+3\sqrt{3}}{9}\overrightarrow{X_3}\right) + 2a\left(\overrightarrow{X_1} + 2\overrightarrow{X_2} - 2\overrightarrow{X_3}\right)$$

$$'' \; '' = a\left(\frac{14}{9}\overrightarrow{X_1} + \frac{37-3\sqrt{3}}{9}\overrightarrow{X_2} - \frac{37+3\sqrt{3}}{9}\overrightarrow{X_3}\right)$$

2

Solid Kinematics

The study of the motion of bodies, whether these be rigid or deformable, liquid or gas, requires an understanding of what characterizes this motion, which is the motion of the particles that constitute them, the duration of the motion and the manner in which it occurs. As a result, the calculation of the trajectories of particles and of velocity and acceleration fields that characterize these along their course is of great importance.

This chapter is, therefore, an essential preliminary reading for the rest of the content of this set of books on the mechanics of rigid solids.

2.1. Generalities on moving solids

2.1.1. *Concept of a rigid material system*

A solid (S) is a rigid material system if, for any particle of matter M_1 and M_2 that belong to it, the distance M_1M_2 is independent of the circumstances, which is written

$$\sqrt{\left(\overrightarrow{M_1M_2}\right)^2} = l \quad \text{such that} \quad \frac{dl}{dt} = 0.$$

As the mechanical system is in motion, even when rigid, vector $\overrightarrow{M_1M_2}$ is able to change to its orientation, and therefore to vary; as a result $\frac{d}{dt}\overrightarrow{M_1M_2} \neq \vec{0}$ as only the vector modulus, which is the length of segment M_1M_2, and does not vary as a function of time.

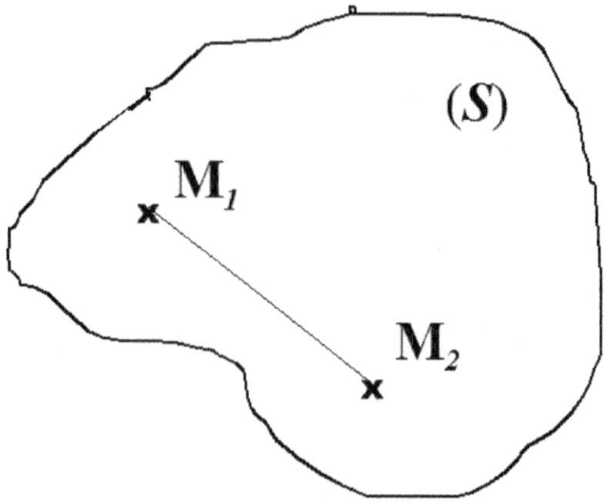

Figure 2.1. *Rigid mechanical system*

The international measurement of length is the meter, as defined in 1983 as *the length of the path traveled by light in a vacuum in 1 / 299,792,458th of a second.* This definition assumes that the definition of the unit of the time that is the second has been previously established.

2.1.2. *Notion of time*

Time is an important value that, although intangible, has a major role in mechanics. This notion deals with the simultaneity of two events, the order in which two events take place, and the length of time that separates them.

Since the 13th General Conference on Weights and Measures, a second is no longer defined in relation to the year, but is rather linked to a property of matter; this base unit of the international system was defined in 1967 as the following: *one second is the duration of 9,192,631,770 periods of the radiation corresponding to the transition between the two hyperfine levels* $F = 3$ *and* $F = 4$ *of the ground state* $S_{1/2}$ *of the cesium 133 atom.* This definition refers to an atom of cesium at rest, at a temperature of 0 K.

2.1.3. *Kinematic components of a solid*

As described previously, identifying a solid and locating it so as to study its motion in an appropriate context requires distinguishing it from a defined point O_S taken as the origin of the solid, and associating it to a frame that will provide information on its situation, as in the diagram below.

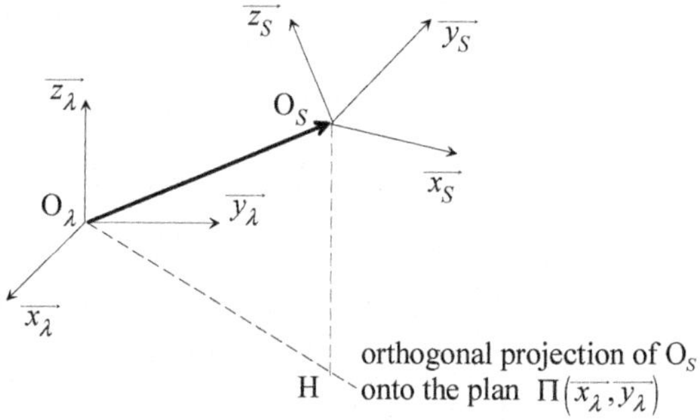

Figure 2.2. *Identification of a solid*

The kinematics of a solid is deduced from this description by using the descriptive elements listed above: the components of the position vector of point O_S in the frame $\langle\lambda\rangle$ and, in relation to that, the orientation of the frame $\langle S\rangle$ linked to this solid.

NOTE 2.1.– the orthogonal projection H of O_S onto the plane $\Pi(\overrightarrow{x_\lambda}, \overrightarrow{y_\lambda})$ results in the expression of the position vector $\overrightarrow{O_\lambda O_S}$ in cylindrical-polar or spherical coordinates.

Point O_S is considered here to be an adimensional geometrical particle without mass, whose kinematics, that is its position, its velocity and its acceleration are intrinsically expressed, without considering anything but its own situation. It is referred to as the *material point* throughout the rest of this work. This term is also used to refer to any point of solid whose kinematics is being described, most notably the center of inertia G, which is encountered in the study of solid kinematics in Chapter 3.

2.2. Kinematics of a material point

2.2.1. *Position vector*

The study of the kinematics of the material point O_s, therefore, relies on the expression of the position vector $\overrightarrow{O_\lambda O_S}$ within the chosen frame, usually ⟨λ⟩, by the most appropriate coordinate system for the problem in question.

2.2.2. *Trajectory of a material point in a reference frame*

Throughout its evolution during the motion studied, a material point M coincides, at a time ***t***, with the point $M_\lambda(t)$ of the frame ⟨λ⟩. All of the points $M_\lambda(t)$ within the time interval $[t_o, t_f]$ during which the motion is studied represent the trajectory $\Gamma_\lambda^{(t_0,t_f)}(M)$ of the material point M in the frame ⟨λ⟩.

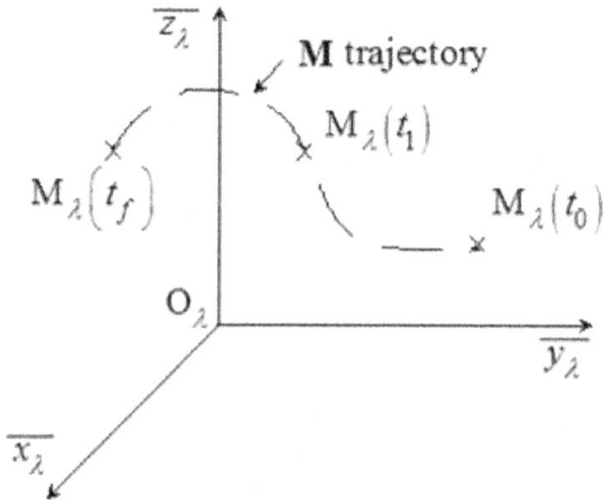

Figure 2.3. *Trajectory of a material point*

The observed trajectory of a material point depends on the situation of the observer.

To illustrate this concept, let us consider the example – albeit a simple one – of a train carriage with a see-through wall, which passes through a station at a constant speed. In the carriage, a child is playing with a ball by bouncing it vertically: for the observer sitting in the carriage and observing the child, the trajectory of the ball is vertical as the train's speed is constant; for an observer sitting in the station watching the train pass by the trajectory of the ball is made of parabolic components. However, the motion is the same, but observed by different observers who are mobile with regards to each other.

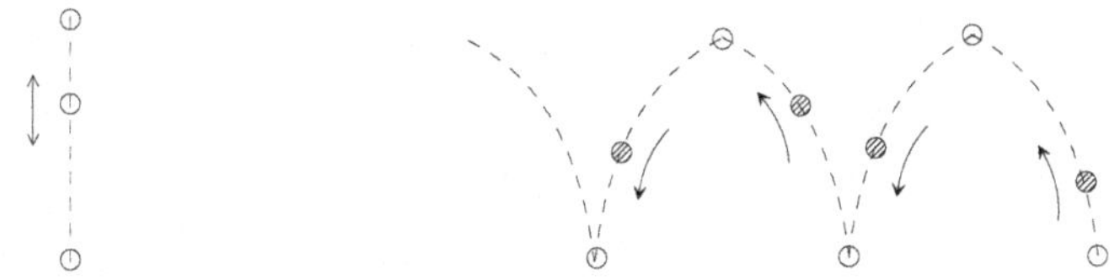

Figure 2.4. *Relative observations of a trajectory*

2.2.3. *Velocity of a material point in a reference frame*

Let us consider the trajectory $\Gamma_{\lambda}^{(t_0,t_f)}(\mathrm{M})$ of the material point M in relation to the frame $\langle\lambda\rangle$ and to point $\mathrm{M}_{\lambda(t)}$ frame, with which M coincide at the time t during its motion.

The position vector $\overrightarrow{\mathrm{O}_{\lambda}\mathrm{M}_{\lambda(t)}}$ is a function of time

$$\overrightarrow{\mathrm{O}_{\lambda}\mathrm{M}_{\lambda(t)}} = \vec{f}(t).$$

The motion $\overrightarrow{D_{\lambda}}^{(t,t^*)}(\mathrm{M})$ of the material point M between two times t and t^* is written

$$\overrightarrow{D_{\lambda}}^{(t,t^*)}(\mathrm{M}) = \overrightarrow{\mathrm{M}_{\lambda(t)}\mathrm{M}_{\lambda(t^*)}} = \overrightarrow{O_{\lambda}\mathrm{M}_{\lambda(t^*)}} - \overrightarrow{O_{\lambda}\mathrm{M}_{\lambda(t)}} = \vec{f}(t^*) - \vec{f}(t),$$

meaning, between the times t and $t+h$

$$\overrightarrow{D_{\lambda}}^{(t,t+h)}(\mathrm{M}) = \vec{f}(t+h) - \vec{f}(t).$$

The *velocity-vector* or velocity of the material point M in relation to the frame ⟨λ⟩, at a time *t*, is the vector written as $\overrightarrow{v^{(\lambda)}}(\mathrm{M}|t)$ and defined by

$$\overrightarrow{v^{(\lambda)}}(\mathrm{M}|t) = \lim_{h \to 0} \frac{1}{h} \overrightarrow{D_\lambda}^{(t,t+h)}(\mathrm{M}),$$

meaning that, by considering the definition of the derivative of a vector,

$$\overrightarrow{v^{(\lambda)}}(\mathrm{M}|t) = \frac{d^{(\lambda)}}{dt}\vec{f}(t) = \frac{d^{(\lambda)}}{dt}\overrightarrow{\mathrm{O}_\lambda \mathrm{M}}.$$

The notation $\frac{d^{(\lambda)}}{dt}$ means that, in the derivation operation the unit vectors of the basis (λ) are considered to be constant, meaning that

$$\frac{d^{(\lambda)}}{dt}\overrightarrow{x_\lambda} = \vec{0} \quad ; \quad \frac{d^{(\lambda)}}{dt}\overrightarrow{y_\lambda} = \vec{0} \quad ; \quad \frac{d^{(\lambda)}}{dt}\overrightarrow{z_\lambda} = \vec{0}.$$

According to this definition, the essential and imperative conditions that characterize the velocity vector of the material point M, in the frame ⟨λ⟩ taken as the reference location for the observer of the motion, meaning the derivative of vector $\overrightarrow{\mathrm{O}_\lambda \mathrm{M}}$ in this frame, are that:

– the origin of the position vector of M in $\langle \lambda \rangle$ is a fixed point of this frame, which here is the point O_λ;

– the unit vectors of the frame $\langle \lambda \rangle$ are fixed and their derivatives in it are equal to zero as a result.

2.2.4. *Components of the velocity vector or velocity*

2.2.4.1. *Velocity in Cartesian coordinates*

$$\overrightarrow{v^{(\lambda)}}(\mathrm{M}|t) = \frac{d^{(\lambda)}}{dt}\overrightarrow{O_\lambda M_{\lambda(t)}} = \frac{d^{(\lambda)}}{dt}\left(x\,\overrightarrow{x_\lambda} + y\,\overrightarrow{y_\lambda} + z\overrightarrow{z_\lambda}\right),$$

$$\overrightarrow{v^{(\lambda)}}(\mathrm{M}|t) = x'\overrightarrow{x_\lambda} + y'\overrightarrow{y_\lambda} + z'\overrightarrow{z_\lambda}.$$

2.2.4.2. *Velocity in the cylindrical-polar coordinates*

$$\overrightarrow{v^{(\lambda)}}(\mathrm{M}|t)=\frac{d^{(\lambda)}}{dt}\overrightarrow{O_\lambda M_{\lambda(t)}}=\frac{d^{(\lambda)}}{dt}\left(r\vec{u}(\alpha)+z\overrightarrow{z_\lambda}\right)$$

$$\text{'' ''}\quad=\frac{dr}{dt}\vec{u}(\alpha)+r\frac{d^{(\lambda)}\vec{u}(\alpha)}{dt}+\frac{dz}{dt}\overrightarrow{z_\lambda}.$$

By definition $\vec{u}(\alpha)=\cos\alpha\,\overrightarrow{x_\lambda}+\sin\alpha\,\overrightarrow{y_\lambda}$, and as a result

$$\frac{d^{(\lambda)}}{dt}\vec{u}(\alpha)=\frac{d\cos\alpha}{dt}\overrightarrow{x_\lambda}+\frac{d\sin\alpha}{dt}\overrightarrow{y_\lambda}=\frac{d\cos\alpha}{d\alpha}\frac{d\alpha}{dt}\overrightarrow{x_\lambda}+\frac{d\sin\alpha}{d\alpha}\frac{d\alpha}{dt}\overrightarrow{y_\lambda},$$

$$\frac{d^{(\lambda)}}{dt}\vec{u}(\alpha)=\alpha'\left(-\sin\alpha\,\overrightarrow{x_\lambda}+\cos\alpha\,\overrightarrow{y_\lambda}\right)$$

$$\text{'' ''}\quad=\alpha'\left[\cos\left(\alpha+\frac{\pi}{2}\right)\overrightarrow{x_\lambda}+\sin\left(\alpha+\frac{\pi}{2}\right)\overrightarrow{y_\lambda}\right],$$

$$\Rightarrow\quad\frac{d^{(\lambda)}}{dt}\vec{u}(\alpha)=\alpha'\vec{u}\left(\alpha+\frac{\pi}{2}\right).$$

This result, which is of little significance in mechanics, will be very useful later as the derivative of the vector $\vec{u}(\alpha)$ will often take place in calculations. Moreover, it can be applied to any more complex form given to α.

Summarily, the expression of the velocity vector of the material point M in the cylindrical-polar coordinates is written

$$\overrightarrow{v^{(\lambda)}}(\mathrm{M}|t)=r'\vec{u}(\alpha)+r\alpha'\vec{u}\left(\alpha+\frac{\pi}{2}\right)+z'\overrightarrow{z_\lambda}.$$

2.2.4.3. *Velocity in the spherical coordinates*

In the spherical coordinates, the position vector of the material point M is expressed as

$$\overrightarrow{\mathrm{O}_\lambda\mathrm{M}_{\lambda(t)}}=R\vec{k}(\alpha,\beta).$$

$$\overrightarrow{v^{(\lambda)}}(\mathrm{M}|t)=\frac{d^{(\lambda)}}{dt}\overrightarrow{\mathrm{O}_\lambda\mathrm{M}}=\frac{dR}{dt}\vec{k}(\alpha,\beta)+R\frac{d^{(\lambda)}}{dt}\vec{k}(\alpha,\beta),$$

where $\frac{d^{(\lambda)}}{dt}\vec{k}(\alpha,\beta)=\frac{\partial^{(\lambda)}}{\partial\alpha}\vec{k}(\alpha,\beta)\times\frac{d\alpha}{dt}+\frac{\partial^{(\lambda)}}{\partial\beta}\vec{k}(\alpha,\beta)\times\frac{d\beta}{dt}$

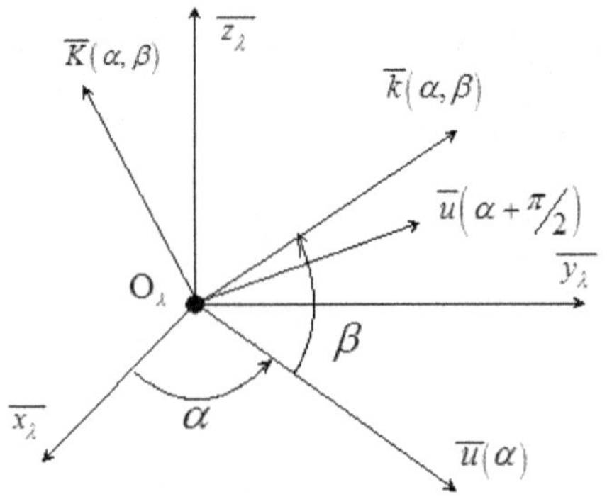

Figure 2.5. *The spherical coordinates*

As $\vec{k}(\alpha,\beta)=\cos\beta\,\vec{u}(\alpha)+\sin\beta\,\overrightarrow{z_\lambda}$, this means that

$$\frac{\partial^{(\lambda)}}{\partial\alpha}\vec{k}(\alpha,\beta)=\cos\beta\,\vec{u}\left(\alpha+\frac{\pi}{2}\right)$$
$$\frac{\partial^{(\lambda)}}{\partial\beta}\vec{k}(\alpha,\beta)=-\sin\beta\,\vec{u}(\alpha)+\cos\beta\,\overrightarrow{z_\lambda},$$

$$\frac{d^{(\lambda)}}{dt}\vec{k}(\alpha,\beta)=-\beta'\sin\beta\,\vec{u}(\alpha)+\alpha'\cos\beta\,\vec{u}\left(\alpha+\frac{\pi}{2}\right)+\beta'\cos\beta\,\overrightarrow{z_\lambda}$$

By inserting the vector

$$\vec{K}(\alpha,\beta)=-\sin\beta\,\vec{u}(\alpha)+\cos\beta\,\overrightarrow{z_\lambda},$$

this provides us, within the basis $\left(\vec{k}(\alpha,\beta)\ \vec{u}\left(\alpha+\frac{\pi}{2}\right)\ \vec{K}(\alpha,\beta)\right)$, with the expression of the velocity vector in the spherical coordinates

$$\overrightarrow{v^{(\lambda)}}(\mathrm{M}|t)=R'\,\vec{k}(\alpha,\beta)+R\alpha'\cos\beta\,\vec{u}\left(\alpha+\frac{\pi}{2}\right)+R\beta'\,\vec{K}(\alpha,\beta).$$

2.2.4.4. *Velocity in generalized coordinates*

With generalized coordinates, there is

$$\overrightarrow{O_\lambda M_{\lambda(t)}}=\vec{f}(Q_1,Q_2,Q_3),$$

meaning the velocity vector is

$$\overrightarrow{v^{(\lambda)}}(\mathrm{M}|t)=\frac{\partial^{(\lambda)}\vec{f}}{\partial Q_1}Q_1'+\frac{\partial^{(\lambda)}\vec{f}}{\partial Q_2}Q_2'+\frac{\partial^{(\lambda)}\vec{f}}{\partial Q_3}Q_3'.$$

2.2.5. *Derivative of a vector in a basis*

Seeing as the derivation of a vectorial function is common practice in mechanics, it is useful here to provide a general approach, explored more deeply in Volume 2 of this set [BOR 16a].

Let us consider two direct orthonormal bases $(e)\equiv\left(\vec{x}\,\vec{y}\,\vec{z}\right)$ and $(E)\equiv\left(\vec{X}\,\vec{Y}\,\vec{Z}\right)$ whose motion in relation to each other is governed by a number of n parameters $q_1,q_2,\ldots,q_n$.

We consider $\vec{f}$ to be a variable vector with regards to the two bases. According to the laws for deriving the vectorial functions (see Volume 2 [BOR 16a]), generally this means that

$$\frac{\partial^{(e)}\vec{f}}{\partial q_i}-\frac{\partial^{(E)}\vec{f}}{\partial q_i}=\overrightarrow{{}^{q_i}\delta}(e,E)\wedge\vec{f},$$

where the vector $\overrightarrow{{}^{q_i}\delta}(e,E)$, called the rate of rotation relative to q_i, is expressed as a projection on the two bases in the following form

$$\overrightarrow{{}^{q_i}\delta}(e,E)=P_i^{(e)}\vec{x}+Q_i^{(e)}\vec{y}+R_i^{(e)}\vec{z}=P_i^{(E)}\vec{X}+Q_i^{(E)}\vec{Y}+R_i^{(E)}\vec{Z},$$

with

$$\begin{cases}P_i^{(e)}=\dfrac{\partial^{(E)}\vec{y}}{\partial q_i}\cdot\vec{z}\\[2ex] Q_i^{(e)}=\dfrac{\partial^{(E)}\vec{z}}{\partial q_i}\cdot\vec{x}\\[2ex] R_i^{(e)}=\dfrac{\partial^{(E)}\vec{x}}{\partial q_i}\cdot\vec{y}\end{cases};\quad\begin{cases}P_i^{(E)}=\dfrac{\partial^{(e)}\vec{Y}}{\partial q_i}\cdot\vec{Z}\\[2ex] Q_i^{(E)}=\dfrac{\partial^{(e)}\vec{Z}}{\partial q_i}\cdot\vec{X}\\[2ex] R_i^{(E)}=\dfrac{\partial^{(e)}\vec{X}}{\partial q_i}\cdot\vec{Y}\end{cases}.$$

The relative situation of the two bases is given in the basis change table

$(e)\|(E)$	$\vec{X}$	$\vec{Y}$	$\vec{Z}$
$\vec{x}$	p_{11}	p_{12}	p_{13}
$\vec{y}$	p_{21}	p_{22}	p_{23}
$\vec{z}$	p_{31}	p_{32}	p_{33}

,

and in the corresponding basis change matrix

$$p(e,E)=\begin{bmatrix} p_{11} & p_{12} & p_{13} \\ p_{21} & p_{22} & p_{23} \\ p_{31} & p_{32} & p_{33} \end{bmatrix},$$

whose terms are functions of the n parameters q_i

$$p_{\alpha\beta} = p_{\alpha\beta}(q_1, q_2, \ldots, q_n).$$

2.2.5.1. *Application to the local basis*

The basis change table between (λ) and $(e) \equiv \left(\vec{u}(\alpha)\ \vec{u}\left(\alpha+\pi/2\right)\ \vec{z}_\lambda\right)$ is written

$\lambda\|(e)$	$\vec{u}(\alpha)$	$\vec{u}\left(\alpha+\frac{\pi}{2}\right)$	$\vec{z}_\lambda$
$\vec{x}_\lambda$	$\cos\alpha$	$-\sin\alpha$	0
$\vec{y}_\lambda$	$\sin\alpha$	$\cos\alpha$	0
$\vec{z}_\lambda$	0	0	1

.

The relative position of the two bases is only tied to the parameter α; as a result

$$\frac{d^{(\lambda)}}{d\alpha}\vec{u}(\alpha)-\frac{d^{(e)}}{d\alpha}\vec{u}(\alpha)={}^{\alpha}\vec{\delta}(\lambda,e)\wedge\vec{u}(\alpha),$$

where $\frac{d^{(e)}}{d\alpha}\vec{u}(\alpha)=\vec{0}$ as $\vec{u}(\alpha)$ is a vector of the basis (*e*), which is fixed in it.

$$\overrightarrow{^{\alpha}\delta}(\lambda,e)=P_{\alpha}\vec{u}(\alpha)+Q_{\alpha}\vec{u}\left(\alpha+\frac{\pi}{2}\right)+R_{\alpha}\overrightarrow{z_{\lambda}},$$

where components P_{α}, Q_{α}, R_{α} are defined as follows:

$$\begin{cases} P_{\alpha}=\frac{\partial^{(\lambda)}}{\partial\alpha}\vec{u}\left(\alpha+\frac{\pi}{2}\right)\cdot\overrightarrow{z_{\lambda}}=\left(-\cos\alpha\,\overrightarrow{x_{\lambda}}-\sin\alpha\,\overrightarrow{y_{\lambda}}\right)\cdot\overrightarrow{z_{\lambda}}=0 \\ Q_{\alpha}=\frac{\partial^{(\lambda)}}{\partial\alpha}\overrightarrow{z_{\lambda}}\cdot\vec{u}(\alpha)=0 \\ R_{\alpha}=\frac{\partial^{(\lambda)}}{\partial\alpha}\vec{u}(\alpha)\cdot\vec{u}\left(\alpha+\frac{\pi}{2}\right) \\ \text{""}=\left(-\sin\alpha\,\overrightarrow{x_{\lambda}}+\cos\alpha\,\overrightarrow{y_{\lambda}}\right)\cdot\left(-\sin\alpha\,\overrightarrow{x_{\lambda}}+\cos\alpha\,\overrightarrow{y_{\lambda}}\right)=1 \end{cases}.$$

As a result $\overrightarrow{^{\alpha}\delta}(\lambda,e)=\overrightarrow{z_{\lambda}}\quad\Rightarrow\quad\frac{d^{(\lambda)}}{d\alpha}\vec{u}(\alpha)=\overrightarrow{z_{\lambda}}\wedge\vec{u}(\alpha)=\vec{u}\left(\alpha+\frac{\pi}{2}\right)$,

$$\Rightarrow\quad\frac{d^{(\lambda)}}{dt}\vec{u}(\alpha)=\frac{d^{(\lambda)}}{d\alpha}\vec{u}(\alpha)\times\frac{d\alpha}{dt}=\alpha'\vec{u}\left(\alpha+\frac{\pi}{2}\right).$$

This result is one that is very useful in constructing equations to deal with mechanical problems. It shall be used extensively hereafter.

2.2.5.2. *Insertion of the rotation rate vector*

Vector $\vec{u}(\alpha)$ belongs to the basis (*e*), and therefore can now be written as

$$\frac{d^{(\lambda)}}{dt}\vec{u}(\alpha)=\alpha'\,\overrightarrow{^{\alpha}\delta}(\lambda,e)\wedge\vec{u}(\alpha)=\alpha'\overrightarrow{z_{\lambda}}\wedge\vec{u}(\alpha)=\overrightarrow{\omega_{e}^{\lambda}}\wedge\vec{u}(\alpha).$$

Vector $\overrightarrow{\omega_{e}^{\lambda}}$, here equal to $\alpha'\overrightarrow{z_{\lambda}}$, is the *rotation rate vector* of the basis (e) in relation to basis (λ); thanks to this notation, the following can be written

$$\frac{d^{(\lambda)}}{dt}\vec{u}\left(\alpha+\frac{\pi}{2}\right)=\overrightarrow{\omega_e^{\lambda}}\wedge\vec{u}\left(\alpha+\frac{\pi}{2}\right)=\alpha'\overrightarrow{z_\lambda}\wedge\vec{u}\left(\alpha+\frac{\pi}{2}\right)=-\alpha'\vec{u}(\alpha),$$

and the following can be verified

$$\frac{d^{(\lambda)}}{dt}\overrightarrow{z_\lambda}=\overrightarrow{\omega_e^{\lambda}}\wedge\overrightarrow{z_\lambda}=\alpha'\overrightarrow{z_\lambda}\wedge\overrightarrow{z_\lambda}=\vec{0},$$

since vector $\overrightarrow{z_\lambda}$ of basis (e) also belongs to the basis (λ).

More generally, the derivation of the vectorial functions[1] highlights the part of the rotation rate vector of the motion in relation to the two bases (e) and (E) in the following expression

$$\frac{d^{(e)}}{dt}\vec{f}-\frac{d^{(E)}}{dt}\vec{f}=\overrightarrow{\omega_E^{e}}\wedge\vec{f},$$

where, if the relative situation of the two bases depends on the n parameters $q_1, q_2,\ldots, q_n$ the rate of rotation is equal to

$$\overrightarrow{\omega_E^{e}}=\sum_{i=1}^{i=n}\overrightarrow{{}^{q_i}\delta(e,E)}q_i',$$

with $\overrightarrow{{}^{q_i}\delta(e,E)}$ as defined and determined above.

2.2.6. *Acceleration vector of a material point in a reference frame*

While it is important to know the velocity field acting on the components of a moving object, this knowledge is incomplete as it does not help us in describing this motion under the effect of the forces that act on the object. Knowledge of the acceleration field in effect is essential; without it the fundamental principal of dynamics cannot be applied. However, it is necessary to go through the velocity field to get to it.

---

1 See the relevant chapter in Volume 2 [BOR 16a].

2.2.6.1. *Definition*

We consider the velocity vector of a material point M at two successive instances t and $t+h$,

$$\overrightarrow{v^{(\lambda)}}(\mathrm{M}|t) \quad ; \quad \overrightarrow{v^{(\lambda)}}(\mathrm{M}|t+h),$$

the *acceleration vector* or acceleration of this material point is defined by the vector

$$\overrightarrow{J^{(\lambda)}}(\mathrm{M}|t) = \lim_{h\to 0}\frac{\overrightarrow{v^{(\lambda)}}(\mathrm{M}|t+h)-\overrightarrow{v^{(\lambda)}}(\mathrm{M}|t)}{h} = \frac{d^{(\lambda)}}{dt}\overrightarrow{v^{(\lambda)}}(\mathrm{M}|t).$$

2.2.6.2. *Acceleration in the Cartesian coordinates*

$$\overrightarrow{v^{(\lambda)}}(\mathrm{M}|t) = x'\overrightarrow{x_\lambda} + y'\overrightarrow{y_\lambda} + z'\overrightarrow{z_\lambda},$$

$$\overrightarrow{J^{(\lambda)}}(\mathrm{M}|t) = \frac{d^{(\lambda)}}{dt}\overrightarrow{v^{(\lambda)}}(\mathrm{M}|t) = x''\overrightarrow{x_\lambda} + y''\overrightarrow{y_\lambda} + z''\overrightarrow{z_\lambda}.$$

2.2.6.3. *Acceleration in the cylindrical-polar coordinates*

$$\vec{v}^{(\lambda)}(\mathrm{M}|t) = r'\vec{u}(\alpha) + r\alpha'\vec{u}\left(\alpha+\frac{\pi}{2}\right) + z'\overrightarrow{z_\lambda},$$

$$\Rightarrow \quad \overrightarrow{J^{(\lambda)}}(\mathrm{M}|t) = \frac{d^{(\lambda)}}{dt}\overrightarrow{v^{(\lambda)}}(\mathrm{M}|t) = r''\vec{u}(\alpha) + r'\frac{d^{(\lambda)}}{dt}\vec{u}(\alpha) + \ldots$$

$$\ldots + (r'\alpha' + \mathrm{r}\alpha'')\vec{u}\left(\alpha+\frac{\pi}{2}\right) + r\alpha'\frac{d^{(\lambda)}}{dt}\vec{u}\left(\alpha+\frac{\pi}{2}\right) + z''\overrightarrow{z_\lambda},$$

with

$$\frac{d^{(\lambda)}}{dt}\vec{u}(\alpha) = \alpha'\vec{u}\left(\alpha+\frac{\pi}{2}\right) \text{ and } \frac{d^{(\lambda)}}{dt}\vec{u}\left(\alpha+\frac{\pi}{2}\right) = -\alpha'\vec{u}(\alpha).$$

As a result,

$$\overrightarrow{J^{(\lambda)}}(\mathrm{M}|t) = (r'' - r\alpha'^2)\vec{u}(\alpha) + (r\alpha'' + 2r'\alpha')\vec{u}\left(\alpha+\frac{\pi}{2}\right) + z''\overrightarrow{z_\lambda}.$$

2.2.6.4. *Acceleration in the spherical coordinates*

$$\overrightarrow{v^{(\lambda)}}(\mathrm{M}|t) = R'\vec{k}(\alpha,\beta) + R\alpha'\cos\beta\,\vec{u}\left(\alpha+\frac{\pi}{2}\right) + R\beta'\vec{K}(\alpha,\beta)$$

$$\overrightarrow{J^{(\lambda)}}(\mathrm{M}|t) = \frac{d^{(\lambda)}}{dt}\overrightarrow{v^{(\lambda)}}(\mathrm{M}|t)$$,

$$\Rightarrow \quad \overrightarrow{J^{(\lambda)}}(\mathrm{M}|t) = R''\vec{k}(\alpha,\beta) + R'\frac{d^{(\lambda)}}{dt}\vec{k}(\alpha,\beta)\ldots$$

$$\ldots + (R'\alpha'\cos\beta + R\alpha''\cos\beta - R\alpha'\beta'\sin\beta)\vec{u}\left(\alpha+\frac{\pi}{2}\right)\ldots$$,

$$\ldots + R\alpha'\cos\beta\frac{d^{(\lambda)}}{dt}\vec{u}\left(\alpha+\frac{\pi}{2}\right)(R'\beta' + R\beta'')\vec{K}(\alpha,\beta) + R\beta'\frac{d^{(\lambda)}}{dt}\vec{K}(\alpha,\beta)$$

with

$$-\quad \frac{d^{(\lambda)}}{dt}\vec{k}(\alpha,\beta) = \frac{\partial^{(\lambda)}}{\partial\alpha}\vec{k}(\alpha,\beta)\times\frac{d\alpha}{dt} + \frac{\partial^{(\lambda)}}{\partial\beta}\vec{k}(\alpha,\beta)\times\frac{d\beta}{dt},$$

$$-\quad \frac{d^{(\lambda)}}{dt}\vec{K}(\alpha,\beta) = \frac{\partial^{(\lambda)}}{\partial\alpha}\vec{K}(\alpha,\beta)\times\frac{d\alpha}{dt} + \frac{\partial^{(\lambda)}}{\partial\beta}\vec{K}(\alpha,\beta)\times\frac{d\beta}{dt}.$$

To calculate the derivatives that act in these two expressions, we must follow the rules of the derivation of vectors, considering the basis

$$(b) = \left(\vec{k}(\alpha,\beta)\ \vec{u}\left(\alpha+\frac{\pi}{2}\right)\ \vec{K}(\alpha,\beta)\right),$$

$$-\quad \frac{\partial^{(\lambda)}}{\partial\alpha}\vec{k}(\alpha,\beta) - \frac{\partial^{(b)}}{\partial\alpha}\vec{k}(\alpha,\beta) = {}^{\alpha}\vec{\delta}(\lambda,b)\wedge\vec{k}(\alpha,\beta)$$

with $\frac{\partial^{(b)}}{\partial\alpha}\vec{k}(\alpha,\beta) = \vec{0}$,

$$-\quad \frac{\partial^{(\lambda)}}{\partial\beta}\vec{k}(\alpha,\beta) - \frac{\partial^{(b)}}{\partial\beta}\vec{k}(\alpha,\beta) = {}^{\beta}\vec{\delta}(\lambda,b)\wedge\vec{k}(\alpha,\beta)$$

with $\frac{\partial^{(b)}}{\partial\beta}\vec{k}(\alpha,\beta) = \vec{0}$,

$$-\quad \frac{\partial^{(\lambda)}}{\partial\alpha}\vec{K}(\alpha,\beta) - \frac{\partial^{(b)}}{\partial\alpha}\vec{K}(\alpha,\beta) = {}^{\alpha}\vec{\delta}(\lambda,b)\wedge\vec{K}(\alpha,\beta)$$

with $\frac{\partial^{(b)}}{\partial\alpha}\vec{K}(\alpha,\beta) = \vec{0}$,

$$-\ \frac{\partial^{(\lambda)}}{\partial\beta}\overrightarrow{K}(\alpha,\beta)-\frac{\partial^{(b)}}{\partial\beta}\overrightarrow{K}(\alpha,\beta)={}^{\beta}\overrightarrow{\delta}(\lambda,b)\wedge\overrightarrow{K}(\alpha,\beta)$$

with $\dfrac{\partial^{(b)}}{\partial\beta}\overrightarrow{K}(\alpha,\beta)=\vec{0}$,

expressions where

$$-\ {}^{\alpha}\overrightarrow{\delta}(\lambda,b)=P_{\alpha}^{(b)}\,\vec{k}(\alpha,\beta)+Q_{\alpha}^{(b)}\,\vec{u}\left(\alpha+\frac{\pi}{2}\right)+R_{\alpha}^{(b)}\,\overrightarrow{K}(\alpha,\beta),$$

$$-\ {}^{\beta}\overrightarrow{\delta}(\lambda,b)=P_{\beta}^{(b)}\,\vec{k}(\alpha,\beta)+Q_{\beta}^{(b)}\,\vec{u}\left(\alpha+\frac{\pi}{2}\right)+R_{\beta}^{(b)}\,\overrightarrow{K}(\alpha,\beta).$$

The coefficients of these two relations are expressed generically as

$$\begin{cases} P_i^{(b)}=\dfrac{\partial^{(\lambda)}\vec{u}\left(\alpha+\dfrac{\pi}{2}\right)}{\partial q_i}\cdot\overrightarrow{K}(\alpha,\beta) \\[2ex] Q_i^{(b)}=\dfrac{\partial^{(\lambda)}\overrightarrow{K}(\alpha,\beta)}{\partial q_i}\cdot\vec{k}(\alpha,\beta) \\[2ex] R_i^{(b)}=\dfrac{\partial^{(\lambda)}\vec{k}(\alpha,\beta)}{\partial q_i}\cdot\vec{u}\left(\alpha+\dfrac{\pi}{2}\right) \end{cases},$$

with the basis change table

$(\lambda)\|(b)$	$\vec{k}(\alpha,\beta)$	$\vec{u}\left(\alpha+\frac{\pi}{2}\right)$	$\overrightarrow{K}(\alpha,\beta)$
$\overrightarrow{x_\lambda}$	$\cos\alpha\cos\beta$	$-\sin\alpha$	$-\cos\alpha\sin\beta$
$\overrightarrow{y_\lambda}$	$\sin\alpha\cos\beta$	$\cos\alpha$	$-\sin\alpha\sin\beta$
$\overrightarrow{z_\lambda}$	$\sin\beta$	0	$\cos\beta$

NOTE 2.2.– To maintain the coherence of this basis change table, the sum of the squares of the terms of a same line or of a same column must always be equal to 1, and the sum of the products of the corresponding terms of two lines or two columns must be equal to 0.

This results in

$$\begin{aligned} P_\alpha^{(b)} &= \sin\beta & P_\beta^{(b)} &= 0 \\ Q_\alpha^{(b)} &= 0 & Q_\beta^{(b)} &= -1 \\ R_\alpha^{(b)} &= \cos\beta & R_\beta^{(b)} &= 0 \end{aligned}$$

meaning

$$\overrightarrow{{}^{\alpha}\delta}(\lambda,S) = \sin\beta\,\vec{k}(\alpha,\beta) + \cos\beta\,\vec{K}(\alpha,\beta)$$

$$\overrightarrow{{}^{\beta}\delta}(\lambda,S) = -\vec{u}\left(\alpha + \frac{\pi}{2}\right)$$

After calculation, the expression of the acceleration vector in the spherical coordinate system is obtained.

$$\overrightarrow{J^{(\lambda)}}(\mathrm{M}|\mathrm{t}) = (R'' - R\alpha'^2\cos^2\beta - R\beta'^2)\,\vec{k}(\alpha,\beta)\ldots$$

$$\ldots + (2R'\alpha'\cos\beta + R\alpha''\cos\beta - 2R\alpha'\beta'\sin\beta)\,\vec{u}\left(\alpha + \frac{\pi}{2}\right)\ldots$$

$$\ldots + (R\beta'' + 2R'\beta' + R\alpha'^2\sin\beta\cos\beta)\,\vec{K}(\alpha,\beta)$$

2.2.6.5. *Acceleration in generalized coordinates*

$$\overrightarrow{v^{(\lambda)}}(\mathrm{M}|t) = \frac{\partial^{(\lambda)}\vec{f}}{\partial Q_1}Q_1' + \frac{\partial^{(\lambda)}\vec{f}}{\partial Q_2}Q_2' + \frac{\partial^{(\lambda)}\vec{f}}{\partial Q_3}Q_3' = \sum_{i=1}^{3}\frac{\partial^{(\lambda)}\vec{f}}{\partial Q_i}Q_i',$$

$$\overrightarrow{J^{(\lambda)}}(\mathrm{M}|t) = \frac{d^{(\lambda)}}{dt}\overrightarrow{v^{(\lambda)}}(\mathrm{M}|t) = \frac{d^{(\lambda)}}{dt}\left[\frac{\partial^{(\lambda)}\vec{f}}{\partial Q_1}Q_1' + \frac{\partial^{(\lambda)}\vec{f}}{\partial Q_2}Q_2' + \frac{\partial^{(\lambda)}\vec{f}}{\partial Q_3}Q_3'\right]$$

$$\overrightarrow{J^{(\lambda)}}(\mathrm{M}|t) = \frac{\partial^{(\lambda)}\vec{f}}{\partial Q_1}Q_1'' + \frac{\partial^{(\lambda)}\vec{f}}{\partial Q_2}Q_2'' + \frac{\partial^{(\lambda)}\vec{f}}{\partial Q_3}Q_3''\ldots$$

$$\ldots + \frac{\partial^{(\lambda)}}{\partial Q_1}\left(\frac{\partial^{(\lambda)}\vec{f}}{\partial Q_1}Q_1' + \frac{\partial^{(\lambda)}\vec{f}}{\partial Q_2}Q_2' + \frac{\partial^{(\lambda)}\vec{f}}{\partial Q_3}Q_3'\right)Q_1'\ldots$$

$$\ldots + \frac{\partial^{(\lambda)}}{\partial Q_2}\left(\frac{\partial^{(\lambda)}\vec{f}}{\partial Q_1}Q_1' + \frac{\partial^{(\lambda)}\vec{f}}{\partial Q_2}Q_2' + \frac{\partial^{(\lambda)}\vec{f}}{\partial Q_3}Q_3'\right)Q_2'\ldots$$

$$\ldots + \frac{\partial^{(\lambda)}}{\partial Q_3}\left(\frac{\partial^{(\lambda)}\vec{f}}{\partial Q_1}Q_1' + \frac{\partial^{(\lambda)}\vec{f}}{\partial Q_2}Q_2' + \frac{\partial^{(\lambda)}\vec{f}}{\partial Q_3}Q_3'\right)Q_3'$$

$$\overrightarrow{J^{(\lambda)}}(\mathrm{M}|t) = \frac{\partial^{(\lambda)}\vec{f}}{\partial Q_1}Q_1'' + \frac{\partial^{(\lambda)}\vec{f}}{\partial Q_2}Q_2'' + \frac{\partial^{(\lambda)}\vec{f}}{\partial Q_3}Q_3'' \ldots$$
$$\ldots + \frac{\partial^{(\lambda)2}\vec{f}}{\partial Q_1^2}Q_1'^2 + \frac{\partial^{(\lambda)2}\vec{f}}{\partial Q_2^2}Q_2'^2 + \frac{\partial^{(\lambda)2}\vec{f}}{\partial Q_3^2}Q_3'^2 \ldots \qquad ,$$
$$\ldots + 2\left(\frac{\partial^{(\lambda)2}\vec{f}}{\partial Q_1 \partial Q_2}Q_1'Q_2' + \frac{\partial^{(\lambda)2}\vec{f}}{\partial Q_2 \partial Q_3}Q_2'Q_3' + \frac{\partial^{(\lambda)2}\vec{f}}{\partial Q_1 \partial Q_3}Q_1'Q_3'\right)$$

or, in a more condensed form,

$$\overrightarrow{J^{(\lambda)}}(\mathrm{M}|t) = \sum_{i=1}^{3}\frac{\partial^{(\lambda)}\vec{f}}{\partial Q_i}Q_i'' + \sum_{i=1}^{3}\sum_{j=1}^{3}\frac{\partial^{(\lambda)2}\vec{f}}{\partial Q_i \partial Q_j}Q_i'Q_j' .$$

2.2.7. *Exercises*

2.2.7.1. *Exercise 11 – Involute of a circle*

The *involute* of a circle (Γ) can be defined as the location of the extremity M of a string rolled around a cylindrical spindle, which is unrolled, having been kept taut, and therefore tangential to the circle at its point of separation K.

In the planar reference frame $\langle 0 \rangle \equiv \langle \mathrm{O}|\vec{x}\,\vec{y} \rangle$ linked to the spindle, the trajectory of M is considered to be an involute of the circle whose center is O and whose radius is α. M is located at M_0 for $\alpha = 0$.

We consider $\alpha = \widehat{\vec{x},\overrightarrow{\mathrm{OK}}}$ where α is a twice derivable function of time t and is increasing.

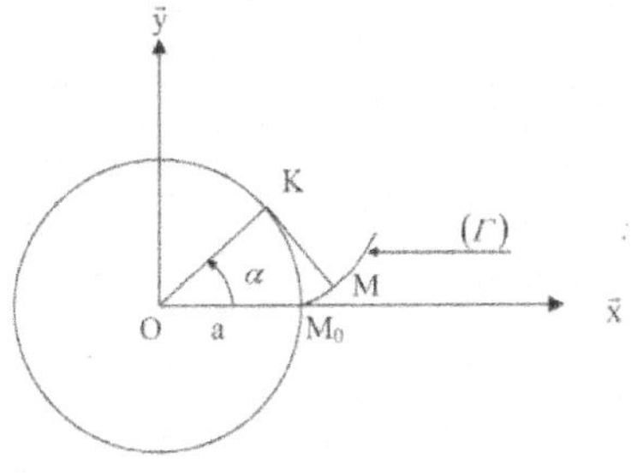

We have: length of the KM segment = length of the arc of the circle KM_0.

– *Question 1*: Determine vector $\overrightarrow{OM} = \vec{f}(\alpha)$:

$$\overrightarrow{OM} = \overrightarrow{OK} + \overrightarrow{KM} = a\vec{u}(\alpha) + a\alpha\vec{u}\left(\alpha - \frac{\pi}{2}\right).$$

– *Question 2*: Determine the velocity vector $\overrightarrow{v^{(0)}}(M)$, the tangential unit vector $\vec{\tau}$ to the involute and the norm v of the velocity vector.

$$\overrightarrow{v^{(0)}}(M) = \frac{d^{(0)}}{dt}\overrightarrow{OM} = a\alpha'\vec{u}\left(\alpha + \frac{\pi}{2}\right) + a\alpha'\vec{u}\left(\alpha - \frac{\pi}{2}\right) + a\alpha\alpha'\vec{u}(\alpha)$$

$$\text{'' ''} \quad = a\alpha\alpha'\vec{u}(\alpha)$$

where $\overrightarrow{v^{(0)}}(M)$ is in the form of $v\vec{\tau}$ with $\alpha' > 0$ since α is an increasing function. This means that:

$$\vec{\tau} = \vec{u}(\alpha) \ , \ v = a\alpha\alpha' > 0.$$

– *Question 3*: Determine the acceleration vector $\overrightarrow{J^{(0)}}(M)$; deduce the normal acceleration vector and the tangential acceleration vector as well as the value of the radius of curvature. Where is the center of curvature?

$$\overrightarrow{J^{(0)}}(M) = \frac{d^{(0)}}{dt}\overrightarrow{v^{(0)}}(M) = a\left(\alpha'^2 + \alpha\alpha''\right)\vec{u}(\alpha) + a\alpha\alpha'^2\vec{u}\left(\alpha + \frac{\pi}{2}\right).$$

If we write this expression as

$$\overrightarrow{J^{(0)}}(M) = J_t\vec{\tau} + J_n\vec{\nu} = J_t\vec{u}(\alpha) + J_n\vec{u}\left(\alpha + \frac{\pi}{2}\right)$$

where J_t is the tangential acceleration and J_n the normal acceleration to the point M. Considering that

$$J_n = \frac{v^2}{R} = \frac{a^2\alpha^2\alpha'^2}{R} = a\alpha\alpha'^2.$$

we can deduce the curvature radius of the trajectory of M

$$R = a\alpha = \text{MK} .$$

The curvature center of trajectory (Γ) of M is the point K, which is the point of separation of the string.

– *Question 4*: Draw trajectory (Γ) of M for $\alpha = \left[0, \frac{3\pi}{2}\right]$.

Seeing as the circle contains the curvature centers of the curve (Γ), it is said to be *involuted*; the curve (Γ) is, therefore, an *involute* of the circle.

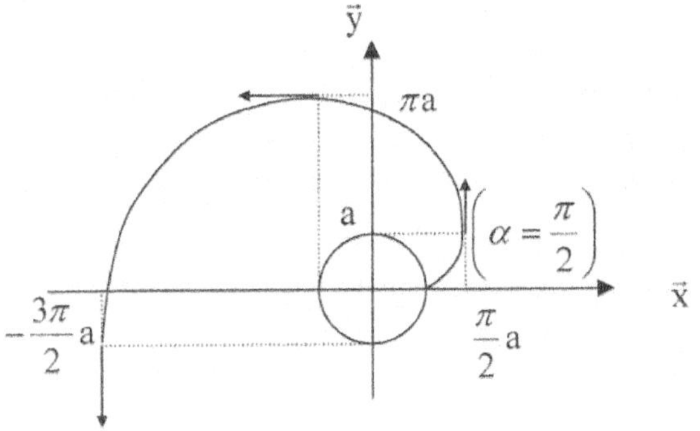

2.2.7.2. *Exercise 12 – Cycloid*

A cycloid is defined as the trajectory of a point M of a circle that rolls without sliding along a line belonging to its plane. For example, let us consider, at a given time, M as the point of contact of the circle with its supporting line; the resulting curve, obtained between this situation and the next after a full rotation of the circle, where the point M is again in contact with the supporting line, is called the *arc* (Γ) of the cycloid.

In the plane of $\Pi\left(\text{O}|\vec{x},\vec{y}\right)$, the circle of center C and of radius a is in contact with the line $\left(\text{O}|\vec{x}\right)$ at point I. We can write $\alpha = \widehat{\overrightarrow{\text{CM}},\overrightarrow{\text{CI}}}$, with $0 < \alpha < 2\pi$.

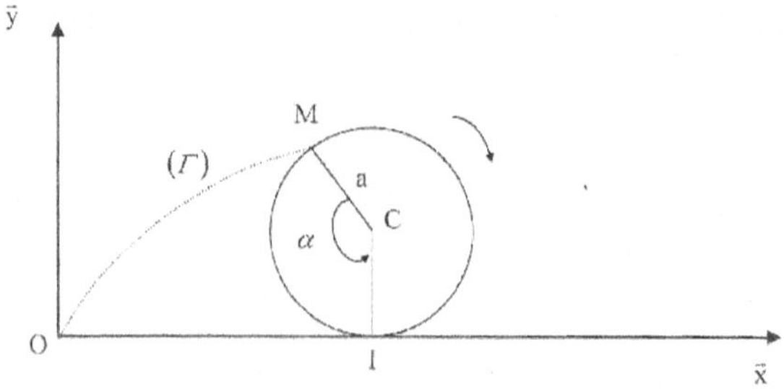

The non-sliding of the circle on the line means that:

length of the circle arc IM = length of segment OI.

– *Question 1*: Determine $\overrightarrow{OM} = \vec{F}(\alpha)$.

$$\overrightarrow{OM} = \overrightarrow{OI} + \overrightarrow{IC} + \overrightarrow{CM} = a\alpha\vec{x} + a\vec{y} + a\alpha\vec{u}\left(\widehat{\vec{x},\overrightarrow{CM}}\right),$$

with $\widehat{\vec{x},\overrightarrow{CM}} = \widehat{\vec{x},\vec{y}} + \widehat{\vec{y},\overrightarrow{CI}} + \widehat{\overrightarrow{CI},\overrightarrow{CM}} = \frac{\pi}{2} + \pi - \alpha$.

$$\begin{aligned} \overrightarrow{OM} &= a\alpha\vec{x} + a\vec{y} + a\alpha\left[-\sin\alpha\vec{x} - \cos\alpha\vec{y}\right] \\ \text{""} &= a(\alpha - \sin\alpha)\vec{x} + a\left(1 - \cos\alpha\vec{y}\right) \end{aligned},$$

$$\Rightarrow \vec{F}(\alpha) = a(\alpha - \sin\alpha)\vec{x} + a\left(1 - \cos\alpha\vec{y}\right).$$

– *Question 2(a)*: Calculate $\vec{F_1} = \frac{d^{(0)}\vec{F}}{d\alpha}$ and deduce:

– $\frac{ds}{d\alpha}$, where s is the curvilinear abscissa of M measured from O;

– $\vec{\tau}$, unit vector tangential to (Γ) at M.

$$\vec{F_1} = \frac{d^{(0)}\vec{F}}{d\alpha} = a(1 - \cos\alpha)\vec{x} + a\sin\alpha\vec{y} = 2a\sin\frac{\alpha}{2}\left(\sin\frac{\alpha}{2}\vec{x} + \cos\frac{\alpha}{2}\vec{y}\right).$$

$$\text{""} = 2a\sin\frac{\alpha}{2}\vec{u}\left(\frac{\pi}{2} - \frac{\alpha}{2}\right)$$

By definition, the curvilinear abscissa, which provides the length of the arc of the curve, is expressed as

$$\left(\frac{d^{(0)}\vec{F}}{ds}\right)^2 = 1 \Rightarrow \left(\frac{d^{(0)}\vec{F}}{ds}\right)^2 = \left(\frac{d^{(0)}\vec{F}}{d\alpha}\frac{d\alpha}{ds}\right)^2 = \left(2a\sin\frac{\alpha}{2}\frac{d\alpha}{ds}\right)^2 = 1,$$

$$\Rightarrow \frac{ds}{d\alpha} = \varepsilon \times 2a\sin\frac{\alpha}{2} \quad \text{with } \varepsilon = \pm 1.$$

Furthermore, $\vec{u}$ is a unit vector and $\vec{F_1}$ is tangential to the curve; as a result the unit vector $\vec{\tau}$ tangential to the curve is

$$\vec{\tau} = \vec{u}\left(\frac{\pi}{2} - \frac{\alpha}{2}\right).$$

– *Question 2(b)*: Calculate the length ℓ of an arc of the cycloid.

The length ℓ of an arc of the cycloid is the integral of ds in an interval of variation of α.

$$\ell = \int_{\alpha_1}^{\alpha_2} 2a\sin\frac{\alpha}{2}d\alpha = \left[-4a\cos\frac{\alpha}{2}\right]_{\alpha_1}^{\alpha_2}.$$

– *Question 2(c)*: Draw the arc of the cycloid obtained for $0 < \alpha < 2\pi$.

$$\ell = \int_0^{2\pi} 2a\sin\frac{\alpha}{2}d\alpha = \left[-4a\cos\frac{\alpha}{2}\right]_0^{2\pi} = 8a.$$

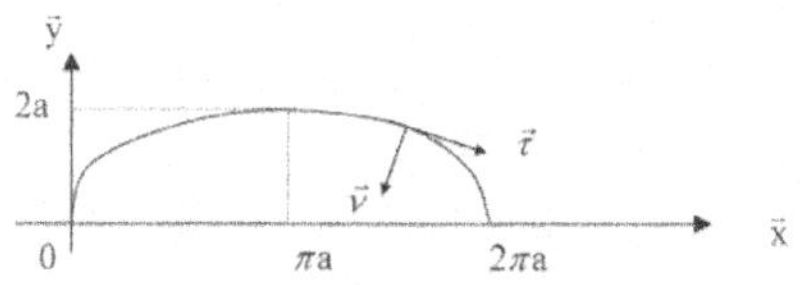

– *Question 3*: Calculate $\frac{d^{(0)}\vec{\tau}}{ds}$ and deduce:

– the curvature radius R;

– the normal unit vector $\vec{\nu}$ to (Γ) at M.

$$\vec{\tau} = \vec{u}\left(\frac{\pi}{2} - \frac{\alpha}{2}\right),$$

$$\Rightarrow \quad \frac{d^{(0)}\vec{\tau}}{ds} = \frac{d^{(0)}}{d\alpha}\vec{u}\left(\frac{\pi}{2} - \frac{\alpha}{2}\right)\frac{d\alpha}{ds}$$

$$\text{" "} = -\frac{1}{2}\vec{u}\left(\pi - \frac{\alpha}{2}\right)\frac{1}{2a\sin\frac{\alpha}{2}} = \frac{1}{2}\vec{u}\left(-\frac{\alpha}{2}\right)\frac{1}{2a\sin\frac{\alpha}{2}} = \frac{\vec{\nu}}{R},$$

$$\Rightarrow \quad R = 4a\sin\frac{\alpha}{2} \quad , \quad \vec{\nu} = \vec{u}\left(-\frac{\alpha}{2}\right).$$

– *Question 4*: α is a twice-derivable function of time *t*. As a function of α, determine the velocity and acceleration vectors of M throughout its trajectory and state the tangential and normal accelerations at this point:

$$- \quad \overrightarrow{v^{(0)}}(\mathrm{M}) = \frac{d^{(0)}\vec{F}}{dt} = \vec{F_1}\frac{d\alpha}{dt} = 2a\alpha' \sin\frac{\alpha}{2}\vec{u}\left(\frac{\pi}{2} - \frac{\alpha}{2}\right);$$

$$- \overrightarrow{J^{(0)}}(\mathrm{M}) = \left(2a\alpha'' \sin\frac{\alpha}{2} + a\alpha'^2 \cos\frac{\alpha}{2}\right)\vec{u}\left(\frac{\pi}{2} - \frac{\alpha}{2}\right) + a\alpha'^2 \sin\frac{\alpha}{2}\vec{u}\left(-\frac{\alpha}{2}\right).$$

Seeing as $\overrightarrow{J^{(0)}}(\mathrm{M}) = J_t\vec{\tau} + J_n\vec{\nu}$, this means that:

$$J_t = a\left(2\alpha'' \sin\frac{\alpha}{2} + \alpha'^2 \cos\frac{\alpha}{2}\right) \quad , \quad J_n = a\alpha'^2 \sin\frac{\alpha}{2}$$

– Question 5: $\alpha = \omega t$, where ω is a positive constant.

What happens to the velocity and acceleration vectors?

State the characteristics of the motion when M describes an arc of the cycloid (accelerated, decelerated or uniform motion).

For $\alpha = \omega t, \alpha' = \omega, \alpha'' = 0$, there is:

– $\overrightarrow{v^{(0)}}(\mathrm{M}) = 2a\omega \sin\frac{\omega t}{2}\vec{u}\left(\frac{\pi}{2} - \frac{\omega t}{2}\right)$;

– $\overrightarrow{J^{(0)}}(\mathrm{M}) = \left(a\omega^2 \cos\frac{\omega t}{2}\right)\vec{u}\left(\frac{\pi}{2} - \frac{\omega t}{2}\right) + a\omega^2 \sin\frac{\omega t}{2}\vec{u}\left(-\frac{\omega t}{2}\right)$.

Depending on the sign of the scalar product:

$$\overrightarrow{v^{(0)}}(\mathrm{M})\cdot\overrightarrow{J^{(0)}}(\mathrm{M}) = a^2\omega^3 \sin\frac{\omega t}{2}\cos\frac{\omega t}{2} = 2a^2\omega^3 \sin \omega t$$

– the motion is accelerated when the sign is positive as the angle between the velocity and acceleration is acute, meaning that the tangential component of the acceleration tends to increase the velocity;

– the motion is slowed when the sign is negative as the angle between the velocity and acceleration is opposed to the direction of the velocity and tends to decrease it.

As a result:

– $0 < \omega t < \pi$, the motion is accelerated;

– $\pi < \omega t < 2\pi$, the motion is slowed.

2.3. Velocity field associated with the motion of the rigid solid

The rigid character of the solid studied in this book would suggest a link between the velocities and accelerations to which its component particles are subjected. This is what this section aims to demonstrate.

2.3.1. *Fundamental formula for the velocity*

Let us consider three particles O_S, M_S and P_S belonging to the same rigid solid (S) attached to the frame $\langle S\rangle \equiv \left\langle \mathrm{O}_S \middle| \overrightarrow{x_S}\ \overrightarrow{y_S}\ \overrightarrow{z_S}\right\rangle$, and whose motion is observed from the frame $\langle \lambda\rangle \equiv \left\langle \mathrm{O}_\lambda \middle| \overrightarrow{x_\lambda}\ \overrightarrow{y_\lambda}\ \overrightarrow{z_\lambda}\right\rangle$. These three particles can be considered as material points and have the following trajectories, respectively, $\Gamma_{\lambda}^{(t_0,t_f)}(\mathrm{O}_S)$, $\Gamma_{\lambda}^{(t_0,t_f)}(\mathrm{M}_S)$ and $\Gamma_{\lambda}^{(t_0,t_f)}(\mathrm{P}_S)$, which are defined from the position vectors $\overrightarrow{\mathrm{O}_{\lambda}\mathrm{O}_S}$, $\overrightarrow{\mathrm{O}_{\lambda}\mathrm{M}_S}$ and $\overrightarrow{\mathrm{O}_{\lambda}\mathrm{P}_S}$.

By definition, the trajectory of a point M_S is the location of the points $M_{\lambda}(t)$ of the affine space with which they coincide at the time t during its motion in the time interval $[t_0,t_f]$; for reasons of simplicity of writing, hereafter we shall combine the material point and its "image" in the affine space. This shall also be the case for all the particles that make up the solid.

Also, by definition, the velocity of these three points in the frame $\langle \lambda \rangle$ is expressed as

$$\overrightarrow{v^{(\lambda)}}(\mathrm{O}_S) = \frac{d^{(\lambda)}}{dt}\overrightarrow{\mathrm{O}_{\lambda}\mathrm{O}_S} \ ; \ \overrightarrow{v^{(\lambda)}}(\mathrm{M}_S) = \frac{d^{(\lambda)}}{dt}\overrightarrow{\mathrm{O}_{\lambda}\mathrm{M}_S} \ ; \ \overrightarrow{v^{(\lambda)}}(\mathrm{P}_S) = \frac{d^{(\lambda)}}{dt}\overrightarrow{\mathrm{O}_{\lambda}\mathrm{P}_S} .$$

Furthermore,

$$\overrightarrow{\mathrm{O}_{\lambda}\mathrm{M}_S} = \overrightarrow{\mathrm{O}_{\lambda}\mathrm{O}_s} + \overrightarrow{\mathrm{O}_S\mathrm{M}_S} = \overrightarrow{\mathrm{O}_{\lambda}\mathrm{O}_s} + x_S(\mathrm{M}_S)\overrightarrow{x_S} + y_S(\mathrm{M}_S)\overrightarrow{y_S} + z_S(\mathrm{M}_S)\overrightarrow{z_S} .$$

The terms $x_S(\mathrm{M}_S)$, $y_S(\mathrm{M}_S)$, $z_S(\mathrm{M}_S)$ are independent of time, since the three points, which belong to the solid, are fixed within the frame $\langle S \rangle$. However, the vector $\overrightarrow{\mathrm{O}_S\mathrm{M}_S}$ (even though its modulus is constant) is variable in relation to $\langle \lambda \rangle$ as the solid, and as a result the system attached to it, both evolve and change within it over time:

$$\begin{aligned}\overrightarrow{v^{(\lambda)}}(\mathrm{M}_S) &= \frac{d^{(\lambda)}}{dt}\overrightarrow{\mathrm{O}_{\lambda}\mathrm{M}_S} = \frac{d^{(\lambda)}}{dt}\overrightarrow{\mathrm{O}_{\lambda}\mathrm{O}_S} + \frac{d^{(\lambda)}}{dt}\overrightarrow{\mathrm{O}_S\mathrm{M}_S} \\ \text{'' ''} \quad &= \overrightarrow{v^{(\lambda)}}(\mathrm{O}_S) + \frac{d^{(\lambda)}}{dt}\overrightarrow{\mathrm{O}_S\mathrm{M}_S}\end{aligned} .$$

According to the derivation of the vectorial functions[2]:

$$\frac{d^{(\lambda)}}{dt}\overrightarrow{\mathrm{O}_S\mathrm{M}_S} - \frac{d^{(S)}}{dt}\overrightarrow{\mathrm{O}_S\mathrm{M}_S} = \overrightarrow{\omega_S^{\lambda}} \wedge \overrightarrow{\mathrm{O}_S\mathrm{M}_S} ,$$

2 See Volume 2 [BOR 16a].

where $\frac{d^{(S)}}{dt}\overrightarrow{\mathrm{O}_S\mathrm{M}_S} = \vec{0}$ since the vector $\overrightarrow{\mathrm{O}_S\mathrm{M}_S}$ is fixed in the frame (*S*), meaning that

$$\overrightarrow{v^{(\lambda)}}(\mathrm{M}_S) = \overrightarrow{v^{(\lambda)}}(\mathrm{O}_S) + \overrightarrow{\omega_S^{\lambda}} \wedge \overrightarrow{\mathrm{O}_S\mathrm{M}_S} .$$

This formula can be applied to any other material point of the solids (*S*), notably at the point P_s:

$$\overrightarrow{v^{(\lambda)}}(\mathrm{P}_S) = \overrightarrow{v^{(\lambda)}}(\mathrm{O}_S) + \overrightarrow{\omega_S^{\lambda}} \wedge \overrightarrow{\mathrm{O}_S\mathrm{P}_S} ,$$

and through the difference between the two previous equations, we obtain

$$\overrightarrow{v^{(\lambda)}}(\mathrm{M}_S) - \overrightarrow{v^{(\lambda)}}(\mathrm{P}_S) = \overrightarrow{\omega_S^{\lambda}} \wedge \overrightarrow{\mathrm{M}_S\mathrm{P}_S} ,$$

which is the general expression of the fundamental formula of the velocity of the rigid solid.

2.3.2. *Use of matrix notation*

In matrix notation, a vector is represented in a basis by the column of its coordinates when it is projected onto it.

Thus, in the basis $(e) \equiv (\vec{x}\,\vec{y}\,\vec{z})$, the matrix representation of the vector $\vec{V} = a\vec{x} + b\vec{y} + c\vec{z}$ is the column

$$[V] = \begin{bmatrix} a \\ b \\ c \end{bmatrix} \text{ of dimension (3,1).}$$

The scalar product of the two vectors

$$\vec{V_1} = a_1\vec{x} + b_1\vec{y} + c_1\vec{z} \quad \text{and} \quad \vec{V_2} = a_2\vec{x} + b_2\vec{y} + c_2\vec{z}$$

is the scalar

$$\vec{V_1} \cdot \vec{V_2} = a_1 a_2 + b_1 b_2 + c_1 c_2 .$$

This operation takes on the form of a matrix

$$\vec{V_1} \cdot \vec{V_2} = \overline{[V_1]}[V_2] = [a_1 \quad b_1 \quad c_1] \begin{bmatrix} a_2 \\ b_2 \\ c_2 \end{bmatrix} = a_1 a_2 + b_1 b_2 + c_1 c_2 ,$$

of dimension (1,1) = (1,3)*(3,1).

The vectorial product of the two previous vectors is written

$$\begin{aligned} \vec{V_1} \wedge \vec{V_2} &= \left(a_1 \vec{x} + b_1 \vec{y} + c_1 \vec{z} \right) \wedge \left(a_2 \vec{x} + b_2 \vec{y} + c_2 \vec{z} \right) \ldots \\ \ldots &= (b_1 c_2 - b_2 c_1) \vec{x} + (c_1 a_2 - c_2 a_1) \vec{y} + (a_1 b_2 - a_2 b_1) \vec{z} \end{aligned} .$$

The result of the operation being a vector, of dimension (3,1), in dimensional terms this operation can only be written as (3,3)*(3,1), meaning that the operator $\left(\vec{V_1} \wedge \right)$ is of dimension (3,3) and is therefore represented as a matrix.

$$\left[\vec{V_1} \wedge \right] = \begin{bmatrix} 0 & -c_1 & b_1 \\ c_1 & 0 & -a_1 \\ -b_1 & a_1 & 0 \end{bmatrix} ,$$

such that $\begin{bmatrix} 0 & -c_1 & b_1 \\ c_1 & 0 & -a_1 \\ -b_1 & a_1 & 0 \end{bmatrix} \begin{bmatrix} a_2 \\ b_2 \\ c_2 \end{bmatrix} = \begin{bmatrix} b_1 c_2 - b_2 c_1 \\ c_1 a_2 - c_2 a_1 \\ a_1 b_2 - a_2 b_1 \end{bmatrix} .$

As a result, if we write

$$\vec{\omega} = \omega_1 \vec{x} + \omega_2 \vec{y} + \omega_3 \vec{z} ,$$

the matrix representation of the operator $\left(\vec{\omega} \wedge \right)$ is written

$$\left[\vec{\omega} \wedge \right] = \begin{bmatrix} 0 & -\omega_3 & \omega_2 \\ \omega_3 & 0 & -\omega_1 \\ -\omega_2 & \omega_1 & 0 \end{bmatrix} .$$

This matrix notation will be used often hereafter.

2.3.3. *Velocity-distributing torsor*

The kinematic situation in $\langle \lambda \rangle$ at any material point M of the solid (S) is, therefore, made apparent as soon as the essential vectors $\overrightarrow{\omega_S^{\lambda}}$ and $\overrightarrow{v^{(\lambda)}}(\mathrm{O}_S)$ are known, where O_S is a particular point of (S) taken as the origin of the frame $\langle S \rangle$ attached to it.

This provides $\overrightarrow{v^{(\lambda)}}(\mathrm{M})$ through the following relation:

$$\overrightarrow{v^{(\lambda)}}(\mathrm{M}) = \overrightarrow{v^{(\lambda)}}(\mathrm{O}_S) + \overrightarrow{\omega_S^{\lambda}} \wedge \overrightarrow{\mathrm{O}_S\mathrm{M}} = \overrightarrow{v^{(\lambda)}}(\mathrm{O}_S) + \overrightarrow{\mathrm{MO}_S} \wedge \overrightarrow{\omega_S^{\lambda}} \ ,$$

whose form, with its last term, is that of a moment relation, such as is used in the definition of the torsor, which is an essential mathematical tool further developed in Volume 2 [BOR 16a], where:

- $\overrightarrow{\omega_S^{\lambda}}$ plays the role of the resultant of this torsor;
- $\overrightarrow{v^{(\lambda)}}(\mathrm{O}_S)$ plays that of the moment at O_S of this same torsor.

They make up the *reduction elements at* O_S of a torsor called the *velocity-distributing* torsor, or *kinematic* torsor of the motion of the rigid solid (S) observed from the frame $\langle \lambda \rangle$. This particular torsor is written as

$$\left\{ {}^{\lambda}_{S} \right\}_{\mathrm{O}_S} = \left[\overrightarrow{\omega_S^{\lambda}} \,\middle|\, \overrightarrow{v^{(\lambda)}}(\mathrm{O}_S) \right].$$

2.3.4. *Partial distributing*

The motion of (S) in $\langle \lambda \rangle$ is generally defined by six parameters $Q_1, Q_2, Q_3, Q_4, Q_5, Q_6$ that are variable as functions of time, and the reduction elements of the velocity-distributing torsor at O_S are written as:

– $\overrightarrow{O_\lambda O_S} = \vec{f}(Q_1, Q_2, Q_3)$,

$$\Rightarrow \overrightarrow{v^{(\lambda)}}(O_S) = \frac{d^{(\lambda)}}{dt}\overrightarrow{O_\lambda O_S} = \sum_{i=1}^{3}\frac{\partial^{(\lambda)}\vec{f}}{\partial Q_i}Q_i' = \sum_{i=1}^{3}\overrightarrow{{}^{i}d}(O_S)Q_i'$$

– $\overrightarrow{\omega_S^\lambda} = \sum_{j=4}^{6}\overrightarrow{{}^{j}\delta}(\lambda, S)Q_j'$.

The velocity-distributing torsor is then written as

$$\left\{{}^{\lambda}_{S}\right\}_{O_S} = \left[\overrightarrow{\omega_S^\lambda}\middle|\overrightarrow{v^{(\lambda)}}(O_S)\right] = \left[\sum_{j=4}^{6}\overrightarrow{{}^{j}\delta}\,Q_j'\middle|\sum_{i=1}^{3}\overrightarrow{{}^{i}d}\,Q_i'\right].$$

These reduction elements appear as linear vectorial expressions whose coefficients are $Q_1',\ldots,Q_6'$; the torsor can then be broken down linearly in the following way:

$$\left\{{}^{\lambda}_{S}\right\}_{O_S} = \sum_{i=1}^{3}\left\{{}^{\lambda}_{S,i}\right\}Q_i' + \sum_{j=4}^{6}\left\{{}^{\lambda}_{S,j}\right\}Q_j' = \sum_{k=1}^{6}\left\{{}^{\lambda}_{S,k}\right\}Q_k',$$

an expression where new torsors $\left\{{}^{\lambda}_{S,k}\right\}$ emerge for $k = 1,\ldots,6$, which group together the parts of $\left\{{}^{\lambda}_{S}\right\}$ that are independent of the Q_k' in their reduction elements, resulting in

– $\left\{{}^{\lambda}_{S,1}\right\}_{O_S} = \left[\vec{0}\middle|\overrightarrow{{}^{1}d}(O_S) = \frac{\partial^{(\lambda)}\vec{f}}{\partial Q_1}\right]$,

– $\left\{{}^{\lambda}_{S,2}\right\}_{O_S} = \left[\vec{0}\middle|\overrightarrow{{}^{2}d}(O_S) = \frac{\partial^{(\lambda)}\vec{f}}{\partial Q_2}\right]$,

– $\left\{{}^{\lambda}_{S,3}\right\}_{O_S} = \left[\vec{0}\middle|\overrightarrow{{}^{3}d}(O_S) = \frac{\partial^{(\lambda)}\vec{f}}{\partial Q_3}\right]$,

– $\left\{{}^{\lambda}_{S,4}\right\}_{O_S} = \left[\overrightarrow{{}^{4}\delta}(\lambda, S)\middle|\vec{0}\right]$,

– $\left\{{}^{\lambda}_{S,5}\right\}_{O_S} = \left[\overrightarrow{{}^{5}\delta}(\lambda, S)\middle|\vec{0}\right]$,

$$-\left\{{}^{\lambda}_{S,6}\right\}_{O_S}=\left[\overrightarrow{{}^{6}\delta(\lambda,S)}\middle|\vec{0}\right].$$

The six torsors are called partial distributing torsors, the first three are coupled-torsors, whose resultant is equal to zero, and are the equivalent of respective moment couples $\overrightarrow{{}^{1}d}(O_S), \overrightarrow{{}^{2}d}(O_S), \overrightarrow{{}^{3}d}(O_S)$; the three others are called sliding torsors and are equivalent to vectors $\overrightarrow{{}^{4}\delta}(O_S), \overrightarrow{{}^{5}\delta}(O_S), \overrightarrow{{}^{6}\delta}(O_S)$ whose supporting line passes through the point O_S and whose moments are, therefore, equal to zero in that point.

NOTE 2.3.– The classification of torsors is presented in detail in Volume 2 [BOR 16a].

2.4. Acceleration field of the rigid solid

Let us consider two material points M_S and P_S that belong to the solids (*S*). The acceleration of P_S is defined by

$$\overrightarrow{J^{[\lambda]}}(P_S)=\frac{d^{(\lambda)}}{dt}\overrightarrow{v^{(\lambda)}}(P_S)=\frac{d^{(\lambda)}}{dt}\left[\overrightarrow{v^{(\lambda)}}(M_S)+\overrightarrow{\omega_S^{\lambda}}\wedge\overrightarrow{M_SP_S}\right]$$

$$\text{'' ''}\quad=\overrightarrow{J^{(\lambda)}}(M_S)+\frac{d^{(\lambda)}}{dt}\overrightarrow{\omega_S^{\lambda}}\wedge\overrightarrow{M_SP_S}+\overrightarrow{\omega_S^{\lambda}}\wedge\frac{d^{(\lambda)}}{dt}\overrightarrow{M_SP_S}.$$

2.4.1. *Derivative in relation to the time of the rate of rotation*

Seeing as the vector $\overrightarrow{\omega_S^{\lambda}}$ is the vectorial function of time, its derivative in relation to this parameter follows the law:

$$\frac{d^{(\lambda)}}{dt}\overrightarrow{\omega_S^{\lambda}}-\frac{d^{(S)}}{dt}\overrightarrow{\omega_S^{\lambda}}=\overrightarrow{\omega_S^{\lambda}}\wedge\overrightarrow{\omega_S^{\lambda}}=\vec{0},$$

which is $\dfrac{d^{(\lambda)}}{dt}\overrightarrow{\omega_S^{\lambda}}=\dfrac{d^{(S)}}{dt}\overrightarrow{\omega_S^{\lambda}}=\left(\overrightarrow{\omega_S^{\lambda}}\right)'$.

In these conditions, the derivation of the rotation rate vector $\overrightarrow{\omega_S^{\lambda}}$ can be done similarly in either of the two bases, (λ) or (S); the choice chiefly depends on the simplicity of the calculation or objective chosen.

2.4.2. *Derivation of a vector of the solid*

If we once again consider the two points M_S and P_S of the solid (S), the vector $\overrightarrow{M_SP_S}$ is a vectorial function of time; its derivative is therefore written

$$\frac{d^{(\lambda)}}{dt}\overrightarrow{M_SP_S} - \frac{d^{(S)}}{dt}\overrightarrow{M_SP_S} = \frac{d^{(\lambda)}}{dt}\overrightarrow{M_SP_S} = \overrightarrow{\omega_S^{\lambda}} \wedge \overrightarrow{M_SP_S} \text{ as } \frac{d^{(S)}}{dt}\overrightarrow{M_SP_S} = \vec{0}.$$

2.4.3. *Fundamental formula of acceleration*

Considering the results obtained above, the acceleration equation that links the two material points M_S and P_S, both belonging to the solid (*S*), is, therefore, written as

$$\overrightarrow{J^{(\lambda)}}(P_S) = \overrightarrow{J^{(\lambda)}}(M_S) + \left(\overrightarrow{\omega_S^{\lambda}}\right)' \wedge \overrightarrow{M_SP_S} + \overrightarrow{\omega_S^{\lambda}} \wedge \left(\overrightarrow{\omega_S^{\lambda}} \wedge \overrightarrow{M_SP_S}\right).$$

NOTE 2.4.– This relation between the accelerations of two material points in a same solid is only applicable if the two points belong to this same solid and are animated by the rate of rotation $\overrightarrow{\omega_S^{\lambda}}$ in relation to the basis (λ).

2.4.4. *Matrix notation of the vectorial product*

As discussed previously, the matrix representation of a matrix is a matrix column (3,1) and that of the vectorial operation $\vec{\omega} \wedge \vec{a}$ is the product of a squared matrix by a column:

$$\left[\vec{a}\right] = \begin{bmatrix} a_1 \\ a_2 \\ a_3 \end{bmatrix} \;;\; \left[\vec{\omega} \wedge \vec{a}\right] = \begin{bmatrix} 0 & -\omega_3 & \omega_2 \\ \omega_3 & 0 & -\omega_1 \\ -\omega_2 & \omega_1 & 0 \end{bmatrix} \begin{bmatrix} a_1 \\ a_2 \\ a_3 \end{bmatrix} = \begin{bmatrix} \omega_2 a_3 - \omega_3 a_2 \\ \omega_3 a_1 - \omega_1 a_3 \\ \omega_1 a_2 - \omega_2 a_1 \end{bmatrix}.$$

Seeing as the operator $\left(\vec{\omega}\wedge\right)$ is represented by the matrix (3,3) above, the operation $\vec{\omega}\wedge\left(\vec{\omega}\wedge\vec{a}\right)$ is then written:

$$\begin{bmatrix} 0 & -\omega_3 & \omega_2 \\ \omega_3 & 0 & -\omega_1 \\ -\omega_2 & \omega_1 & 0 \end{bmatrix}\begin{bmatrix} 0 & -\omega_3 & \omega_2 \\ \omega_3 & 0 & -\omega_1 \\ -\omega_2 & \omega_1 & 0 \end{bmatrix}\begin{bmatrix} a_1 \\ a_2 \\ a_3 \end{bmatrix}$$

$$'' \quad '' \quad = \begin{bmatrix} -\omega_2^2-\omega_3^2 & \omega_1\omega_2 & \omega_3\omega_1 \\ \omega_1\omega_2 & -\omega_3^2-\omega_1^2 & \omega_2\omega_3 \\ \omega_3\omega_1 & \omega_2\omega_3 & -\omega_1^2-\omega_2^2 \end{bmatrix}\begin{bmatrix} a_1 \\ a_2 \\ a_3 \end{bmatrix}.$$

This provides the matrix elements that are involved in the expression of acceleration.

2.4.5. *Exercises*

2.4.5.1. *Exercise 13 – Kinematics of an articulated solid*

In frame $\langle\lambda\rangle \equiv \left\langle \mathrm{A}\middle|\overrightarrow{x_\lambda}\,\overrightarrow{y_\lambda}\,\overrightarrow{z_\lambda}\right\rangle$, we consider an articulated parallelogram ABCD made of four rigid bars. Its situation is governed by a single parameter $\gamma(t)$.

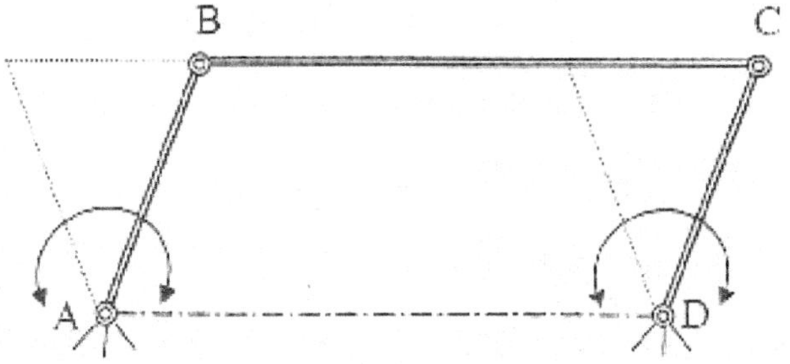

The situation of the articulated device is described in the following figure:

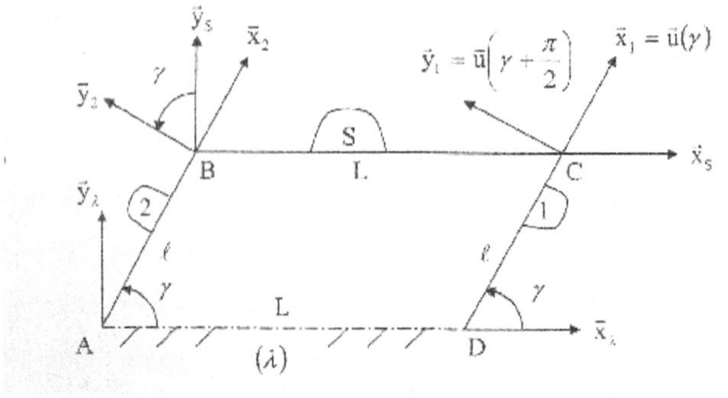

– *Question 1*: Bar BC, whose length is L, is identified as the solid (S). Write out the velocity-distributing torsor $\left\{{}^{\lambda}_{S}\right\}_B$. What are the trajectories in $\langle\lambda\rangle$ of the different points of the bar BC? Characterize the motion of (S) in relation to $\langle\lambda\rangle$.

Bar BC, to which the frame $\langle S\rangle \equiv \left\langle B|\vec{x_S}\ \vec{y_S}\ \vec{z_S}\right\rangle$ is affected, with $\vec{z_S} \equiv \vec{z_\lambda}$, and which remains parallel to the device represented by the segment AD, experiences translational motion (rate of rotation equal to zero). Its extremity B is located in $\langle\lambda\rangle$ by the position vector

$$\overrightarrow{AB} = \ell\vec{u}(\gamma),$$

$$\vec{v}(B) = \frac{d^{(\lambda)}}{dt}\overrightarrow{AB} = \ell\gamma'\vec{u}\left(\gamma+\frac{\pi}{2}\right) = \vec{v},$$

which means that the velocity-distributing torsor $\left\{{}^{\lambda}_{S}\right\}$ at B, which is a couple-torsor is expressed as

$$\left\{{}^{\lambda}_{S}\right\}_B = \left[\vec{0}\middle|\ell\gamma'\vec{u}\left(\gamma+\frac{\pi}{2}\right)\right].$$

$$\forall M_S \in (S),\ \overrightarrow{AM_S} = \overrightarrow{AB} + \overrightarrow{BM_S} = \ell\vec{u}(\gamma) + \mu\vec{x_S} = (\ell\cos\gamma + \mu)\vec{x_\lambda} + \ell\sin\gamma\vec{y_\lambda}$$

with $\mu < L$.

$$\overrightarrow{AM_S}^2 = (\ell\cos\gamma + \mu)^2 + \ell^2\sin^2\gamma.$$

This expression is the equation of a circle with a radius of ℓ and a center Q_{M_S} such that

$$\overrightarrow{AQ_{M_S}} = \mu \overrightarrow{x_\lambda} .$$

This verifies that

$$\overrightarrow{Q_{M_S} M_S}^2 = \left(\overrightarrow{AM_S} - \overrightarrow{AQ_{M_S}} \right)^2 = \ell^2 .$$

Each point M_S of the bar BC, therefore, moves when γ varies with time, on a circle of radius ℓ, which is unique to it and whose center Q_{M_S} is located on the fixed AD.

Bar (S) experiences a motion of circular translation in relation to $\langle \lambda \rangle$.

– *Question 2*: Bar DC is identified as the solid (1). Write out the torsor $\left\{ {}^\lambda_1 \right\}_D$ and determine $\overrightarrow{M_C}\left\{ {}^\lambda_1 \right\}$. What are the trajectories in $\langle \lambda \rangle$ of the different points of the bar DC? Characterize the motion of (1) in relation to $\langle \lambda \rangle$.

The bar DC is linked to the frame $\langle 1 \rangle \equiv \left\langle C | \overrightarrow{x_1}\, \overrightarrow{y_1}\, \overrightarrow{z_1} \right\rangle$ with the following equation:

$$\overrightarrow{x_1} = \vec{u}(\gamma) \quad , \quad \overrightarrow{y_1} = \vec{u}\left(\gamma + \frac{\pi}{2} \right) \quad , \quad \overrightarrow{z_1} \equiv \overrightarrow{z_\lambda} .$$

This bar undergoes a motion of rotation around the axis $\left(D | \overrightarrow{z_\lambda} \right)$

$$\Rightarrow \quad \overrightarrow{\omega_1^\lambda} = \gamma' \overrightarrow{z_\lambda} .$$

D is fixed in relation to $\langle \lambda \rangle$ and $\langle 1 \rangle$ $\Rightarrow$ $\overrightarrow{\mathscr{M}_D}\left\{ {}^\lambda_1 \right\} = \vec{0}$. As a result

$$\left\{ {}^\lambda_1 \right\}_D = \left[\gamma' \overrightarrow{z_\lambda} \middle| \vec{0} \right] .$$

According to the base property of the velocities of the rigid solid

$$\overrightarrow{\mathcal{M}_C}\left\{{}^{\lambda}_{1}\right\} = \overrightarrow{v^{(\lambda)}}(C) = \overrightarrow{v^{(\lambda)}}(D) + \overrightarrow{\omega_1^{\lambda}} \wedge \overrightarrow{DC}$$

$$\text{'' ''} \quad = \gamma' \overrightarrow{z_\lambda} \wedge \ell \vec{u}(\gamma) = \ell \gamma' \vec{u}\left(\gamma + \frac{\pi}{2}\right)$$

$$\forall M_1 \in (1),\ \overrightarrow{DM_1} = \nu \overrightarrow{x_1} = \nu \cos\gamma \overrightarrow{X_1} + \nu \sin\gamma \overrightarrow{Y_1}\ , \text{ with } \nu < \ell$$

$$\Rightarrow\ \overrightarrow{DM_1}^{2} = \nu^2 , \forall \gamma .$$

Furthermore,

$$\overrightarrow{v^{(\lambda)}}(M_1) = \overrightarrow{v^{(\lambda)}}(D) + \overrightarrow{\omega_1^{\lambda}} \wedge \overrightarrow{DM_1} = \gamma' \overrightarrow{z_\lambda} \wedge \nu \vec{u}(\gamma) = \nu \gamma' \vec{u}\left(\gamma + \frac{\pi}{2}\right),$$

meaning that the velocity vector of M_1 is always orthogonal to the radius vector $\overrightarrow{DM_1}$.

Point M_1 is, therefore, located on a circle of center D and of radius $\nu = \left\|\overrightarrow{DM_1}\right\|$. Solid (1) undergoes a motion of rotation around a fixed axis $\left(D|\overrightarrow{z_\lambda}\right)$ in relation to $\langle\lambda\rangle$.

– *Question 3*: Bar AB is identified as Bolid (2). Write out the torsors $\left\{{}^{\lambda}_{2}\right\}_A$ and $\left\{{}^{\lambda}_{2}\right\}_B$. Characterize the motion of (2) in relation to $\langle\lambda\rangle$.

Bar AB is linked to the frame $\langle 2\rangle \equiv \left\langle B|\overrightarrow{x_2}\,\overrightarrow{y_2}\,\overrightarrow{z_2}\right\rangle$ with

$$\overrightarrow{x_2} = \vec{u}(\gamma)\ ,\ \overrightarrow{y_2} = \vec{u}\left(\gamma + \frac{\pi}{2}\right)\ ,\ \overrightarrow{z_2} = \overrightarrow{z_\lambda} .$$

This bar undergoes a motion of rotation around the axis $\left(A|\overrightarrow{z_\lambda}\right)$

$$\Rightarrow\ \overrightarrow{\omega_2^{\lambda}} = \gamma' \overrightarrow{z_\lambda} .$$

A is fixed in relation to $\langle\lambda\rangle$ and $\langle 2\rangle$ $\Rightarrow$ $\overrightarrow{\mathcal{M}_A}\left\{ {}^{\lambda}_{2} \right\} = \vec{0}$. As a result

$$\left\{ {}^{\lambda}_{2} \right\}_A = \left[\gamma' \overrightarrow{z_\lambda} \middle| \vec{0} \right].$$

According to the base property on the velocities of the rigid solid

$$\begin{aligned} \overrightarrow{\mathcal{M}_B}\left\{ {}^{\lambda}_{2} \right\} &= \overrightarrow{v^{(\lambda)}}(B) = \overrightarrow{v^{(\lambda)}}(A) + \overrightarrow{\omega_2^\lambda} \wedge \overrightarrow{AB} \\ \text{" "} \quad &= \gamma' \overrightarrow{z_\lambda} \wedge \ell \vec{u}(\gamma) = \ell \gamma' \vec{u}\left(\gamma + \frac{\pi}{2} \right) \end{aligned}.$$

$$\forall M_2 \in (2),\ \overrightarrow{AM_2} = \eta \overrightarrow{x_1} = \eta \cos\gamma \overrightarrow{X_1} + \eta \sin\gamma \overrightarrow{Y_1}\ ,\ \text{with } \eta < \ell$$

$$\Rightarrow\ \overrightarrow{AM_2}^2 = \eta^2\ , \forall \gamma.$$

Furthermore,

$$\overrightarrow{v^{(\lambda)}}(M_2) = \overrightarrow{v^{(\lambda)}}(A) + \overrightarrow{\omega_2^\lambda} \wedge \overrightarrow{AM_2} = \gamma' \overrightarrow{z_\lambda} \wedge \eta \vec{u}(\gamma) = \eta \gamma' \vec{u}\left(\gamma + \frac{\pi}{2} \right),$$

meaning that the velocity vector of M_2 is always orthogonal to the radius vector $\overrightarrow{AM_2}$. Point M_2 is, therefore, located on a circle of center A and of radius $\eta = \left\| \overrightarrow{AM_2} \right\|$. Solid (2) undergoes a motion of rotation around a fixed axis $\left(A \middle| \overrightarrow{z_\lambda} \right)$ in relation to $\langle\lambda\rangle$.

– *Question 4*: Locate bar (1) in relation to bar (2). Write out the torsor $\left\{ {}^{2}_{1} \right\}_D$. Characterize the motion of (1) in relation to (2).

$$\left\{ {}^{2}_{1} \right\} = \left\{ {}^{\lambda}_{1} \right\} + \left\{ {}^{2}_{\lambda} \right\} = \left\{ {}^{\lambda}_{1} \right\} - \left\{ {}^{\lambda}_{2} \right\} \ \Rightarrow\ \overrightarrow{\omega_1^2} = \overrightarrow{\omega_1^\lambda} - \overrightarrow{\omega_2^\lambda} = \vec{0},$$

$$\begin{aligned} \overrightarrow{\mathcal{M}_D}\left\{ {}^{2}_{1} \right\} &= \overrightarrow{\mathcal{M}_D}\left\{ {}^{\lambda}_{1} \right\} - \overrightarrow{\mathcal{M}_D}\left\{ {}^{\lambda}_{2} \right\} = \vec{0} - \overrightarrow{\mathcal{M}_A}\left\{ {}^{\lambda}_{2} \right\} - \overrightarrow{DA} \wedge \vec{s}\left\{ {}^{\lambda}_{2} \right\} \\ \text{" "} \quad &= -L \overrightarrow{x_\lambda} \wedge \gamma' \overrightarrow{z_\lambda} = -L\gamma' \overrightarrow{y_\lambda} \end{aligned},$$

$$\left\{ {}^{2}_{1} \right\}_D = \left[\vec{0} \middle| -L\gamma' \overrightarrow{y_\lambda} \right].$$

The motion of (1) in relation to (2) is a circular translation.

2.4.5.2. *Exercise 14 – Kinematics of an articulated system*

The following mechanical system is made up of five solids. Bars (S_1), (S_2), (S_3), and (S_4) make up an articulated lozenge, whose sides are equal to *l*.

We write $\overrightarrow{OA} = \ell \vec{u}(\gamma)$, where γ is a twice-derivable function, and the study is limited to $0 < \gamma < \pi/2$.

The whole system is loaded onto a device (0) to which the frame $\langle 0 \rangle \equiv \langle O|\vec{x}\,\vec{y}\,\vec{z} \rangle$ is linked. The solid (S_5) is guided in its motion by a rectilinear slide linked to the device, with an axis of $(O\,|\,\vec{y})$.

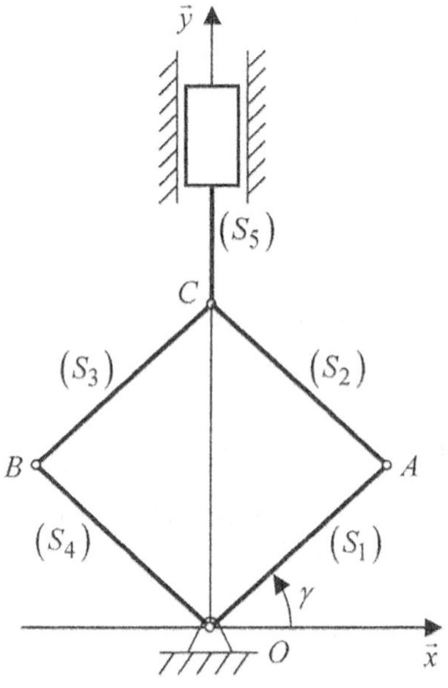

We write $\{{}^j_i\}$, where i and t $j = 0,1,\ldots,5$, as the velocity-distributing torsor of (S_i) in its motion in relation to (S_j).

– *Question 1(a)*: Determine the following velocity-distributing torsors through their reduction elements at the points stated:

$$\left\{ {}^{0}_{1} \right\}_{/O} \;,\; \left\{ {}^{0}_{2} \right\}_{/A} \;,\; \left\{ {}^{0}_{4} \right\}_{/O} \;,\; \left\{ {}^{0}_{3} \right\}_{/B} \;,\; \left\{ {}^{0}_{5} \right\}.$$

Bar (S_1) makes an angle of γ with an axis $\vec{x}$. This results in

$$\left\{ {}^{0}_{1} \right\}_{/O} = \left[\dot{\gamma}\vec{z} \middle| \vec{0} \right].$$

For bar (S_2):

– $\widehat{\vec{x}, \overrightarrow{AC}} = \pi - \gamma \quad \Rightarrow \quad \vec{\omega}_2^0 = -\dot{\gamma}\vec{z}$;

– $\overrightarrow{v^{(0)}}(A) = \dfrac{d^{(0)}}{dt}\overrightarrow{OA} = \dfrac{d^{(0)}}{dt}\ell\vec{u}(\gamma) = \ell\dot{\gamma}\vec{u}\left(\gamma + \dfrac{\pi}{2}\right)$;

$$\Rightarrow \left\{ {}^{0}_{2} \right\}_{/A} = \left[-\dot{\gamma}\vec{z} \middle| \ell\dot{\gamma}\vec{u}\left(\gamma + \frac{\pi}{2}\right) \right].$$

For Bar (S_3):

– $\widehat{\vec{x}, \overrightarrow{BC}} = \gamma \quad \Rightarrow \quad \vec{\omega}_3^0 = \dot{\gamma}\vec{z}$;

– $$\begin{aligned} \overrightarrow{v^{(0)}}(B) &= \frac{d^{(0)}}{dt}\overrightarrow{OB} = \frac{d^{(0)}}{dt}\ell\vec{u}(\pi - \gamma) \\ \text{'' ''} \quad &= -\ell\dot{\gamma}\vec{u}\left(\frac{3\pi}{2} - \gamma\right) = \ell\dot{\gamma}\vec{u}\left(\frac{\pi}{2} - \gamma\right) \end{aligned};$$

$$\Rightarrow \left\{ {}^{0}_{3} \right\}_{/B} = \left[\dot{\gamma}\vec{z} \middle| \ell\dot{\gamma}\vec{u}\left(\frac{\pi}{2} - \gamma\right) \right].$$

NOTE 2.5.–

$$\begin{aligned} \vec{u}\left(\frac{3\pi}{2} - \gamma\right) &= \cos\left(\frac{3\pi}{2} - \gamma\right)\vec{x} + \sin\left(\frac{3\pi}{2} - \gamma\right)\vec{y} \\ \text{'' ''} \quad &= -\cos\left(\frac{\pi}{2} - \gamma\right)\vec{x} - \sin\left(\frac{\pi}{2} - \gamma\right)\vec{y} = -\vec{u}\left(\frac{\pi}{2} - \gamma\right) \end{aligned}.$$

For Bar (S_4):

$$\widehat{\vec{x},\overrightarrow{OB}} = \pi - \gamma \quad \Rightarrow \vec{\omega}_2^0 = -\dot{\gamma}\vec{z}, \Rightarrow \left\{ {}^0_4 \right\}_{/O} = \left[-\dot{\gamma}\vec{z} \middle| \vec{0} \right].$$

For solid (S_5) carried on the axis $\left(O|\vec{y}\right)$:

$$\overrightarrow{OC} = \overrightarrow{OA} + \overrightarrow{AC} = \ell\vec{u}(\gamma) + \ell\vec{u}(\pi - \gamma),$$

$$\Rightarrow \overrightarrow{OC} = \ell\cos\gamma\vec{x} + \ell\sin\gamma\vec{y} - \ell\cos\gamma\vec{x} + \ell\sin\gamma\vec{y} = 2\ell\sin\gamma\vec{y},$$

$$\Rightarrow \left\{ {}^0_5 \right\} = \left[\vec{0} \middle| 2l\dot{\gamma}\cos\gamma\vec{y} \right].$$

– *Question 1(b)*: For each torsor, state the point of intersection of its central axis, when it exists, with the plane $\Pi\left(O|\vec{x},\vec{y}\right)$.

The axis of the torsor $\left\{ {}^0_1 \right\}$ is defined by

$$-\ \overrightarrow{OQ_{10}} = \frac{\vec{s}\left\{ {}^0_1 \right\} \wedge \overrightarrow{\mathrm{M}_O}\left\{ {}^0_1 \right\}}{\vec{s}^{\,2}\left\{ {}^0_1 \right\}} = \vec{0};$$

$$-\ \overrightarrow{AQ_{20}} = \frac{\vec{s}\left\{ {}^0_2 \right\} \wedge \overrightarrow{\mathrm{M}_A}\left\{ {}^0_2 \right\}}{\vec{s}^{\,2}\left\{ {}^0_2 \right\}} = \frac{-\dot{\gamma}\vec{z} \wedge \ell\dot{\gamma}\vec{u}\left(\gamma + \frac{\pi}{2}\right)}{\dot{\gamma}^2} = \ell\vec{u}(\gamma),$$

$$\Rightarrow \quad \overrightarrow{OQ_{20}} = \overrightarrow{OA} + \overrightarrow{AQ_{20}} = 2\ell\overrightarrow{u(\gamma)};$$

$$-\ \overrightarrow{BQ_{30}} = \frac{\vec{s}\left\{ {}^0_3 \right\} \wedge \overrightarrow{\mathrm{M}_A}\left\{ {}^0_3 \right\}}{\vec{s}^{\,2}\left\{ {}^0_3 \right\}} = \frac{\dot{\gamma}\vec{z} \wedge \ell\dot{\gamma}\vec{u}\left(\frac{\pi}{2} - \gamma\right)}{\dot{\gamma}^2} = \ell\vec{u}(\pi - \gamma),$$

$$\Rightarrow \quad \overrightarrow{OQ_{30}} = \overrightarrow{OB} + \overrightarrow{BQ_{30}} = 2\ell\vec{u}(\pi - \gamma);$$

$$-\ \overrightarrow{OQ_{40}} = \frac{\vec{s}\left\{ {}^0_4 \right\} \wedge \overrightarrow{\mathrm{M}_O}\left\{ {}^0_4 \right\}}{\vec{s}^{\,2}\left\{ {}^0_4 \right\}} = \vec{0}.$$

Torsor $\left\{\begin{smallmatrix}0\\5\end{smallmatrix}\right\}$ is a couple-torsor; it does not present a central axis.

– *Question 2*: Answer the previous question with the following torsors:

$$\left\{\begin{smallmatrix}2\\1\end{smallmatrix}\right\}_{/A}, \quad \left\{\begin{smallmatrix}5\\2\end{smallmatrix}\right\}_{/A}, \quad \left\{\begin{smallmatrix}1\\5\end{smallmatrix}\right\}_{/O}.$$

– $\left\{\begin{smallmatrix}2\\1\end{smallmatrix}\right\}=\left\{\begin{smallmatrix}0\\1\end{smallmatrix}\right\}-\left\{\begin{smallmatrix}0\\2\end{smallmatrix}\right\}$ with

$$\overrightarrow{\mathcal{M}_A}\left\{\begin{smallmatrix}0\\1\end{smallmatrix}\right\}=\overrightarrow{\mathcal{M}_O}\left\{\begin{smallmatrix}0\\1\end{smallmatrix}\right\}+\vec{s}\left\{\begin{smallmatrix}0\\1\end{smallmatrix}\right\}\wedge\overrightarrow{OA}=\gamma'\vec{z}\wedge\ell\vec{u}(\gamma)=\ell\gamma'\vec{u}\left(\gamma+\frac{\pi}{2}\right),$$

$$\Rightarrow \left\{\begin{smallmatrix}1\\2\end{smallmatrix}\right\}_{/A}=\left[2\gamma'\vec{z}\,\middle|\,\vec{0}\right];$$

– $\left\{\begin{smallmatrix}5\\2\end{smallmatrix}\right\}=\left\{\begin{smallmatrix}0\\2\end{smallmatrix}\right\}-\left\{\begin{smallmatrix}0\\5\end{smallmatrix}\right\}$ with

$$\overrightarrow{\mathcal{M}_A}\left\{\begin{smallmatrix}5\\2\end{smallmatrix}\right\}=\overrightarrow{\mathcal{M}_A}\left\{\begin{smallmatrix}0\\2\end{smallmatrix}\right\}-\overrightarrow{\mathcal{M}_A}\left\{\begin{smallmatrix}0\\5\end{smallmatrix}\right\}=\ell\gamma'\vec{u}\left(\gamma+\frac{\pi}{2}\right)-2\ell\gamma'\cos\gamma\vec{y}$$

$$\text{'' ''} \quad =\ell\gamma'\left(-\sin\gamma\vec{x}-\cos\gamma\vec{y}\right)=\ell\gamma'\vec{u}\left(\frac{3\pi}{2}-\gamma\right),$$

$$\Rightarrow \left\{\begin{smallmatrix}5\\2\end{smallmatrix}\right\}_{A}=\left[-\gamma'\vec{z}\,\middle|\,\ell\gamma'\vec{u}\left(\frac{3\pi}{2}-\gamma\right)\right];$$

– $\left\{\begin{smallmatrix}1\\5\end{smallmatrix}\right\}=\left\{\begin{smallmatrix}0\\5\end{smallmatrix}\right\}-\left\{\begin{smallmatrix}0\\1\end{smallmatrix}\right\} \Rightarrow \left\{\begin{smallmatrix}1\\5\end{smallmatrix}\right\}_{/O}=\left[-\gamma'\vec{z}\,\middle|\,2\ell\gamma'\cos\gamma\vec{y}\right].$

– *Question 3*: What are the trajectories in $\langle 0\rangle$ of C and of a point M of AC [bar (S_2)] assuming $\overrightarrow{AM}=\lambda\overrightarrow{AC}$, with $0<\lambda<1$:

$$\overrightarrow{OA}=\ell\vec{u}(\gamma)=\ell\cos\gamma\vec{x}+\ell\sin\gamma\vec{y}=x\vec{x}+y\vec{y}.$$

The coordinates of the point A verify the relation: $x^2+y^2=\ell^2$, which is the equation of a circle of center O and of radius l:

$$\overrightarrow{OC}=2\ell\sin\gamma\vec{y}.$$

The location of C is a segment with an origin at O and a length of 2ℓ on the line $\left(O|\vec{y}\right)$.

For a point M of (S_2) defined by $\overrightarrow{AM} = \lambda\overrightarrow{AC} = \lambda\ell\vec{u}(\pi-\gamma)$, we write

$$\begin{aligned}\overrightarrow{OM} &= \overrightarrow{OA} + \overrightarrow{AM} = \ell\left(\cos\gamma\vec{x} + \sin\gamma\vec{y}\right) + \lambda\ell\left(-\cos\gamma\vec{x} + \sin\gamma\vec{y}\right)\\ \text{""} &= \ell(1-\lambda)\cos\gamma\vec{x} + \ell(1+\lambda)\sin\gamma\vec{y} = x\vec{x} + y\vec{y}\end{aligned},$$

meaning

$$\frac{x^2}{\ell^2(1-\lambda)^2} + \frac{y^2}{\ell^2(1+\lambda)^2} = 1,$$

which is the equation of an ellipsis of center O and with demi-axes $\ell(1-\gamma)$ and $\ell(1+\gamma)$, which is in fact a quarter ellipsis according to the diagram below.

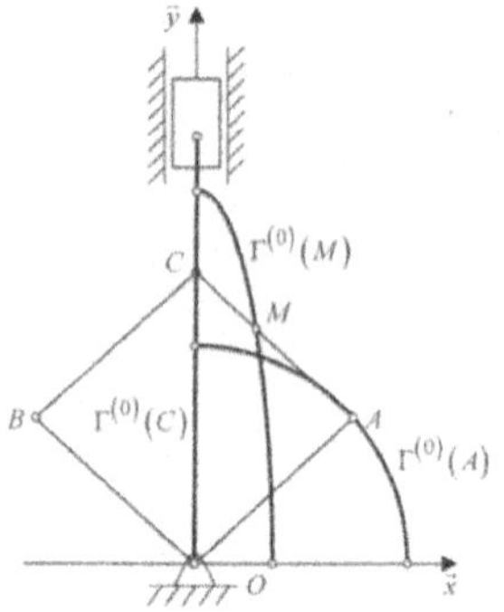

2.5. Motion with fixed plane

2.5.1. *Position of the problem*

Here, we consider the case of the motion of a solid (S), characterized by the shifting of a plane, real or virtual, fixed to the solid, which is geometrically coincident with a plane linked to the frame $\langle\lambda\rangle$.

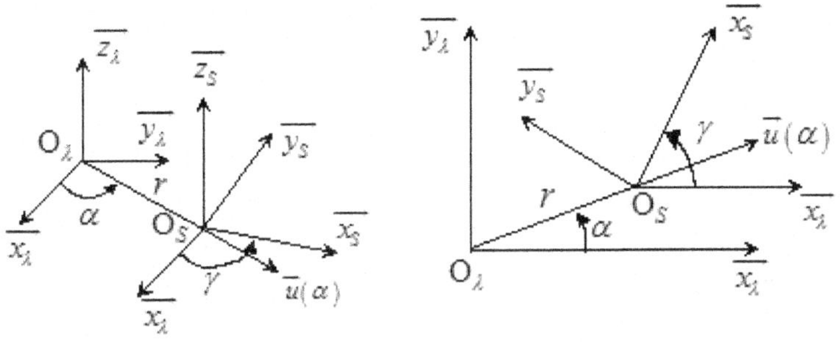

Figure 2.6. *Motion with fixed plane*

To deal with the subject easily, we assume that the plane $\Pi_S \equiv \Pi\left(\mathrm{O}_S|\overrightarrow{x_S}\,\overrightarrow{y_S}\right)$ of the solid (S) remains coincident with the plane $\Pi_\lambda \equiv \left(\mathrm{O}_\lambda|\overrightarrow{x_\lambda}\,\overrightarrow{y_\lambda}\right)$. The axes $\overrightarrow{z_\lambda}$ and $\overrightarrow{z_S}$, therefore remain collinear during the motion.

The values of the problem are the following:

- $\overrightarrow{\mathrm{O}_\lambda\mathrm{O}_S} = r\vec{u}(\alpha)$;
- $\alpha = \widehat{\overrightarrow{x_\lambda},\overrightarrow{\mathrm{O}_\lambda\mathrm{O}_S}}$;
- $\gamma = \widehat{\overrightarrow{x_\lambda},\overrightarrow{x_S}}$.

If we consider a point M_S of the solid and its projection P_S onto the plane Π_S, with $\overrightarrow{\mathrm{O}_S\mathrm{P}_S} = a\overrightarrow{x_S} + b\overrightarrow{y_S}$ and $\overrightarrow{\mathrm{P}_S\mathrm{M}_S} = c\overrightarrow{z_S} = c\overrightarrow{z_\lambda}$, where a,b,c are independent of time since the two points M_S and P_S both belong to the solid (S), this results in

$$\begin{aligned}\overrightarrow{\mathrm{O}_\lambda\mathrm{M}_S} &= \overrightarrow{\mathrm{O}_\lambda\mathrm{O}_S} + \overrightarrow{\mathrm{O}_S\mathrm{M}_S} = \overrightarrow{\mathrm{O}_\lambda\mathrm{O}_S} + \overrightarrow{\mathrm{O}_S\mathrm{P}_S} + \overrightarrow{\mathrm{P}_S\mathrm{M}_S} \\ \text{'' ''} &= r\vec{u}(\alpha) + a\overrightarrow{x_S} + b\overrightarrow{y_S} + c\overrightarrow{z_\lambda}\end{aligned},$$

$$\begin{aligned}\overrightarrow{v^{(\lambda)}}(\mathrm{M}_S) &= \overrightarrow{v^{(\lambda)}}(\mathrm{O}_S) + \frac{d^{(\lambda)}}{dt}\overrightarrow{\mathrm{O}_S\mathrm{M}_S} \\ \text{''''} &= r'\vec{u}(\alpha) + r\alpha'\vec{u}\left(\alpha + \frac{\pi}{2}\right) + a\frac{d^{(\lambda)}}{dt}\overrightarrow{x_S} + b\frac{d^{(\lambda)}}{dt}\overrightarrow{y_S}\end{aligned},$$

with

$$\frac{d^{(\lambda)}}{dt}\overrightarrow{x_S} = \frac{d^{(S)}}{dt}\overrightarrow{x_S} + \overrightarrow{\omega_S^{\lambda}} \wedge \overrightarrow{x_S}, \quad \frac{d^{(\lambda)}}{dt}\overrightarrow{y_S} = \frac{d^{(S)}}{dt}\overrightarrow{y_S} + \overrightarrow{\omega_S^{\lambda}} \wedge \overrightarrow{y_S},$$

and

$$\overrightarrow{\omega_S^{\lambda}} = \gamma'\overrightarrow{z_\lambda},$$

as the rotation of angle γ happens around the axis $\overrightarrow{z_\lambda}$, meaning that

$$\overrightarrow{v^{(\lambda)}}(M_S) = r'\vec{u}(\alpha) + r\alpha'\vec{u}\left(\alpha + \frac{\pi}{2}\right) + a\gamma'\overrightarrow{y_S} - b\gamma'\overrightarrow{x_S}.$$

Furthermore, this means that

$$\begin{aligned}\overrightarrow{v^{(\lambda)}}(M_S) &= \overrightarrow{v^{(\lambda)}}(O_S) + \frac{d^{(\lambda)}}{dt}\left(\overrightarrow{O_SP_S} + \overrightarrow{P_SM_S}\right) \\ &= \overrightarrow{v^{(\lambda)}}(O_S) + \gamma'\overrightarrow{z_\lambda} \wedge \left(\overrightarrow{O_SP_S} + c\overrightarrow{z_\lambda}\right) \\ &= \overrightarrow{v^{(\lambda)}}(O_S) + \gamma'\overrightarrow{z_\lambda} \wedge \overrightarrow{O_SP_S} = \overrightarrow{v^{(\lambda)}}(P_S)\end{aligned}.$$

In conclusion, the trajectory $\Gamma_\lambda(M_S)$ is deduced from the trajectory of P_S, $\Gamma_\lambda(P_S)$, through the translation of the vector $c\,\overrightarrow{z_\lambda}$. The two points M_S and P_S describe their respective trajectories with the same velocity, which warrants the planar representation of the motion of the solids (one or several) with a fixed plane motion.

2.5.2. *Instantaneous rotation center*

The velocity-distributing torsor of (S) in relation to $\langle\lambda\rangle$ is written as:

$$\left\{{}^{\lambda}_{S}\right\}_{O_S} = \left[\gamma'\overrightarrow{z_\lambda} \middle| \overrightarrow{v^{(\lambda)}}(O_S)\right]$$

and has as its *scalar invariant*[3] the quantity

$$\mathcal{I} = \vec{s}\{{}^{\lambda}_{S}\} \cdot \overrightarrow{\mathcal{M}_{O_S}}\{{}^{\lambda}_{S}\} = \gamma' \overrightarrow{z_\lambda} \cdot \overrightarrow{v^{(\lambda)}}(O_S) = 0 .$$

A torsor represents all the sets of vectors that have the same reduction elements at the same point; this is called an equivalence class. All the sets of a same class, which are therefore represented by the same torsor, have the same properties, which are those that are established from the reduction elements of the torsor.

The scalar invariant is one of these indicators. When it is equal to zero, this means that at any moment the resultant of the torsor is orthogonal to its moment in a point or that the moment of the torsor is equal to zero. Like in the case of the motion studied here, the rate of rotation has a fixed direction and the velocity is not also equal to zero, meaning that this velocity, in any point of the solid, is always applied in a plane that is perpendicular to this direction; as a result the motion is planar.

Another fundamental characteristic of torsors is the existence of what is called the central axis. Torsor theory, developed in Volume 2 [BOR 16a], shows that for every torsor {T} whose resultant is not equal to zero, $\vec{s}\{\mathcal{T}\} \neq \vec{0}$, it is associated with an axis (Δ) called the central axis of the torsor, such that

$$\forall Q \in (\Delta), \ \overrightarrow{O_S Q} = \frac{\vec{s}\{T\} \wedge \overrightarrow{M_{O_S}}\{T\}}{\left(\vec{s}\{T\}\right)^2} + k\,\vec{s}\{T\} \quad \text{with} \quad k \in \mathbb{R} .$$

In the current case

$$\overrightarrow{O_S Q} = \frac{\gamma' \overrightarrow{z_\lambda} \wedge \left(r' \vec{u}(\alpha) + r\alpha' \vec{u}\left(\alpha + \frac{\pi}{2}\right)\right)}{\gamma'^2} + k\gamma' \overrightarrow{z_\lambda} = \overrightarrow{O_S I} + k\gamma' \overrightarrow{z_\lambda} .$$

3 See Volume 2 [BOR 16a].

We note that the vector

$$\overrightarrow{O_S I} = \frac{-r\alpha' \vec{u}(\alpha) + r' \vec{u}\left(\alpha + \frac{\pi}{2}\right)}{\gamma'} = \vec{z_\lambda} \wedge \frac{\overrightarrow{v^{(\lambda)}}(O_s)}{\gamma'},$$

– is orthogonal to the axis $\vec{z_\lambda}$;

– is contained within the plane Π_S;

– is orthogonal to $\overrightarrow{v^{(\lambda)}}(O_S)$ as if it is the radius of an arc of the circle to which $\overrightarrow{v^{(\lambda)}}(O_S)$ would be tangential, as in the case of an instantaneous rotation.

Point I is, therefore, the trace in the plane Π_S of an axis (Δ), collinear to $\vec{z_\lambda}$, around which a rotation would take place at the time t considered. This point I is the *instantaneous center of rotation* of the solid (S) in a motion with fixed plane in relation to the frame $\langle\lambda\rangle$.

Motion with fixed plane is, therefore, considered as a continuous succession of rotations around an axis $\vec{z_\lambda}$ whose center I moves throughout the motion.

2.5.3. *Fixed and mobile centroids of the motion*

We now look at the motion of the instantaneous center of rotation I throughout the motion; it comes to coincide with a point of plane Π_S and with a point of plane Π_λ at each instant.

The trajectory $\Gamma_{(S)}(\mathrm{I})$ of the instantaneous center of rotation in the plane Π_S is governed by the equation

$$\overrightarrow{O_S I} = \vec{z_\lambda} \wedge \frac{\overrightarrow{v^{(\lambda)}}(O_S)}{\gamma'} = X_S(t)\overrightarrow{x_S} + Y_S(t)\overrightarrow{y_S}.$$

It is called the *mobile centroid* of the motion with a fixed plane.

The trajectory $\Gamma_{(\lambda)}(\mathrm{I})$ of the instantaneous center of rotation in the plane Π_λ is governed by the equation

$$\overrightarrow{\mathrm{O}_\lambda \mathrm{I}} = \overrightarrow{\mathrm{O}_\lambda \mathrm{O}_S} + \overrightarrow{\mathrm{O}_S \mathrm{I}} = r\vec{u}(\alpha) + \vec{z_\lambda} \wedge \frac{\overrightarrow{v^{(\lambda)}}(\mathrm{O}_S)}{\gamma'} = X_\lambda(t)\vec{x_\lambda} + Y_\lambda(t)\vec{y_\lambda}.$$

It is called the *fixed centroid* of the motion with a fixed plane.

2.5.4. *The instantaneous center of rotation on the fixed centroid and on the movable centroid*

The situation of the instantaneous center of rotation I is, therefore, defined by the two vectors:

$$\overrightarrow{\mathrm{O}_S \mathrm{I}} = \vec{z_\lambda} \wedge \frac{\overrightarrow{v^{(\lambda)}}(\mathrm{O}_S)}{\gamma'} \text{ and } \overrightarrow{\mathrm{O}_\lambda \mathrm{I}} = \overrightarrow{\mathrm{O}_\lambda \mathrm{O}_S} + \vec{z_\lambda} \wedge \frac{\overrightarrow{v^{(\lambda)}}(\mathrm{O}_S)}{\gamma'}.$$

The evolution velocities of this point on each of the two curves is given by:

$$\overrightarrow{v^{(S)}}(\mathrm{I}) = \frac{d^{(S)}}{dt}\overrightarrow{\mathrm{O}_S \mathrm{I}} = \vec{z_\lambda} \wedge \frac{d^{(S)}}{dt}\left[\frac{\overrightarrow{v^{(\lambda)}}(\mathrm{O}_S)}{\gamma'}\right],$$

$$\text{" "} \quad = \vec{z_\lambda} \wedge \frac{\gamma'\frac{d^{(S)}}{dt}\overrightarrow{v^{(\lambda)}}(\mathrm{O}_S) - \gamma''\overrightarrow{v^{(\lambda)}}(\mathrm{O}_S)}{\gamma'^2}$$

$$\overrightarrow{v^{(S)}}(\mathrm{I}) = \vec{z_\lambda} \wedge \frac{\gamma'\left[\frac{d^{(\lambda)}}{dt}\overrightarrow{v^{(\lambda)}}(\mathrm{O}_S) + \vec{\omega_\lambda^S} \wedge \overrightarrow{v^{(\lambda)}}(\mathrm{O}_S)\right] - \gamma''\overrightarrow{v^{(\lambda)}}(\mathrm{O}_S)}{\gamma'^2}.$$

$$\text{" "} \quad = \vec{z_\lambda} \wedge \frac{\overrightarrow{J^{(\lambda)}}(\mathrm{O}_S)}{\gamma'} + \frac{\vec{z_\lambda} \wedge \left[\vec{\omega_\lambda^S} \wedge \overrightarrow{v^{(\lambda)}}(\mathrm{O}_S)\right]}{\gamma'} - \vec{z_\lambda} \wedge \frac{\gamma''}{\gamma'^2}\overrightarrow{v^{(\lambda)}}(\mathrm{O}_S)$$

Seeing as the double vectorial product[4] verifies $\vec{a} \wedge (\vec{b} \wedge \vec{c}) = (\vec{a} \cdot \vec{c})\vec{b} - (\vec{a} \cdot \vec{b})\vec{c}$, this results in:

4 See the relevant chapter of Volume 2 [BOR 16a].

$$\overrightarrow{z_\lambda} \wedge \left[\overrightarrow{\omega_\lambda^S} \wedge \overrightarrow{v^{(\lambda)}}(O_S) \right] = \left[\overrightarrow{z_\lambda} \cdot \overrightarrow{v^{(\lambda)}}(O_S) \right] \overrightarrow{\omega_\lambda^S} - \left(\overrightarrow{z_\lambda} \cdot \overrightarrow{\omega_\lambda^S} \right) \overrightarrow{v^{(\lambda)}}(O_S)$$

$$'''' \qquad = \left(\overrightarrow{z_\lambda} \cdot \overrightarrow{\omega_S^\lambda} \right) \overrightarrow{v^{(\lambda)}}(O_S) = \gamma' \overrightarrow{v^{(\lambda)}}(O_S) ,$$

and, finally

$$\overrightarrow{v^{(S)}}(I) = \overrightarrow{z_\lambda} \wedge \frac{\overrightarrow{J^{(\lambda)}(O_S)}}{\gamma'} + \overrightarrow{v^{(\lambda)}}(O_S) - \overrightarrow{z_\lambda} \wedge \frac{\gamma''}{\gamma'^2} \overrightarrow{v^{(\lambda)}}(O_S) .$$

Seeing as

$$\overrightarrow{v^{(\lambda)}}(I) = \frac{d^{(\lambda)}}{dt} \overrightarrow{O_\lambda O_S} + \overrightarrow{z_\lambda} \wedge \frac{d^{(\lambda)}}{dt} \left[\frac{\overrightarrow{v^{(\lambda)}}(O_S)}{\gamma'} \right] ,$$

$$\overrightarrow{v^{(\lambda)}}(I) = \overrightarrow{v^{(\lambda)}}(O_S) + \overrightarrow{z_\lambda} \wedge \frac{\gamma' \overrightarrow{J^{(\lambda)}}(O_S) - \gamma'' \overrightarrow{v^{(\lambda)}}(O_S)}{\gamma'^2}$$

$$'''' \qquad = \overrightarrow{v^{(\lambda)}}(O_S) + \overrightarrow{z_\lambda} \wedge \frac{\overrightarrow{J^{(\lambda)}}(O_S)}{\gamma'} - \overrightarrow{z_\lambda} \wedge \frac{\gamma''}{\gamma'^2} \overrightarrow{v^{(\lambda)}}(O_S) ,$$

we can see that the velocity vectors of I on the fixed and mobile centroids are the same; this means that both the fixed and the mobile centroid are tangential at the instantaneous center of rotation and that the mobile centroid rolls without sliding over the fixed centroid since the instantaneous motion of this point along the two curves is the same.

2.5.5. *Physical interpretation of the notions of fixed centroid and mobile centroid*

To materially represent these two notions, let us imagine the motion with a fixed plane of a solid on a flat surface that follows a given law. In order to obey this law, the motion is carried out using a guide on the surface, a sort of rack for example, which the solid follows using a crenellated or cogged slide while maintaining planar contact with the surface. The guide represents the fixed centroid; the slide attached to the solid is the mobile centroid. These two notions are, therefore, of real material use.

2.5.6. *Exercises*

2.5.6.1. *Exercise 15 – Motion of a solid with a fixed plane*

A solid (*S*), attached to a frame $\langle S \rangle \equiv \left\langle \mathrm{O}_S \,|\, \overrightarrow{x_S}\ \overrightarrow{y_S}\ \overrightarrow{z_S} \right\rangle$, moves with a fixed plane in relation to a device linked to the frame $\langle 0 \rangle \equiv \left\langle \mathrm{O}_0 \,|\, \overrightarrow{x_0}\ \overrightarrow{y_0}\ \overrightarrow{z_0} \right\rangle$. Its motion is defined as follows:

– $\overrightarrow{z_S} \equiv \overrightarrow{z_0} \equiv \vec{z}$; $\gamma(t) = \widehat{\overrightarrow{x_0}, \overrightarrow{x_S}} = \widehat{\overrightarrow{y_0}, \overrightarrow{y_S}}$;

– O_S describes a segment of the line $\left\langle \mathrm{O}_0 \,|\, \overrightarrow{x_0} \right\rangle$;

– the edge $\left(\mathrm{O}_S \,|\, \overrightarrow{y_S} \right)$ of the solid rests on the edge $\left\langle \mathrm{B}_0 \,|\, \vec{z} \right\rangle$ of the device.

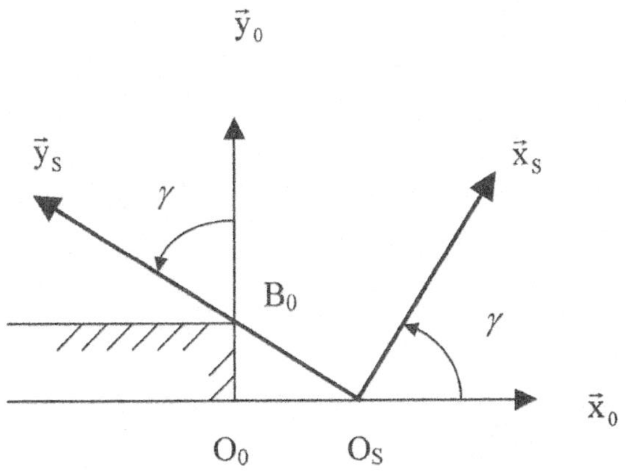

We give: $\overrightarrow{\mathrm{O}_0\mathrm{B}_0} = h\,\overrightarrow{y_0}$, where h is a positive constant.

This means that

$$\overrightarrow{\mathrm{O}_0\mathrm{O}_S} = h \tan \gamma\, \overrightarrow{x_0},$$

and the position of (S) in $\langle 0 \rangle$ is defined using the single parameter $\gamma \in \left[0, \dfrac{\pi}{2} \right]$.

Note that:

$$\frac{d\tan\gamma}{d\gamma} = 1+\tan\gamma = \frac{1}{\cos^2\gamma}$$

– *Question 1*: Determine the reduction elements at O_S of the velocity-distributing torsor $\{ {}^0_S \}$ as a function of h, γ, γ':

$$\overrightarrow{\omega_S^0} = \gamma' \vec{z}\,,\ \overrightarrow{O_0O_S} = h\tan\gamma\,\vec{x_0} \ \Rightarrow\ \overrightarrow{v^{(0)}}(O_S) = \frac{h\gamma'}{\cos^2\gamma}\vec{x_0}\,,$$

$$\Rightarrow\ \{ {}^0_S \}_{/O_S} = \left[\gamma'\vec{z} \left| \frac{h\gamma'}{\cos^2}\vec{x_0} \right. \right].$$

– *Question 2*: Characterize this torsor and determine its central axis (Δ). I is the point of intersection of (Δ) and of the plane $\Pi(O_0|\vec{x_0},\vec{y_0})$.

The scalar invariant of the torsor $\{ {}^0_S \}$ is $I = \vec{s}\{ {}^0_S \} \cdot \overrightarrow{\mathcal{M}_{O_S}}\{ {}^0_S \} = 0$.

This is, therefore, a sliding-torsor whose central axis (Δ) is collinear to $\vec{z}$ and passes through the point I, which is the intersection with the plane $\Pi(O_0|\vec{x_0},\vec{y_0})$, given by:

$$\overrightarrow{O_SI} = \frac{\vec{s}\{ {}^0_S \} \wedge \overrightarrow{\mathcal{M}_{O_S}}\{ {}^0_S \}}{\vec{s}\{ {}^0_S \}^2} = \frac{h}{\cos^2\gamma}\vec{y_0}\,.$$

– *Question 3*: Study of the mobile centroid $\Gamma^{(S)}(I)$, the trajectory of point I in the frame $\langle S \rangle$. For this we give: $\overrightarrow{O_SI} = X(\gamma)\vec{x_S} + Y(\gamma)\vec{y_S}$.

Establish the parametric equations of this curve, and deduce the Cartesian equation and the velocity vector $\overrightarrow{v^{(S)}}(I)$.

The mobile centroid is the curve represented by the following vectorial equation in the frame $\langle S \rangle$:

$$\overrightarrow{O_SI} = X(\gamma)\vec{x_S} + Y(\gamma)\vec{y_S} = \frac{h}{\cos^2\gamma}\left(\sin\gamma\,\vec{x_S} + \cos\gamma\,\vec{y_S}\right),$$

hence, its parametric representation:

$$X(\gamma)=\frac{h\sin\gamma}{\cos^2\gamma}\ ,\ Y(\gamma)=\frac{h}{\cos\gamma}\geq h\ \Rightarrow\ \cos\gamma=\frac{h}{Y},$$

and its Cartesian representation

$$X^2=h^2\frac{1-\cos^2\gamma}{\cos^4\gamma}=h^2\frac{1-\frac{h^2}{Y^2}}{\frac{h^4}{Y^4}}=\frac{Y^4}{h^2}\left(1-\frac{h^2}{Y^2}\right)=Y^2\left(\frac{Y^2}{h^2}-1\right),$$

$$\Rightarrow\ X=\pm Y\sqrt{\frac{Y^2}{h^2}-1}\,.$$

The speed of change of the point I on the movable centroid is

$$\overrightarrow{v^{(S)}}(\mathrm{I})=\frac{d^{(S)}}{dt}\overrightarrow{\mathrm{O}_S\mathrm{I}}=h\gamma'\frac{\cos^3\gamma+2\sin^2\gamma\cos\gamma}{\cos^4\gamma}\overrightarrow{x_S}+h\gamma'\frac{\sin\gamma}{\cos^2\gamma}\overrightarrow{y_S}$$

$$\text{'' ''}\quad=\frac{h\gamma'}{\cos^2\gamma}\left(\frac{1+\sin^2\gamma}{\cos\gamma}\overrightarrow{x_S}+\sin\gamma\,\overrightarrow{y_S}\right)$$

– *Question 4*: Study of the fixed centroid $\Gamma^{(0)}(\mathrm{I})$, trajectory of I in $\langle 0\rangle$.

We give $\overrightarrow{\mathrm{O}_0\mathrm{I}}=x(\gamma)\overrightarrow{x_0}+y(\gamma)\overrightarrow{y_0}$.

Establish the parametric equations of this curve, and deduce its Cartesian equation and the velocity vector $\overrightarrow{v^{(0)}}(\mathrm{I})$.

The fixed centroid is the curve represented by the following vectorial equation in the frame $\langle 0\rangle$:

$$\overrightarrow{\mathrm{O}_0\mathrm{I}}=\overrightarrow{\mathrm{O}_0\mathrm{O}_S}+\overrightarrow{\mathrm{O}_S\mathrm{I}}=h\tan\gamma\overrightarrow{x_0}+h\left(1+\tan^2\gamma\right)\overrightarrow{y_0}=x(\gamma)\overrightarrow{x_0}+y(\gamma)\overrightarrow{y_0}\,,$$

hence its parametric representation

$$x = h\tan\gamma \ , \quad y = h\left(1+\tan^2 y\right) \Rightarrow y = h\left(1+\frac{x^2}{h^2}\right)$$

and its Cartesian representation

$$y = h\left(1+\frac{x^2}{h^2}\right) \quad \text{or} \quad x^2 = h(y-h)$$
,

which is the equation of a parabola. Furthermore, this means that

$$\overrightarrow{v^{(0)}}(\mathrm{I}) = \frac{d^{(0)}}{dt}\overrightarrow{\mathrm{O_0 I}} = \frac{h\gamma'}{\cos^2\gamma}\overrightarrow{x_0} + \frac{2h\gamma'\tan\gamma}{\cos^2\gamma}\overrightarrow{y_0} = \frac{h}{\cos^2\gamma}\left(\overrightarrow{x_0} + 2\tan\gamma\,\overrightarrow{y_0}\right)$$
.

– *Question 5*: Compare $\overrightarrow{v^{(S)}}(\mathrm{I})$ and $\overrightarrow{v^{(0)}}(\mathrm{I})$.

$$\overrightarrow{x_0} + 2\tan\gamma\,\overrightarrow{y_0} = \left(\cos\gamma\,\overrightarrow{x_S} - \sin\gamma\,\overrightarrow{y_S}\right) + 2\tan\gamma\left(\sin\gamma\,\overrightarrow{x_S} + \cos\gamma\,\overrightarrow{y_S}\right)$$

$$\text{" "} \quad = \left(\cos\gamma + \frac{2\sin^2\gamma}{\cos\gamma}\right)\overrightarrow{x_S} + \sin\gamma\,\overrightarrow{y_S} = \frac{1+\sin^2\gamma}{\cos\gamma}\overrightarrow{x_S} + \sin\gamma\,\overrightarrow{y_S}$$

$$\overrightarrow{v^{(0)}}(\mathrm{I}) = \frac{h\gamma'}{\cos^2\gamma}\left(\frac{1+\sin^2\gamma}{\cos\gamma}\overrightarrow{x_S} + \sin\gamma\,\overrightarrow{y_S}\right) = \overrightarrow{v^{(S)}}(\mathrm{I}).$$

Here, the mobile centroid is said to be rolling over of the fixed centroid without sliding over the course of the motion, like a system of cogs and gears.

2.5.6.2. *Exercise 16 – Motion of a system with a fixed plane*

We consider a device linked to the frame $\langle 0\rangle \equiv \left\langle \mathrm{O} \middle| \overrightarrow{x_0}\ \overrightarrow{y_0}\ \overrightarrow{z} \right\rangle$ onto which the mechanical system made up of two bars is installed.

Bar (1), of extremities O and A, of length L and to which the frame $\langle 1 \rangle \equiv \left\langle \mathrm{O} | \vec{x_1}\, \vec{y_1}\, \vec{z} \right\rangle$ is attached, is linked as a pivot of axis $\left(\mathrm{O}|\vec{z}\right)$ to the device. Its position is given by the angle $\alpha = \widehat{\vec{x_0}, \vec{x_1}} = \widehat{\vec{y_0}, \vec{y_1}}$.

This means: $\overrightarrow{\mathrm{OA}} = L\vec{u}(\alpha) = L\vec{x_1}$, $\vec{y_1} = \vec{u}\left(\alpha + \dfrac{\pi}{2}\right)$.

Bar (2), of extremities A and B, of length L and to which the frame $\langle 2 \rangle \equiv \left\langle \mathrm{A} | \vec{x_2}\, \vec{y_2}\, \vec{z} \right\rangle$ is attached, is linked as a pivot of axis $\left(\mathrm{A}|\vec{z}\right)$ to bar (1); it is also linked to the device, meaning that:

$$\overrightarrow{\mathrm{AB}} = L\vec{u}(3\alpha) = L\vec{x_2} \quad \text{where } 3\alpha = \widehat{\vec{x_0}, \vec{x_2}} = \widehat{\vec{y_0}, \vec{y_2}}.$$

α is a twice-derivable function of time.

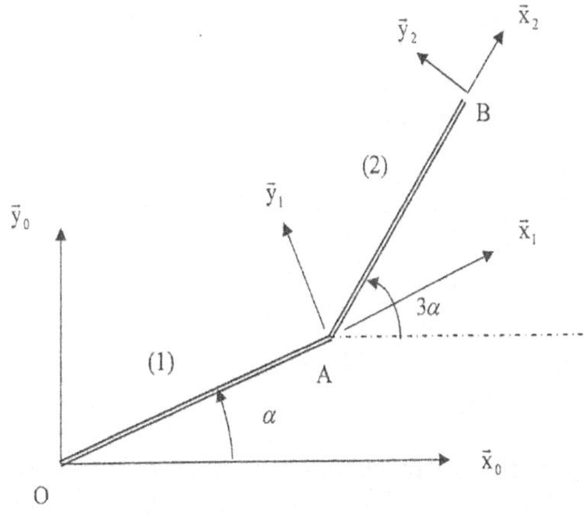

– *Question 1:* Express the vectors $\overrightarrow{\omega_1^0}, \overrightarrow{\omega_2^0}, \overrightarrow{\omega_2^1}$:

$$\overrightarrow{\omega_1^0} = \alpha'\vec{z} \;;\; \overrightarrow{\omega_2^0} = 3\alpha'\vec{z} \;;\; \overrightarrow{\omega_2^1} = \vec{s}\left\{ {}^{1}_{2} \right\} = \vec{s}\left\{ {}^{0}_{2} \right\} - \vec{s}\left\{ {}^{0}_{1} \right\} = 2\alpha'\vec{z}.$$

– *Question 2*: Using the reduction elements expressed in the basis $\left(\vec{u}(\alpha)\vec{u}\left(\alpha+\frac{\pi}{2}\right)\vec{z}\right)$, determine the expression of the velocity-distributing torsors:

– $\left\{ {0 \atop 1} \right\}_{/O}$, $\left\{ {0 \atop 1} \right\}_{/A}$, $\left\{ {1 \atop 2} \right\}_{/A}$, $\left\{ {0 \atop 2} \right\}_{/A}$, $\left\{ {0 \atop 2} \right\}_{/B}$;

– $\left\{ {0 \atop 1} \right\}_{/O} = \left[\alpha'\vec{z}\middle|\vec{0}\right]$, sliding-torsor of axis $\left(O\middle|\vec{z}\right)$;

– $\overrightarrow{\mathcal{M}_A}\left\{ {0 \atop 1} \right\} = \overrightarrow{\mathcal{M}_O}\left\{ {0 \atop 1} \right\} + \vec{s}\left\{ {0 \atop 1} \right\} \wedge \overrightarrow{OA} = \alpha'\vec{z} \wedge L\vec{u}(\alpha)$

$$\Rightarrow \left\{ {0 \atop 1} \right\}_{/A} = \left[\alpha'\vec{z}\middle|L\alpha'\vec{u}\left(\alpha+\frac{\pi}{2}\right)\right];$$

– $\left\{ {1 \atop 2} \right\}_{/A} = \left[2\alpha'\vec{z}\middle|\vec{0}\right]$, sliding-torsor of axis $\left(A\middle|\vec{z}\right)$;

– $\overrightarrow{\mathcal{M}_A}\left\{ {0 \atop 2} \right\} = \overrightarrow{\mathcal{M}_A}\left\{ {1 \atop 2} \right\} + \overrightarrow{\mathcal{M}_A}\left\{ {0 \atop 1} \right\} = L\alpha'\vec{u}\left(\alpha+\frac{\pi}{2}\right)$;

$$\Rightarrow \left\{ {0 \atop 2} \right\}_{/A} = \left[3\alpha'\vec{z}\middle|L\alpha'\vec{u}\left(\alpha+\frac{\pi}{2}\right)\right];$$

– $\overrightarrow{\mathcal{M}_B}\left\{ {0 \atop 2} \right\} = \overrightarrow{\mathcal{M}_A}\left\{ {0 \atop 2} \right\} + \vec{s}\left\{ {0 \atop 2} \right\} \wedge \overrightarrow{AB}$

" " $= L\alpha'\vec{u}\left(\alpha+\frac{\pi}{2}\right) + 3\alpha'\vec{z} \wedge L\vec{u}(3\alpha)$

$$\Rightarrow \left\{ {0 \atop 2} \right\}_{/B} = \left[3\alpha'\vec{z}\middle|L\alpha'\left(\vec{u}\left(\alpha+\frac{\pi}{2}\right)+3\vec{u}\left(3\alpha+\frac{\pi}{2}\right)\right)\right].$$

– *Question 3*: Determine the instantaneous rotation centers I_1^0, I_2^1, I_2^0:

$$\left\{ {0 \atop 1} \right\}_{/O} = \left[\alpha'\vec{z}\middle|\vec{0}\right] \Rightarrow \overrightarrow{OI_1^0} = \frac{\vec{s}\left\{ {0 \atop 1} \right\} \wedge \overrightarrow{\mathcal{M}_O}\left\{ {0 \atop 1} \right\}}{\vec{s}^{\,2}\left\{ {0 \atop 1} \right\}} = \vec{0} \Rightarrow I_1^0 = O,$$

$$\left\{ {1 \atop 2} \right\}_{/A} = \left[2\alpha'\vec{z}\middle|\vec{0}\right] \Rightarrow \overrightarrow{AI_2^1} = \frac{\vec{s}\left\{ {1 \atop 2} \right\} \wedge \overrightarrow{\mathcal{M}_O}\left\{ {1 \atop 2} \right\}}{\vec{s}^{\,2}\left\{ {1 \atop 2} \right\}} = \vec{0} \Rightarrow I_2^1 = A,$$

$$\left\{{}^0_2\right\}_{/A} = \left[3\alpha'\vec{z} \middle| L\alpha'\vec{u}\left(\alpha + \frac{\pi}{2}\right)\right] \Rightarrow I = \vec{s}\cdot\overrightarrow{\mathcal{M}_A}\left\{{}^0_2\right\} = 0 .$$

The torsor $\left\{{}^0_2\right\}$ is a sliding-torsor; its central axis (Δ) passes through the point I^0_2 defined by

$$\overrightarrow{AI^0_2} = \frac{\vec{s}\left\{{}^0_2\right\}\wedge\overrightarrow{\mathcal{M}_A}\left\{{}^0_2\right\}}{\vec{s}^{\,2}\left\{{}^0_2\right\}} = -\frac{L}{3}\vec{u}(\alpha).$$

The three instantaneous centers of rotation I^0_1, I^1_2, I^0_2 are always aligned on bar (1).

– *Question 4*: Express the vectors that define the fixed centroid and the mobile centroid of the motion of (2) in relation to (0). Characterize these curves:

– the fixed centroid for the motion of (2) in relation to (0) is expressed by the vectorial function $\overrightarrow{OI^0_2}$ as a projection onto the fixed centroid (0).

$$\overrightarrow{OI^0_2} = \overrightarrow{OA} + \overrightarrow{AI^0_2} = \frac{2L}{3}\left(\cos\alpha\,\overrightarrow{x_0} + \sin\alpha\,\overrightarrow{y_0}\right).$$

The fixed centroid is the circle of center O and of radius $\frac{2L}{3}$;

– the mobile centroid for the motion of (2) in relation to (0) is expressed by the vectorial function $\overrightarrow{AI^0_2} = -\frac{L}{3}\vec{u}(\alpha)$ which expresses a circle of center A and of radius $\frac{L}{3}$.

– *Question 5*: Determine the velocity vectors of the instantaneous center of rotation I^0_2 : $\overrightarrow{v^{(0)}}\left(I^0_2\right)$ and $\overrightarrow{v^{(2)}}\left(I^0_2\right)$. Compare these vectors.

A vectorial basis to consider is the basis $(2) \equiv \left(\vec{u}(3\alpha)\,\vec{u}\left(3\alpha+\frac{\pi}{2}\right)\vec{z} \right)$:

– $\overrightarrow{v^{(0)}}(\mathrm{I}_2^0) = \frac{d^{(0)}}{dt}\overrightarrow{\mathrm{OI}_2^0} = \frac{d^{(0)}}{dt}\left[\frac{2L}{3}\vec{u}(\alpha)\right] = \frac{2L}{3}\alpha'\vec{u}\left(\alpha+\frac{\pi}{2}\right)$;

– $\overrightarrow{v^{(2)}}(\mathrm{I}_2^0) = \frac{d^{(2)}}{dt}\overrightarrow{\mathrm{AI}_2^0}$

$$\overrightarrow{v^{(2)}}(\mathrm{I}_2^0) = -\frac{L}{3}\frac{d^{(2)}}{dt}\left[\cos 2\alpha\,\vec{u}(3\alpha) - \sin 2\alpha\,\vec{u}\left(3\alpha+\frac{\pi}{2}\right)\right]$$

$$\text{'' ''} \quad = \frac{2L}{3}\alpha'\left[\sin 2\alpha\,\vec{u}(3\alpha) + \cos 2\alpha\,\vec{u}\left(3\alpha+\frac{\pi}{2}\right)\right] = \frac{2L}{3}\alpha'\vec{u}\left(\alpha+\frac{\pi}{2}\right).$$

– *Question 6*: Determine the acceleration vectors of the instantaneous center of rotation I_2^0: $\overrightarrow{J^{(0)}}(\mathrm{I}_2^0)$ and $\overrightarrow{J^{(2)}}(\mathrm{I}_2^0)$. Compare these vectors.

$$\overrightarrow{J^{(0)}}(\mathrm{I}_2^0) = \frac{d^{(0)}}{dt}\overrightarrow{v^{(0)}}(\mathrm{I}_2^0) = \frac{2L}{3}\left[\alpha''\vec{u}\left(\alpha+\frac{\pi}{2}\right) - \alpha'^2\vec{u}(\alpha)\right].$$

$$\overrightarrow{J^{(2)}}(\mathrm{I}_2^0) = \frac{d^{(2)}}{dt}\overrightarrow{v^{(2)}}(\mathrm{I}_2^0),$$

$$\overrightarrow{J^{(2)}}(\mathrm{I}_2^0) = \frac{2L}{3}\left[\alpha''\vec{u}\left(\alpha+\frac{\pi}{2}\right)\ldots\right.$$

$$\left.\ldots + 2\alpha'^2\left(\cos 2\alpha\,\vec{u}(3\alpha) - \sin 2\alpha\,\vec{u}\left(3\alpha+\frac{\pi}{2}\right)\right)\right].$$

$$\text{'' ''} \quad = \frac{2L}{3}\left[2\alpha'^2\vec{u}(\alpha) + \alpha''\vec{u}\left(\alpha+\frac{\pi}{2}\right)\right]$$

If we decompose these accelerations into tangential components collinear to the velocity vectors, we can see that the components are equal to each other: however, the normal components of the acceleration of the instantaneous center of rotation I_2^0 on the fixed centroid and on the mobile centroid are different.

2.6. Combining motions within a mobile frame of reference

As soon as the mechanical problem involves the motion of several solids in relation to each other, the question of their respective situations and the issue of adequately formulating their kinematics become crucial.

The mechanics of the systems of the rigid solids is the subject of Volume 3 [BOR 16b]; however, from a simply kinematic perspective, it is important to be able to express the velocity field of a solid in conditions that are most favorable for satisfactory formulation. Often this involves using frames of reference that are themselves mobile. As a result, the combination of the motions of these different bodies requires a particular development to determine the trajectories, the velocity and acceleration fields of a material point depending on the frame from which the motion is being observed.

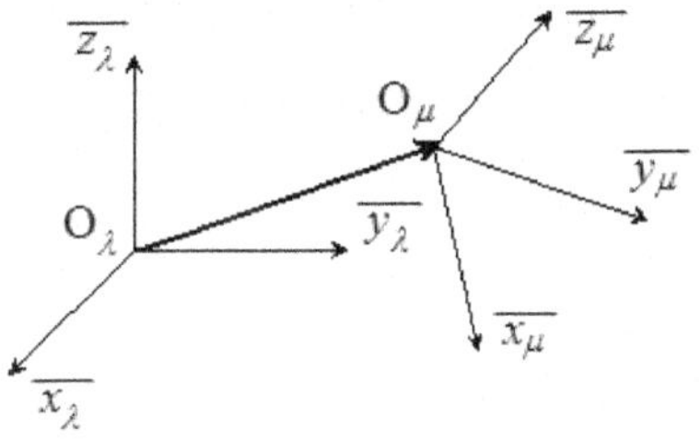

Figure 2.7. *Relative situation of the two systems*

2.6.1. *Position of the problem*

Let us consider two frames $\langle\lambda\rangle \equiv \left\langle O_\lambda \middle| \overrightarrow{x_\lambda}\ \overrightarrow{y_\lambda}\ \overrightarrow{z_\lambda} \right\rangle$ and $\langle\mu\rangle \equiv \left\langle O_\mu \middle| \overrightarrow{x_\mu}\ \overrightarrow{y_\mu}\ \overrightarrow{z_\mu} \right\rangle$ which are mobile in relation to each other and for which the motion of $\langle\mu\rangle$ in relation to $\langle\lambda\rangle$ can be expressed as:

– the position vector $\overrightarrow{O_\lambda O_\mu} = a(t)\overrightarrow{x_\lambda} + b(t)\overrightarrow{y_\lambda} + c(t)\overrightarrow{z_\lambda}$;

– the basis change table:

$(\lambda) \backslash (\mu)$	$\overrightarrow{x_\mu}$	$\overrightarrow{y_\mu}$	$\overrightarrow{z_\mu}$
$\overrightarrow{x_\lambda}$	$p_{11}(t)$	$p_{12}(t)$	$p_{13}(t)$
$\overrightarrow{y_\lambda}$	$p_{21}(t)$	$p_{22}(t)$	$p_{23}(t)$
$\overrightarrow{z_\lambda}$	$p_{31}(t)$	$p_{32}(t)$	$p_{33}(t)$

2.6.2. *Trajectory of a material point in the different frames*

Over the course of the time interval $[t_i, t_f]$, the trajectory of a material point M in $\langle \mu \rangle$ is $\Gamma_{(\mu)}^{(t_i,t_f)}(\mathrm{M})$, defined using the position vector

$$\overrightarrow{\mathrm{O}_\mu \mathrm{M}} = x_\mu(t)\overrightarrow{x_\mu} + y_\mu(t)\overrightarrow{y_\mu} + z_\mu(t)\overrightarrow{z_\mu}.$$

In the same time interval, the trajectory of this point in $\langle \lambda \rangle$, $\Gamma_{(\lambda)}^{(t_i,t_f)}(\mathrm{M})$, is defined by the position vector

$$\overrightarrow{\mathrm{O}_\lambda \mathrm{M}} = \overrightarrow{\mathrm{O}_\lambda \mathrm{O}_\mu} + \overrightarrow{\mathrm{O}_\mu \mathrm{M}} = x_\lambda(t)\overrightarrow{x_\lambda} + y_\lambda(t)\overrightarrow{y_\lambda} + z_\lambda(t)\overrightarrow{z_\lambda},$$

where $\begin{cases} x_\lambda(t) = a + p_{11}\,x_\mu + p_{12}\,y_\mu + p_{13}\,z_\mu \\ y_\lambda(t) = b + p_{21}\,x_\mu + p_{22}\,y_\mu + p_{23}\,z_\mu \\ z_\lambda(t) = c + p_{31}\,x_\mu + p_{32}\,y_\mu + p_{33}\,z_\mu \end{cases}$.

2.6.3. *Combination of velocities*

We know the motion of $\langle \mu \rangle$ in relation to $\langle \lambda \rangle$, which we can represent using the velocity-distributing torsor:

$$\left\{ {}^{\lambda}_{\mu} \right\}_{\mathrm{O}_\mu} = \left[\overrightarrow{\omega_\mu^\lambda} \,\middle|\, \overrightarrow{v^{(\lambda)}}(\mathrm{O}_\mu) \right].$$

We consider a material point M whose velocity is calculated in $\langle\lambda\rangle$:

$$\overrightarrow{v^{(\lambda)}}(\mathrm{M})=\frac{d^{(\lambda)}}{dt}\overrightarrow{\mathrm{O}_\lambda\mathrm{M}}=\frac{d^{(\lambda)}}{dt}\overrightarrow{\mathrm{O}_\lambda\mathrm{O}_\mu}+\frac{d^{(\lambda)}}{dt}\overrightarrow{\mathrm{O}_\mu\mathrm{M}}=\overrightarrow{v^{(\lambda)}}(\mathrm{O}_\mu)+\frac{d^{(\lambda)}}{dt}\overrightarrow{\mathrm{O}_\mu\mathrm{M}}$$

where $\frac{d^{(\lambda)}}{dt}\overrightarrow{\mathrm{O}_\mu\mathrm{M}}-\frac{d^{(\mu)}}{dt}\overrightarrow{\mathrm{O}_\mu\mathrm{M}}=\overrightarrow{\omega_\mu^\lambda}\wedge\overrightarrow{\mathrm{O}_\mu\mathrm{M}}$, according to the rules of the derivation of vectorial functions,[5] which is:

$$\frac{d^{(\lambda)}}{dt}\overrightarrow{\mathrm{O}_\mu\mathrm{M}}=\overrightarrow{v^{(\mu)}}(\mathrm{M})+\overrightarrow{\omega_\mu^\lambda}\wedge\overrightarrow{\mathrm{O}_\mu\mathrm{M}}.$$

As a result,

$$\overrightarrow{v^{(\lambda)}}(\mathrm{M})=\overrightarrow{v^{(\mu)}}(\mathrm{M})+\overrightarrow{v^{(\lambda)}}(\mathrm{O}_\mu)+\overrightarrow{\omega_\mu^\lambda}\wedge\overrightarrow{\mathrm{O}_\mu\mathrm{M}}.$$

2.6.3.1. *General formulation of the combination of velocities*

If we consider the terms of the expression below, we remark that

$$\overrightarrow{v^{(\lambda)}}(\mathrm{O}_\mu)+\overrightarrow{\omega_\mu^\lambda}\wedge\overrightarrow{\mathrm{O}_\mu\mathrm{M}},$$

is the motion at M of a torsor whose reduction elements at O_μ are

$$\overrightarrow{\omega_\mu^\lambda} \text{ and } \overrightarrow{v^{(\lambda)}}(\mathrm{O}_\mu),$$

or in other words the relative velocity-distributing torsor $\{{}^\lambda_\mu\}$ of the motion of frame $\langle\mu\rangle$ in relation to $\langle\lambda\rangle$.

The vector

$$\overrightarrow{\mathcal{M}_{\mathrm{M}}}\{{}^\lambda_\mu\}=\overrightarrow{v^{(\lambda)}}(\mathrm{O}_\mu)+\overrightarrow{\omega_\mu^\lambda}\wedge\overrightarrow{\mathrm{O}_\mu\mathrm{M}},$$

corresponds to the velocity that the material point M would have in the frame $\langle\lambda\rangle$ if it were fixed in the frame $\langle\mu\rangle$. This is called the

5 See Volume 2 [BOR 16a].

drive velocity of M at a time t during the motion of $\langle\mu\rangle$ relative to $\langle\lambda\rangle$; it is written $\overrightarrow{v_e}\left(\mathrm{M}\big|_{\mu}^{\lambda}\right)$:

$$\overrightarrow{\mathcal{M}_{\mathrm{M}}\left\{\begin{smallmatrix}\lambda\\\mu\end{smallmatrix}\right\}}=\overrightarrow{v^{(\lambda)}}(\mathrm{O}_\mu)+\overrightarrow{\omega_\mu^\lambda}\wedge\overrightarrow{\mathrm{O}_\mu\mathrm{M}}=\overrightarrow{v_e}\left(\mathrm{M}\big|_{\mu}^{\lambda}\right).$$

The other term in the above expression of $\overrightarrow{v^{(\lambda)}}(\mathrm{M})$ leads us to the general formulation of the law of the combination of velocities in the relative motion of frames $\langle\lambda\rangle$ and $\langle\mu\rangle$:

$$\overrightarrow{v^{(\lambda)}}(\mathrm{M})-\overrightarrow{v^{(\mu)}}(\mathrm{M})=\overrightarrow{\mathcal{M}_{\mathrm{M}}\left\{\begin{smallmatrix}\lambda\\\mu\end{smallmatrix}\right\}}.$$

2.6.3.2. *Properties of the relative velocity-distributing torsor* $\left\{\begin{smallmatrix}\lambda\\\mu\end{smallmatrix}\right\}$

2.6.3.2.1. *Opposed relative torsors*

Inversely, looking at the motion of $\langle\lambda\rangle$ relative to $\langle\mu\rangle$ we can write

$$\overrightarrow{v^{(\mu)}}(\mathrm{M})-\overrightarrow{v^{(\lambda)}}(\mathrm{M})=\overrightarrow{\mathcal{M}_{\mathrm{M}}\left\{\begin{smallmatrix}\mu\\\lambda\end{smallmatrix}\right\}}.$$

By combining this equation with the previous one, we obtain

$$\overrightarrow{\mathcal{M}_{\mathrm{M}}\left\{\begin{smallmatrix}\lambda\\\mu\end{smallmatrix}\right\}}+\overrightarrow{\mathcal{M}_{\mathrm{M}}\left\{\begin{smallmatrix}\mu\\\lambda\end{smallmatrix}\right\}}=\vec{0}.$$

Seeing as this result is applicable for any point M, at another point P there is

$$\overrightarrow{\mathcal{M}_{\mathrm{P}}\left\{\begin{smallmatrix}\lambda\\\mu\end{smallmatrix}\right\}}+\overrightarrow{\mathcal{M}_{\mathrm{P}}\left\{\begin{smallmatrix}\mu\\\lambda\end{smallmatrix}\right\}}=\vec{0},$$

$$\text{with}\quad\begin{cases}\overrightarrow{\mathcal{M}_{\mathrm{P}}\left\{\begin{smallmatrix}\lambda\\\mu\end{smallmatrix}\right\}}=\overrightarrow{\mathcal{M}_{\mathrm{M}}\left\{\begin{smallmatrix}\lambda\\\mu\end{smallmatrix}\right\}}+\overrightarrow{\omega_\mu^\lambda}\wedge\overrightarrow{\mathrm{MP}}\\\overrightarrow{\mathcal{M}_{\mathrm{P}}\left\{\begin{smallmatrix}\mu\\\lambda\end{smallmatrix}\right\}}=\overrightarrow{\mathcal{M}_{\mathrm{M}}\left\{\begin{smallmatrix}\mu\\\lambda\end{smallmatrix}\right\}}+\overrightarrow{\omega_\lambda^\mu}\wedge\overrightarrow{\mathrm{MP}}\end{cases}.$$

By summing the terms of the two equations, term by term, and considering the results seen above, this means that

$$\left(\overrightarrow{\omega_\mu^\lambda}+\overrightarrow{\omega_\lambda^\mu}\right)\wedge\overrightarrow{\mathrm{MP}}=\vec{0}.$$

This result can be applied for any point M or P, which leads to this result

$$\overrightarrow{\omega_{\mu}^{\lambda}} + \overrightarrow{\omega_{\lambda}^{\mu}} = \vec{0} .$$

The respective reduction elements of the two velocity-distributing torsors $\{{}^{\lambda}_{\mu}\}$ and $\{{}^{\mu}_{\lambda}\}$ are opposed, and as a result the two torsors verify the following relation

$$\{{}^{\lambda}_{\mu}\} + \{{}^{\mu}_{\lambda}\} = \{0\} .$$

2.6.3.2.2. Inversion of motions

The motion of $\langle\lambda\rangle$ relative to $\langle\mu\rangle$ is the inverse of the motion of $\langle\mu\rangle$ relative to $\langle\lambda\rangle$. The link between these two motions relies on these three equations

$$\begin{cases} \overrightarrow{\mathcal{M}_{\mathrm{M}}}\{{}^{\mu}_{\lambda}\} = -\overrightarrow{\mathcal{M}_{\mathrm{M}}}\{{}^{\lambda}_{\mu}\} \\ \overrightarrow{\omega_{\lambda}^{\mu}} = -\overrightarrow{\omega_{\mu}^{\lambda}} \\ \{{}^{\mu}_{\lambda}\} = -\{{}^{\lambda}_{\mu}\} \end{cases} ,$$

which mean that the drive velocities, the rates of rotation and the velocity-distributing torsors are opposed.

The motion of $\langle\mu\rangle$ relative to $\langle\lambda\rangle$ is represented by the velocity-distributing torsor $\{{}^{\lambda}_{\mu}\}_{\mathrm{O}_{\mu}} = \left[\overrightarrow{\omega_{\mu}^{\lambda}} \,\middle|\, \overrightarrow{v^{(\lambda)}}(\mathrm{O}_{\mu})\right]$ which is expressed at a point Q by

$$\{{}^{\lambda}_{\mu}\}_{\mathrm{Q}} = \left[\overrightarrow{\omega_{\mu}^{\lambda}} \,\middle|\, \overrightarrow{v^{(\lambda)}}(\mathrm{O}_{\mu}) + \overrightarrow{\omega_{\mu}^{\lambda}} \wedge \overrightarrow{\mathrm{O}_{\mu}\mathrm{Q}}\right].$$

In the same way, the velocity-distributing torsor

$$\{{}^{\mu}_{\lambda}\}_{\mathrm{O}_{\lambda}} = \left[\overrightarrow{\omega_{\lambda}^{\mu}} \,\middle|\, \overrightarrow{v^{(\mu)}}(\mathrm{O}_{\lambda})\right]$$

of the motion of $\langle\lambda\rangle$ in relation to $\langle\mu\rangle$ at the same point Q is written

$$\left\{ {}^{\mu}_{\lambda} \right\}_{\mathrm{Q}} = \left[\overrightarrow{\omega_{\lambda}^{\mu}} \,\middle|\, \overrightarrow{v^{(\mu)}}(\mathrm{O}_{\lambda}) + \overrightarrow{\omega_{\lambda}^{\mu}} \wedge \overrightarrow{\mathrm{O}_{\lambda}\mathrm{Q}} \right].$$

The drive velocities of the point Q during the motion of one of these systems relative to the other are written, respectively, as

$$\begin{cases} \overrightarrow{\mathcal{M}_{\mathrm{Q}}}\left\{ {}^{\lambda}_{\mu} \right\} = \overrightarrow{v^{(\lambda)}}(\mathrm{O}_{\mu}) + \overrightarrow{\omega_{\mu}^{\lambda}} \wedge \overrightarrow{\mathrm{O}_{\mu}\mathrm{Q}} \\ \overrightarrow{\mathcal{M}_{\mathrm{Q}}}\left\{ {}^{\mu}_{\lambda} \right\} = \overrightarrow{v^{(\mu)}}(\mathrm{O}_{\lambda}) + \overrightarrow{\omega_{\lambda}^{\mu}} \wedge \overrightarrow{\mathrm{O}_{\lambda}\mathrm{Q}} \end{cases},$$

and verify the equation

$$\overrightarrow{\mathcal{M}_{\mathrm{Q}}}\left\{ {}^{\lambda}_{\mu} \right\} + \overrightarrow{\mathcal{M}_{\mathrm{Q}}}\left\{ {}^{\mu}_{\lambda} \right\} = \vec{0}.$$

As a result the two velocities $\overrightarrow{v^{(\lambda)}}(\mathrm{O}_{\mu})$ and $\overrightarrow{v^{(\mu)}}(\mathrm{O}_{\lambda})$ are linked by the equation

$$\overrightarrow{v^{(\lambda)}}(\mathrm{O}_{\mu}) + \overrightarrow{v^{(\mu)}}(\mathrm{O}_{\lambda}) = \overrightarrow{\omega_{\mu}^{\lambda}} \wedge \overrightarrow{\mathrm{O}_{\lambda}\mathrm{O}_{\mu}} = \overrightarrow{\omega_{\lambda}^{\mu}} \wedge \overrightarrow{\mathrm{O}_{\mu}\mathrm{O}_{\lambda}}.$$

2.6.3.3. *Combination of velocities in a succession of frames*

We consider three frames $\langle\lambda\rangle$, $\langle\mu\rangle$ and $\langle\nu\rangle$, the motion of each relative to the others is represented by the three formulae for the combination of their motions

$$\begin{cases} \overrightarrow{v^{(\lambda)}}(\mathrm{M}) = \overrightarrow{v^{(\mu)}}(\mathrm{M}) + \overrightarrow{\mathcal{M}_{\mathrm{M}}}\left\{ {}^{\lambda}_{\mu} \right\} \\ \overrightarrow{v^{(\mu)}}(\mathrm{M}) = \overrightarrow{v^{(\nu)}}(\mathrm{M}) + \overrightarrow{\mathcal{M}_{\mathrm{M}}}\left\{ {}^{\mu}_{\nu} \right\} \\ \overrightarrow{v^{(\nu)}}(\mathrm{M}) = \overrightarrow{v^{(\lambda)}}(\mathrm{M}) + \overrightarrow{\mathcal{M}_{\mathrm{M}}}\left\{ {}^{\nu}_{\lambda} \right\} \end{cases},$$

the sum of which, term by term, leads to this equation

$$\overrightarrow{\mathcal{M}_{\mathrm{M}}}\left\{ {}^{\lambda}_{\mu} \right\} + \overrightarrow{\mathcal{M}_{\mathrm{M}}}\left\{ {}^{\mu}_{\nu} \right\} + \overrightarrow{\mathcal{M}_{\mathrm{M}}}\left\{ {}^{\nu}_{\lambda} \right\} = \vec{0}.$$

By developing in the following way each of the three terms of the drive velocity of the equation below

$$\overrightarrow{\mathcal{M}_{\mathrm{M}}}\left\{{\lambda \atop \mu}\right\}=\overrightarrow{v^{(\lambda)}}(\mathrm{O}_{\mu})+\overrightarrow{\omega_{\mu}^{\lambda}}\wedge\overrightarrow{\mathrm{O}_{\mu}\mathrm{M}}=\overrightarrow{v^{(\lambda)}}(\mathrm{O}_{\mu})+\overrightarrow{\omega_{\mu}^{\lambda}}\wedge\left(\overrightarrow{\mathrm{O}_{\lambda}\mathrm{M}}-\overrightarrow{\mathrm{O}_{\lambda}\mathrm{O}_{\mu}}\right),$$

and by using the formulae for the combination of velocities seen above, we obtain the following equation

$$\left(\overrightarrow{\omega_{\mu}^{\lambda}}+\overrightarrow{\omega_{\nu}^{\mu}}+\overrightarrow{\omega_{\lambda}^{\nu}}\right)\wedge\overrightarrow{\mathrm{O}_{\lambda}\mathrm{M}}=\vec{0},$$

which is verified for any point M considered, and as a result

$$\overrightarrow{\omega_{\mu}^{\lambda}}+\overrightarrow{\omega_{\nu}^{\mu}}+\overrightarrow{\omega_{\lambda}^{\nu}}=\vec{0}.$$

We can deduce the two following formulae for the combination of motions applied to the three relative velocity-distributing torsors

$$\left\{{\lambda \atop \mu}\right\}+\left\{{\mu \atop \nu}\right\}+\left\{{\nu \atop \lambda}\right\}=\{0\} \text{ or even } \left\{{\lambda \atop \mu}\right\}=\left\{{\nu \atop \mu}\right\}+\left\{{\lambda \atop \nu}\right\}.$$

These torsor relations enable the calculation of the rate of rotation in the determination of coordinates using the Euler angles

$$\overrightarrow{\omega_{S}^{\lambda}}=\overrightarrow{\omega_{1}^{\lambda}}+\overrightarrow{\omega_{2}^{1}}+\overrightarrow{\omega_{S}^{2}}=\psi'\overrightarrow{z_{\lambda}}+\theta'\vec{n}+\varphi'\overrightarrow{z_{S}}.$$

2.6.4. *Combination of accelerations*

When the motion of frame $\langle\mu\rangle$ relative to the frame $\langle\lambda\rangle$ is known, meaning that the velocity-distributing torsor of this motion is $\left\{{\lambda \atop \mu}\right\}_{\mathrm{O}_{\mu}}=\left[\overrightarrow{\omega_{\mu}^{\lambda}}\,\middle|\,\overrightarrow{v^{(\lambda)}}(\mathrm{O}_{\mu})\right]$ and, as a result, $\left(\overrightarrow{\omega_{\mu}^{\lambda}}\right)'$ and $\overrightarrow{J^{(\lambda)}}(\mathrm{O}_{\mu})$ are known, the derivation relative to time of the formula of the combination of velocities is

$$\overrightarrow{v^{(\lambda)}}(\mathrm{M})=\overrightarrow{v^{(\mu)}}(\mathrm{M})+\overrightarrow{v^{(\lambda)}}(\mathrm{O}_{\mu})+\overrightarrow{\omega_{\mu}^{\lambda}}\wedge\overrightarrow{\mathrm{O}_{\mu}\mathrm{M}}.$$

In $\langle\lambda\rangle$, this means that

$$\frac{d^{(\lambda)}}{dt}\overrightarrow{v^{(\lambda)}}(\mathrm{M})=\frac{d^{(\lambda)}}{dt}\overrightarrow{v^{(\mu)}}(\mathrm{M})+\frac{d^{(\lambda)}}{dt}\overrightarrow{v^{(\lambda)}}(\mathrm{O}_\mu)+\frac{d^{(\lambda)}}{dt}\left(\overrightarrow{\omega_\mu^\lambda}\wedge\overrightarrow{\mathrm{O}_\mu\mathrm{M}}\right),$$

an equation whose different terms are developed below

– $\frac{d^{(\lambda)}}{dt}\overrightarrow{v^{(\lambda)}}(\mathrm{M})=\overrightarrow{J^{(\lambda)}}(\mathrm{M})$;

– $$\begin{aligned}\frac{d^{(\lambda)}}{dt}\overrightarrow{v^{(\mu)}}(\mathrm{M})&=\frac{d^{(\mu)}}{dt}\overrightarrow{v^{(\mu)}}(\mathrm{M})+\overrightarrow{\omega_\mu^\lambda}\wedge\overrightarrow{v^{(\mu)}}(\mathrm{M})\\ \text{" "}\quad&=\overrightarrow{J^{(\mu)}}(\mathrm{M})+\overrightarrow{\omega_\mu^\lambda}\wedge\overrightarrow{v^{(\mu)}}(\mathrm{M})\quad;\end{aligned}$$

– $\frac{d^{(\lambda)}}{dt}\overrightarrow{v^{(\lambda)}}(\mathrm{O}_\mu)=\overrightarrow{J^{(\lambda)}}(\mathrm{O}_\mu)$;

– $$\begin{aligned}\frac{d^{(\lambda)}}{dt}\left(\overrightarrow{\omega_\mu^\lambda}\wedge\overrightarrow{\mathrm{O}_\mu\mathrm{M}}\right)&=\left(\frac{d^{(\lambda)}}{dt}\overrightarrow{\omega_\mu^\lambda}\right)\wedge\overrightarrow{\mathrm{O}_\mu\mathrm{M}}+\overrightarrow{\omega_\mu^\lambda}\wedge\frac{d^{(\lambda)}}{dt}\overrightarrow{\mathrm{O}_\mu\mathrm{M}}\\ \text{" "}\quad&=\left(\overrightarrow{\omega_\mu^\lambda}\right)'\wedge\overrightarrow{\mathrm{O}_\mu\mathrm{M}}+\overrightarrow{\omega_\mu^\lambda}\wedge\left[\frac{d^{(\mu)}}{dt}\overrightarrow{\mathrm{O}_\mu\mathrm{M}}+\overrightarrow{\omega_\mu^\lambda}\wedge\overrightarrow{\mathrm{O}_\mu\mathrm{M}}\right]\\ \text{" "}\quad&=\left(\overrightarrow{\omega_\mu^\lambda}\right)'\wedge\overrightarrow{\mathrm{O}_\mu\mathrm{M}}+\overrightarrow{\omega_\mu^\lambda}\wedge\overrightarrow{v^{(\mu)}}(\mathrm{M})+\overrightarrow{\omega_\mu^\lambda}\wedge\left(\overrightarrow{\omega_\mu^\lambda}\wedge\overrightarrow{\mathrm{O}_\mu\mathrm{M}}\right)\end{aligned}.$$

2.6.4.1. *General formulation*

We obtain the formula for the combination of accelerations

$$\begin{aligned}\overrightarrow{J^{(\lambda)}}(\mathrm{M})-\overrightarrow{J^{(\mu)}}(\mathrm{M})&=\overrightarrow{\omega_\mu^\lambda}\wedge\overrightarrow{v^{(\mu)}}(\mathrm{M})+\overrightarrow{J^{(\lambda)}}(\mathrm{O}_\mu)\ldots\\ \text{" "}\quad&\ldots+\left(\overrightarrow{\omega_\mu^\lambda}\right)'\wedge\overrightarrow{\mathrm{O}_\mu\mathrm{M}}+\overrightarrow{\omega_\mu^\lambda}\wedge\left(\overrightarrow{\omega_\mu^\lambda}\wedge\overrightarrow{\mathrm{O}_\mu\mathrm{M}}\right)\end{aligned}.$$

2.6.4.2. *Identification of the different terms of acceleration*

In this expression, on the right side we can distinguish two terms that express the different aspects of the acceleration of M at a time t during the motion of the frame $\langle\mu\rangle$ relative to the frame $\langle\lambda\rangle$:

– $\vec{J}_c\left(\mathrm{M}|_\mu^\lambda\right)=2\,\overrightarrow{\omega_\mu^\lambda}\wedge\overrightarrow{v^{(\mu)}}(\mathrm{M})$, also known as Coriolis acceleration;

$$- \ \overrightarrow{J_e}\left(\mathrm{M}|_{\mu}^{\lambda}\right) = \overrightarrow{J^{(\lambda)}}\left(\mathrm{O}_{\mu}\right) + \left(\overrightarrow{\omega_{\mu}^{\lambda}}\right)' \wedge \overrightarrow{\mathrm{O}_{\mu}\mathrm{M}} + \overrightarrow{\omega_{\mu}^{\lambda}} \wedge \left(\overrightarrow{\omega_{\mu}^{\lambda}} \wedge \overrightarrow{\mathrm{O}_{\mu}\mathrm{M}}\right),$$

called the *drive acceleration*, meaning that the formula for the combination of accelerations during the relative motion of the two frames $\langle \lambda \rangle$ and $\langle \mu \rangle$ can be written synthetically as:

$$\overrightarrow{J^{(\lambda)}}(\mathrm{M}) - \overrightarrow{J^{(\mu)}}(\mathrm{M}) = \overrightarrow{J_c}\left(\mathrm{M}|_{\mu}^{\lambda}\right) + \overrightarrow{J_e}\left(\mathrm{M}|_{\mu}^{\lambda}\right).$$

2.6.4.3. *Interpretation of the Coriolis acceleration*

In the expression of the Coriolis acceleration, we can note that the corresponding vector is orthogonal to the plane made by the rotation rate vector $\overrightarrow{\omega_{\mu}^{\lambda}}$ and the velocity vector $\overrightarrow{v^{(\mu)}}(\mathrm{M})$. Furthermore, the existence of an acceleration is tied to existence of a force as a result of the fundamental principle of dynamics, as seen in Volume 3 [BOR 16b]. In the trajectory in $\langle \lambda \rangle$ of a body evolving in the frame $\langle \mu \rangle$, this term of acceleration is, therefore, seen as an effect exerted orthogonally on the plane $\Pi\left(\overrightarrow{\omega_{\mu}^{\lambda}}, \overrightarrow{v^{(\mu)}}(\mathrm{M})\right)$.

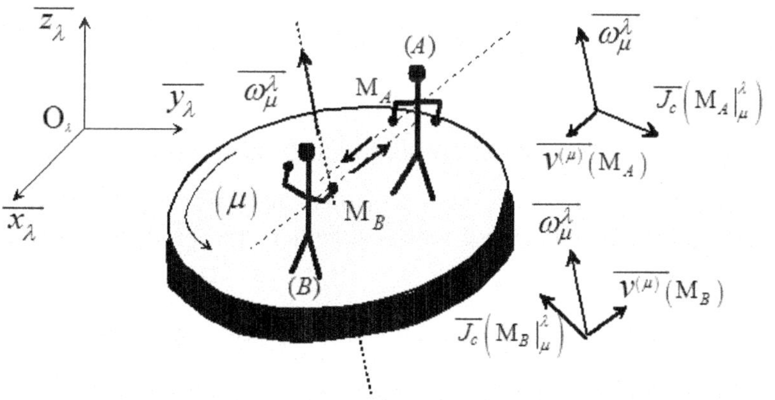

Figure 2.8. *Illustration of the effect of Coriolis acceleration*

To illustrate the significance of this notion, let us take the following example – although it may seem a bit outlandish – of a boxing match taking place on a rotating platform, as it is easy to understand.

When person (A) moves their fist $\mathrm{M_A}$ toward (B) with a velocity of $\overrightarrow{v^{(\mu)}}(\mathrm{M}_A)$ relative to the platform (μ), the rotation of the latter, with a rate of rotation $\overrightarrow{\omega_\mu^\lambda}$ in the frame of reference $\langle\lambda\rangle$, results in a Coriolis acceleration $\overrightarrow{J_c}\left(\mathrm{M}_A|_\mu^\lambda\right)$, which, considering the situation of the three vectors, results in the trajectory of the fist becoming deviated toward the outside of the platform, as if under the effect of a centrifugal force.

Inversely, if person (B) is aiming to reach their opponent with their fist M_B with a velocity of $\overrightarrow{v^{(\mu)}}(\mathrm{M}_B)$, the Coriolis acceleration $\overrightarrow{J_c}\left(\mathrm{M}_B|_\mu^\lambda\right)$ would deviate its trajectory toward the center of the platform as if under the effect of centripetal force.

With regards to the drive acceleration $\overrightarrow{J_e}\left(\mathrm{M}|_\mu^\lambda\right)$ of the material point M $\left(\mathrm{M}_A \text{ or } \mathrm{M}_B\right)$ in the motion of $\langle\mu\rangle$ relative to $\langle\lambda\rangle$, it is the acceleration that would be present in this point M in $\langle\lambda\rangle$ if it were fixed in $\langle\mu\rangle$.

2.6.4.4. *Properties of the combination of accelerations*

The terms of drive and Coriolis that are involved in the combination of motions work together to produce particular properties.

2.6.4.4.1. Derivation in relation to time of the drive velocity

$$\overrightarrow{\mathcal{M}_{\mathrm{M}}}\left\{\begin{smallmatrix}\lambda\\ \mu\end{smallmatrix}\right\} = \overrightarrow{v^{(\lambda)}}(\mathrm{O}_\mu) + \overrightarrow{\omega_\mu^\lambda} \wedge \overrightarrow{\mathrm{O}_\mu\mathrm{M}}.$$

$$\frac{d^{(\lambda)}}{dt}\overrightarrow{\mathcal{M}_{\mathrm{M}}}\{{}^{\lambda}_{\mu}\} = \frac{d^{(\lambda)}}{dt}\overrightarrow{v^{(\lambda)}}(\mathrm{O}_\mu) + \left(\frac{d^{(\lambda)}}{dt}\overrightarrow{\omega_\mu^\lambda}\right)\wedge\overrightarrow{\mathrm{O}_\mu\mathrm{M}} + \overrightarrow{\omega_\mu^\lambda}\wedge\frac{d^{(\lambda)}}{dt}\overrightarrow{\mathrm{O}_\mu\mathrm{M}}$$

$$\text{'' ''} \quad = \overrightarrow{J^{(\lambda)}}(\mathrm{O}_\mu) + \left(\overrightarrow{\omega_\mu^\lambda}\right)'\wedge\overrightarrow{\mathrm{O}_\mu\mathrm{M}}\ldots$$

$$\ldots + \overrightarrow{\omega_\mu^\lambda}\wedge\left[\frac{d^{(\mu)}}{dt}\overrightarrow{\mathrm{O}_\mu\mathrm{M}} + \overrightarrow{\omega_\mu^\lambda}\wedge\overrightarrow{\mathrm{O}_\mu\mathrm{M}}\right],$$

$$\text{'' ''} \quad = \overrightarrow{\omega_\mu^\lambda}\wedge\overrightarrow{v^{(\mu)}}(\mathrm{M}) + \overrightarrow{J^{(\lambda)}}(\mathrm{O}_\mu)\ldots$$

$$\ldots + \left(\overrightarrow{\omega_\mu^\lambda}\right)'\wedge\overrightarrow{\mathrm{O}_\mu\mathrm{M}} + \overrightarrow{\omega_\mu^\lambda}\wedge\left(\overrightarrow{\omega_\mu^\lambda}\wedge\overrightarrow{\mathrm{O}_\mu\mathrm{M}}\right)$$

which gives the expression of this derivative

$$\frac{d^{(\lambda)}}{dt}\overrightarrow{\mathcal{M}_{\mathrm{M}}}\{{}^{\lambda}_{\mu}\} = \frac{1}{2}\vec{J}_c\left(\mathrm{M}|_\mu^\lambda\right) + \vec{J}_e\left(\mathrm{M}|_\mu^\lambda\right).$$

2.6.4.4.2. Inversion of the motions

For the two motions, which are the inverse of each other, of the two frames, $\langle\lambda\rangle$ and $\langle\mu\rangle$, the equations for the combination of the accelerations are, respectively, written as

$$\begin{aligned}\overrightarrow{J^{(\lambda)}}(\mathrm{M}) &= \overrightarrow{J^{(\mu)}}(\mathrm{M}) + \vec{J}_c\left(\mathrm{M}|_\mu^\lambda\right) + \vec{J}_e\left(\mathrm{M}|_\mu^\lambda\right) \\ \overrightarrow{J^{(\mu)}}(\mathrm{M}) &= \overrightarrow{J^{(\lambda)}}(\mathrm{M}) + \vec{J}_c\left(\mathrm{M}|_\lambda^\mu\right) + \vec{J}_e\left(\mathrm{M}|_\lambda^\mu\right)\end{aligned}.$$

The sum of these two relations, term by term, results in

$$\vec{J}_c\left(\mathrm{M}|_\mu^\lambda\right) + \vec{J}_c\left(\mathrm{M}|_\lambda^\mu\right) + \vec{J}_e\left(\mathrm{M}|_\mu^\lambda\right) + \vec{J}_e\left(\mathrm{M}|_\lambda^\mu\right) = \vec{0},$$

with $\vec{J}_c\left(\mathrm{M}|_\mu^\lambda\right) + \vec{J}_c\left(\mathrm{M}|_\lambda^\mu\right) = 2\overrightarrow{\omega_\mu^\lambda}\wedge\overrightarrow{v^{(\mu)}}(\mathrm{M}) + 2\overrightarrow{\omega_\lambda^\mu}\wedge\overrightarrow{v^{(\lambda)}}(\mathrm{M})$,

$$\Rightarrow \quad \vec{J}_c\left(\mathrm{M}|_\mu^\lambda\right) + \vec{J}_c\left(\mathrm{M}|_\lambda^\mu\right) = 2\overrightarrow{\omega_\mu^\lambda}\wedge\left[\overrightarrow{v^{(\mu)}}(\mathrm{M}) - \overrightarrow{v^{(\lambda)}}(\mathrm{M})\right] = -2\overrightarrow{\omega_\mu^\lambda}\wedge\overrightarrow{\mathrm{M}_{\mathrm{M}}}\{{}^{\lambda}_{\mu}\}.$$

As a result, we obtain

$$\vec{J}_e\left(\mathrm{M}|_\mu^\lambda\right) + \vec{J}_e\left(\mathrm{M}|_\lambda^\mu\right) = 2\overrightarrow{\omega_\mu^\lambda}\wedge\overrightarrow{\mathcal{M}_{\mathrm{M}}}\{{}^{\lambda}_{\mu}\}.$$

2.6.5. *Exercises*

2.6.5.1. *Exercise 17 – Combination of motions*

Two revolving cylinders (S_1) and (S_2), of respective radii $2R_1$ and $2R_2$ $(R_2 > R_1)$, are linked as the pivots of axis $\left(\mathrm{O}|\vec{z}\right)$ to a device tied to the frame $\langle 0 \rangle \equiv \left\langle \mathrm{O}|\vec{x}\,\vec{y}\,\vec{z} \right\rangle$.

Cylinder (S_2) is hollow, and as such a right circular cylinder (S_3), of radius $R_2 - R_1$ and of axis $\left(\mathrm{O}_3|\vec{z}\right)$, can roll without sliding inside (S_2) and outside (S_1).

We give $\overrightarrow{\omega_1^0} = \omega_1 \vec{z}$, $\overrightarrow{\omega_2^0} = \omega_2 \vec{z}$,

and write $\overrightarrow{\omega_3^0} = \omega_3 \vec{z}$, $\dfrac{\overrightarrow{\mathrm{OO}_3}}{\left\|\overrightarrow{\mathrm{OO}_3}\right\|} = \vec{u}$, $\vec{z} \wedge \vec{u} = \vec{v}$.

We seek to determine ω_3 and $\overrightarrow{v^{(0)}}(\mathrm{O}_3)$, which define the kinematics of (S_3) relative to $\langle 0 \rangle$. (S_3) is in contact with (S_1) and (S_2) following two variable generative lines $\left(\mathrm{M}_1|\vec{z}\right)$ and $\left(\mathrm{M}_2|\vec{z}\right)$.

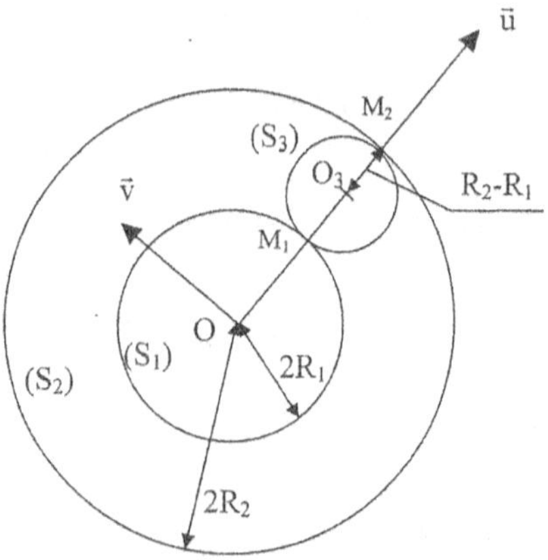

– *Question 1(a)*: In function of $\omega_1, \omega_2, \omega_3, R_1, R_2$, determine the reduction elements of the velocity-distributing torsors at the successive points indicated

$$\left\{ {}^{1}_{2} \right\}_{/O}, \quad \left\{ {}^{1}_{3} \right\}_{/M_1}, \quad \left\{ {}^{2}_{3} \right\}_{/M_2},$$

by using the combination of motions in relation to $\langle 0 \rangle$.

First, we must consider the three initial torsors corresponding to the intrinsic motion of each cylinder:

$$\left\{ {}^{0}_{1} \right\}_{/O} = \left[\omega_1 \vec{z} \,\middle|\, \vec{0} \right] \; ; \; \left\{ {}^{0}_{2} \right\}_{/O} = \left[\omega_2 \vec{z} \,\middle|\, \vec{0} \right] \; ; \; \left\{ {}^{0}_{3} \right\}_{/O_3} = \left[\omega_3 \vec{z} \,\middle|\, \vec{0} \right]$$

$$\left\{ {}^{1}_{2} \right\} = \left\{ {}^{0}_{2} \right\} - \left\{ {}^{0}_{1} \right\} \Rightarrow \left\{ {}^{1}_{2} \right\}_{/O} = \left[(\omega_2 - \omega_1) \vec{z} \,\middle|\, \vec{0} \right].$$

– Seeing as there is a non-sliding roll taking place at M_1 between (S_1) and (S_3), the velocity of this point in relation to each of these two solids is the same:

$$\overrightarrow{v^{(1)}}(M_1) - \overrightarrow{v^{(3)}}(M_1) = \overrightarrow{\mathcal{M}_{M_1}}\left\{ {}^{1}_{3} \right\} = \vec{0}.$$

Furthermore, $\vec{s}\left\{ {}^{1}_{3} \right\} = \vec{s}\left\{ {}^{0}_{3} \right\} - \vec{s}\left\{ {}^{0}_{1} \right\} = \omega_3 \vec{z} - \omega_1 \vec{z}$

$$\Rightarrow \left\{ {}^{1}_{3} \right\}_{/M_1} = \left[(\omega_3 - \omega_1) \vec{z} \,\middle|\, \vec{0} \right].$$

– A non-sliding roll is taking place at M_2 between (S_2) and (S_3). In the same manner we obtain:

$$\Rightarrow \left\{ {}^{2}_{3} \right\}_{/M_2} = \left[(\omega_3 - \omega_2) \vec{z} \,\middle|\, \vec{0} \right].$$

– *Question 1(b)*: Deduce the position of the instantaneous centers of rotation $I_1^0, I_2^0, I_2^1, I_3^1, I_3^2$.

The three previous torsors have scalar invariants that are equal to zero; they are sliding torsors whose central axes are collinear to $\vec{z}$.

$$-\left\{\begin{smallmatrix}0\\1\end{smallmatrix}\right\}_{/O}=\left[\omega_1\vec{z}\,\middle|\,\vec{0}\right]\Rightarrow\overrightarrow{OI_1^0}=\frac{\vec{s}\left\{\begin{smallmatrix}0\\1\end{smallmatrix}\right\}\wedge\overrightarrow{\mathcal{M}_O}\left\{\begin{smallmatrix}0\\1\end{smallmatrix}\right\}}{\vec{s}^{\,2}\left\{\begin{smallmatrix}0\\1\end{smallmatrix}\right\}}=\vec{0}\Rightarrow I_1^0\equiv O;$$

$$-\left\{\begin{smallmatrix}0\\2\end{smallmatrix}\right\}_{/O}=\left[\omega_2\vec{z}\,\middle|\,\vec{0}\right]\Rightarrow I_2^0\equiv O;$$

$$-\left\{\begin{smallmatrix}1\\2\end{smallmatrix}\right\}_{/O}=\left[(\omega_2-\omega_1)\vec{z}\,\middle|\,\vec{0}\right]\Rightarrow I_2^1\equiv O;$$

$$-\left\{\begin{smallmatrix}1\\3\end{smallmatrix}\right\}_{/M_1}=\left[(\omega_3-\omega_1)\vec{z}\,\middle|\,\vec{0}\right]\Rightarrow\overrightarrow{M_1I_3^1}=\vec{0}\Rightarrow I_3^1\equiv M_1;$$

$$-\left\{\begin{smallmatrix}2\\3\end{smallmatrix}\right\}_{/M_2}=\left[(\omega_3-\omega_2)\vec{z}\,\middle|\,\vec{0}\right]\Rightarrow\overrightarrow{M_2I_3^2}=\vec{0}\Rightarrow I_3^2\equiv M_2.$$

– *Question 2(a)*: In function of $\omega_1,\omega_2,\omega_3,R_1,R_2$, determine at the respective points indicated the moments of the velocity-distributing torsors $\overrightarrow{\mathcal{M}_{O_3}}\left\{\begin{smallmatrix}1\\3\end{smallmatrix}\right\}$, $\overrightarrow{\mathcal{M}_{O_3}}\left\{\begin{smallmatrix}2\\3\end{smallmatrix}\right\}$:

$$-\ \overrightarrow{\mathcal{M}_{O_3}}\left\{\begin{smallmatrix}1\\3\end{smallmatrix}\right\}=\overrightarrow{\mathcal{M}_{M_1}}\left\{\begin{smallmatrix}1\\3\end{smallmatrix}\right\}+\overrightarrow{O_3M_1}\wedge\vec{s}\left\{\begin{smallmatrix}1\\3\end{smallmatrix}\right\}=-(R_2-R_1)\vec{u}\wedge(\omega_3-\omega_1)\vec{z},$$

$$\Rightarrow\ \overrightarrow{\mathcal{M}_{O_3}}\left\{\begin{smallmatrix}1\\3\end{smallmatrix}\right\}=(R_2-R_1)(\omega_3-\omega_1)\vec{v};$$

$$-\ \overrightarrow{\mathcal{M}_{O_3}}\left\{\begin{smallmatrix}2\\3\end{smallmatrix}\right\}=\overrightarrow{\mathcal{M}_{M_2}}\left\{\begin{smallmatrix}2\\3\end{smallmatrix}\right\}+\overrightarrow{O_3M_2}\wedge\vec{s}\left\{\begin{smallmatrix}2\\3\end{smallmatrix}\right\}=(R_2-R_1)\vec{u}\wedge(\omega_3-\omega_2)\vec{z},$$

$$\Rightarrow\ \overrightarrow{\mathcal{M}_{O_3}}\left\{\begin{smallmatrix}2\\3\end{smallmatrix}\right\}=-(R_2-R_1)(\omega_3-\omega_2)\vec{v}.$$

– *Question 2(b)*: In function of $\omega_1,\omega_2,\omega_3,R_1,R_2$, determine at the respective points indicated the moments of the velocity-distributing elements $\overrightarrow{\mathcal{M}_{O_3}}\left\{\begin{smallmatrix}0\\1\end{smallmatrix}\right\}$, $\overrightarrow{\mathcal{M}_{O_3}}\left\{\begin{smallmatrix}0\\2\end{smallmatrix}\right\}$:

$$-\ \overrightarrow{\mathcal{M}_{O_3}}\left\{\begin{smallmatrix}0\\1\end{smallmatrix}\right\}=\overrightarrow{\mathcal{M}_O}\left\{\begin{smallmatrix}0\\1\end{smallmatrix}\right\}+\vec{s}\left\{\begin{smallmatrix}0\\1\end{smallmatrix}\right\}\wedge\overrightarrow{OO_3}$$

$$\text{'' ''}\quad=\vec{z}\wedge(R_1+R_2)\vec{u}=\omega_1(R_1+R_2)\vec{v},$$

$$-\ \overrightarrow{\mathcal{M}_{O_3}}\left\{\begin{smallmatrix}0\\2\end{smallmatrix}\right\}=\overrightarrow{\mathcal{M}_O}\left\{\begin{smallmatrix}0\\2\end{smallmatrix}\right\}+\vec{s}\left\{\begin{smallmatrix}0\\2\end{smallmatrix}\right\}\wedge\overrightarrow{OO_3}=\omega_2(R_1+R_2)\vec{v}.$$

– *Question 3(a)*: Determine vector $\overrightarrow{\mathcal{M}_{O_3}}\left\{\begin{smallmatrix}0\\3\end{smallmatrix}\right\}$ through its components in the basis $\left(\vec{u}\ \vec{v}\ \vec{z}\right)$, using the combination of motions relative to (S_1):

$$\overrightarrow{\mathcal{M}_{O_3}}\left\{\begin{smallmatrix}0\\3\end{smallmatrix}\right\} = \overrightarrow{\mathcal{M}_{O_3}}\left\{\begin{smallmatrix}1\\3\end{smallmatrix}\right\} + \overrightarrow{\mathcal{M}_{O_3}}\left\{\begin{smallmatrix}0\\1\end{smallmatrix}\right\}$$

$$'' \quad '' \quad = [(R_2 - R_1)(\omega_3 - \omega_1) + (R_1 + R_2)\omega_1]\vec{v},$$

$$\Rightarrow \overrightarrow{\mathcal{M}_{O_3}}\left\{\begin{smallmatrix}0\\3\end{smallmatrix}\right\} = [(R_2 - R_1)\omega_3 + 2R_1\omega_1]\vec{v}.$$

– *Question 3(b)*: Determine vector $\overrightarrow{\mathcal{M}_{O_3}}\left\{\begin{smallmatrix}0\\3\end{smallmatrix}\right\}$ through its components in the basis $(\vec{u}\,\vec{v}\,\vec{z})$, using the combination of motions relative to (S_2):

$$\overrightarrow{\mathcal{M}_{O_3}}\left\{\begin{smallmatrix}0\\3\end{smallmatrix}\right\} = \overrightarrow{\mathcal{M}_{O_3}}\left\{\begin{smallmatrix}2\\3\end{smallmatrix}\right\} + \overrightarrow{\mathcal{M}_{O_3}}\left\{\begin{smallmatrix}0\\2\end{smallmatrix}\right\}$$

$$'' \quad '' \quad = \left[-(R_2 - R_1)(\omega_3 - \omega_2)\vec{v} + (R_1 + R_2)\omega_2\right]\vec{v},$$

$$\Rightarrow \overrightarrow{\mathcal{M}_{O_3}}\left\{\begin{smallmatrix}0\\3\end{smallmatrix}\right\} = [-(R_2 - R_1)\omega_3 + 2R_2\omega_2]\vec{v}.$$

– *Question 4*: In function of $\omega_1, \omega_2, R_1, R_2$, determine the rotation rate ω_3, the velocity vector $\overrightarrow{v^{(0)}}(O_3)$ and the position of the instantaneous center of rotation I_3^0.

From the expressions of $\overrightarrow{\mathcal{M}_{O_3}}\left\{\begin{smallmatrix}0\\3\end{smallmatrix}\right\}$ above, we can deduce $\omega_3 = \dfrac{R_2\omega_2 - R_1\omega_1}{R_2 - R_1}$.

Furthermore, $\overrightarrow{v^{(0)}}(O_3) = \overrightarrow{\mathcal{M}_{O_3}}\left\{\begin{smallmatrix}0\\3\end{smallmatrix}\right\} = (R_2\omega_2 + R_1\omega_1)\vec{v}$.

The scalar invariant of the torsor $\left\{\begin{smallmatrix}0\\3\end{smallmatrix}\right\}$: $\vec{s}\left\{\begin{smallmatrix}0\\3\end{smallmatrix}\right\} \cdot \overrightarrow{\mathcal{M}_{O_3}}\left\{\begin{smallmatrix}0\\3\end{smallmatrix}\right\} = 0$. It is, therefore, a sliding torsor whose central axis, collinear to $\vec{z}$, crosses the plane passing through O and orthogonally to $\vec{z}$ at the point I_3^0 defined by

$$\overrightarrow{O_3I_3^0} = \frac{\vec{s}\left\{\begin{smallmatrix}0\\3\end{smallmatrix}\right\} \wedge \overrightarrow{M_{O_3}}}{\vec{s}^{\,2}\left\{\begin{smallmatrix}0\\3\end{smallmatrix}\right\}} = \frac{\omega_3\vec{z} \wedge (R_2\omega_2 + R_1\omega_1)\vec{v}}{\omega_3^2} = -\frac{R_2\omega_2 + R_1\omega_1}{\omega_3}\vec{u}.$$

– *Question 5(a)*: Characterize the motion of (S_3) when $\omega_1 = \omega_2$

$$\omega_3 = \omega_1 = \omega_2 \; ; \; \overrightarrow{v^{(0)}}(\mathrm{O}_3) = (R_2 + R_1)\omega_1 \vec{v} \; ; \; \overrightarrow{\mathrm{O}_3\mathrm{I}_3^0} = -(R_2 + R_1)\vec{u} \, ,$$

$$\Rightarrow \left\{ {}^1_2 \right\} = \left\{ {}^1_3 \right\} = \left\{ {}^2_3 \right\} = \{0\} \, .$$

There is no relative motion of the three cylinders relative to each other. There is an overall rotation of the group with a rotation rate of $\omega_1 \vec{z}$.

– *Question 5(b)*: Characterize the motion of (S_3) when $\omega_1 = \dfrac{R_2}{R_1}\omega_2$

$$R_2\omega_2 - R_1\omega_1 = 0 \;\Rightarrow\; \omega_3 = 0 \, .$$

Cylinder (3) does not experience a rotation relative to $\langle 0 \rangle$; it provides a drag force without sliding for cylinder (S_2) when (S_1) is rotating.

2.6.5.2. *Exercise 18 – Combination of motions*

In a device representing the frame $\langle 0 \rangle \equiv \left\langle \mathrm{O} \middle| \vec{x}\,\vec{y}\,\vec{z} \right\rangle$, we consider the following mechanical group, made up of three solids.

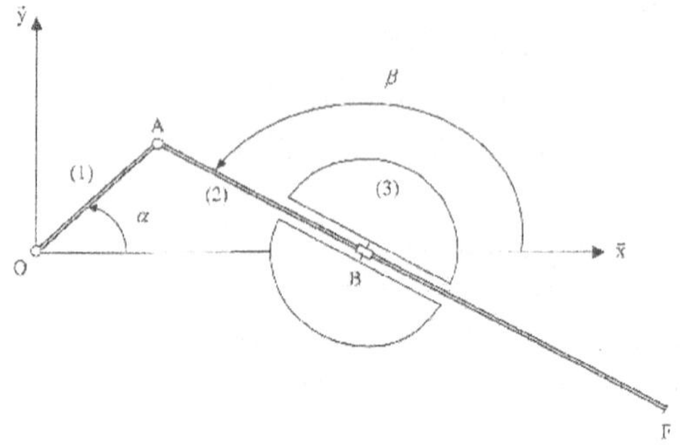

Solid (1) is the bar OA, with a length of R, articulated by the pivot (pivot linkage) of axis $\left(\mathrm{O}\middle|\vec{z}\right)$ with the device (0). There is:

$$\overrightarrow{\mathrm{OA}} = R\vec{u}(\alpha) \text{ with } \alpha = \widehat{\vec{x},\overrightarrow{\mathrm{OA}}} \text{ measured on } \vec{z}$$

where $\alpha(t)$ is a derivable function of the time *t*, and its derivative is α'.

Solid (2) is the bar AF with the pivot linkage of axis $\left(\mathrm{A}|\vec{z}\right)$ with (1). Moreover, it is affected by solid (3) through a slide (sliding linkage).

Solid (3) has the pivot linkage of axis $\left(\mathrm{B}|\vec{z}\right)$ with the device (0). There is:

$$\overrightarrow{\mathrm{OB}} = a\vec{x}$$

where a is a constant length such that $a > R$.

The position of (2) is defined by:

$$\beta = \widehat{\vec{x}, \overrightarrow{\mathrm{FA}}} \text{ measured on } \vec{z} \text{ and } \overrightarrow{\mathrm{BA}} = r\vec{u}(\beta) \text{ with } r > 0 \text{ , } \forall t \text{ ,}$$

where $r(t)$ and $\beta(t)$ are derivable functions of the time *t*, with derivatives called r' and β'.

– *Question 1(a)*: Project onto $\vec{x}$ and $\vec{y}$ the geometrical locking relation: $\overrightarrow{\mathrm{OA}} + \overrightarrow{\mathrm{AB}} = \overrightarrow{\mathrm{OB}}$. Express r and $\tan\beta$ in function of *R*, a and of the trigonometric functions of α :

$$\overrightarrow{\mathrm{OA}} + \overrightarrow{\mathrm{AB}} = \overrightarrow{\mathrm{OB}} \text{ meaning } R\vec{u}(\alpha) - r\vec{u}(\beta) = a\vec{x} \text{ .}$$

By projection onto $\vec{x}$ and $\vec{y}$, we get obtain

$$\begin{cases} R\cos\alpha - r\cos\beta = a \\ R\sin\alpha - r\sin\beta = 0 \end{cases} \Rightarrow \tan\beta = \frac{R\sin\alpha}{R\cos\alpha - a} \text{ ,}$$

$$r^2 = (R\cos\alpha - a)^2 + R^2\sin^2\alpha = R^2 - 2aR\cos\alpha + a^2 \geq (a - R)^2 \text{ ,}$$

$$r = \sqrt{R^2 - 2aR\cos\alpha + a^2} > 0 \text{ .}$$

The previous expression of r and β show that actually the motion frame described above only depends on the single parameter α ; the position of the crank (1) affects that of the bar (2) and of the slide-disc (3).

Question 1(b): In function of R and of a, determine the extreme values of r.

The two extreme values of r are provided by the extreme situations of the system when point A is on the axis $\left(\mathrm{O}|\vec{x}\right)$, which is when it is located either between O and B or along the extension of BO:

$$\overrightarrow{\mathrm{OA}} = \pm R\vec{x} \text{ .}$$

$\overrightarrow{\mathrm{OA}} = R\vec{x} \Rightarrow r = a - R$ meaning $\cos\alpha = 1$ and $\cos\beta = -1$,

$\Rightarrow r$ is minimal when $\alpha = 0 + 2k\pi$, $\beta = n$,

$\overrightarrow{\mathrm{OA}} = -R\vec{x} \Rightarrow r = a + R$ meaning $\cos = -1$ and $\cos\beta = -1$,

$\Rightarrow r$ is maximal when $\alpha = \pi + 2k\pi$ and $\beta = \pi$.

Question 1(c): Determine β'/α' as a function of R, a and $\cos\alpha$. Deduce that the extreme values of β are reached when OA is perpendicular to AB. Give the expressions of $\beta_{\min}$ and of $\beta_{\max}$

$$\tan\beta = \frac{R\sin\alpha}{R\cos\alpha - a} \quad \Rightarrow \quad \left(1 + \tan^2\beta\right)\beta' = \frac{R\left(R - a\cos\alpha\right)}{\left(R\cos\alpha - a\right)^2}\alpha',$$

with $1 + \tan^2\beta = \dfrac{R^2 - 2aR\cos\alpha + a^2}{\left(R\cos\alpha - a\right)^2}$.

As a result, there is

$$\frac{\beta'}{\alpha'} = R\frac{R - a\cos\alpha}{R^2 - 2aR\cos\alpha + a^2} \text{ .}$$

The extreme values of β as a function of α are reached for

$$\frac{d\beta}{d\alpha} = \frac{\beta'}{\alpha'} = 0 \quad \Rightarrow \quad \cos\alpha_1 = \frac{R}{a} < 1,$$

which means that, at the positions where β is at an extreme, triangle OAB is a right-angled triangle in A. These extreme values are, therefore, reached when OA is orthogonal to AB.

For this value of *cos* α, there is

$$\sin\alpha_1 = \pm\frac{1}{a}\sqrt{a^2 - R^2} < 1 \quad \Rightarrow \quad \tan\beta_m = \pm\frac{R\sqrt{a^2 - R^2}}{R^2 - a^2} < 0 \,.$$

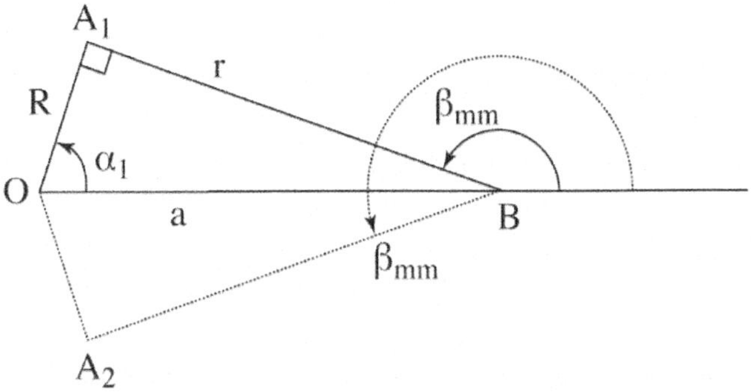

The two configurations of $\beta_{\min}$ and $\beta_{\max}$ are illustrated in the figure above. We can see that

$$\begin{cases} \alpha_1 = \arccos\dfrac{R}{a} & , \sin(\pi - \beta_{\min}) = \dfrac{R}{a} \\ \alpha_1 = 2\pi - \arccos\dfrac{R}{a} & , \sin(\beta_{\max} - \pi) = \dfrac{R}{a} \end{cases} \Rightarrow \begin{cases} \beta_{\min} = \pi - \arcsin\dfrac{R}{a} \\ \beta_{\max} = \pi + \arcsin\dfrac{R}{a} \end{cases}.$$

– *Question 2(a)*: Determine $\left\{ {}^0_1 \right\}_O$, $\overrightarrow{\mathcal{M}_A}\left\{ {}^0_1 \right\}$ and the instantaneous center of rotation $I\left\{ {}^0_1 \right\}$.

The motion of the crank $\overline{\mathrm{OA}}$ takes place in the plane $\Pi\left(\mathrm{O}|\vec{x},\vec{y}\right)$ and presents as an angular rotation perpendicular to $\vec{z}$. This means that

$$\overrightarrow{\omega_1^0} = \alpha'\vec{z}.$$

During the motion of the crank, its rotation point O is fixed in relation to the frame $\langle 0\rangle$; as a result

$$\overrightarrow{v^{(0)}}(\mathrm{O}) = \vec{0},$$

$$\Rightarrow \quad \left\{{}^0_1\right\}_{\mathrm{O}} = \left[\overrightarrow{\omega_1^0}\middle|\overrightarrow{v^{(0)}}(\mathrm{O})\right] = \left[\alpha'\vec{z}\middle|\vec{0}\right].$$

$$\overrightarrow{\mathcal{M}_{\mathrm{A}}}\left\{{}^0_1\right\} = \overrightarrow{\mathcal{M}_{\mathrm{O}}}\left\{{}^0_1\right\} + \vec{s}\left\{{}^0_1\right\}\wedge\overrightarrow{\mathrm{OA}} = \alpha'\vec{z}\wedge R\vec{u}(\alpha) = R\alpha'\vec{u}\left(\alpha+\frac{\pi}{2}\right),$$

and, seeing as O and A both belong to the solid (1)

$$\overrightarrow{v^{(0)}}(\mathrm{A}) = \overrightarrow{\mathcal{M}_{\mathrm{A}}}\left\{{}^0_1\right\} = R\alpha'\vec{u}\left(\alpha+\frac{\pi}{2}\right).$$

The moment of the torsor $\left\{{}^0_1\right\}$ is known at point O, and as a result the position of the instantaneous center of rotation $\mathrm{I}\left\{{}^0_1\right\}$ is given by the equation

$$\overrightarrow{\mathrm{OI}\left\{{}^0_1\right\}} = \frac{\vec{s}\left\{{}^0_1\right\}\wedge\overrightarrow{\mathcal{M}_{\mathrm{O}}}\left\{{}^0_1\right\}}{\vec{s}\left\{{}^0_1\right\}^2} = \frac{\alpha'\vec{z}\wedge\vec{0}}{\alpha'^2} = \vec{0}.$$

The instantaneous center of rotation $\mathrm{I}\left\{{}^0_1\right\}$ is at O.

– *Question 2.(b)*: Determine $\left\{{}^0_2\right\}_{\mathrm{A}}$, $\overrightarrow{\mathcal{M}_{\mathrm{B}}}\left\{{}^0_2\right\}$ and the vector $\overrightarrow{\mathrm{AI}\left\{{}^0_2\right\}}$

Rod ABF, called solid (2), undergoes a motion of rotation of angle β in the plane $\Pi(\mathrm{O}|\vec{x},\vec{y})$ and slides at B along a slide in the same plane. As a result

$$\vec{s}\left\{{}^{0}_{2}\right\} = \overrightarrow{\omega_2^0} = \beta'\vec{z}\,.$$

Point A is shared by the two solids (1) and (2), and consequently

$$\overrightarrow{v^{(0)}}(\mathrm{A}) = \overrightarrow{\mathcal{M}_{\mathrm{A}}}\left\{{}^{0}_{2}\right\} = R\alpha'\vec{u}\left(\alpha+\frac{\pi}{2}\right),$$

$$\left\{{}^{0}_{2}\right\}_{\mathrm{A}} = \left[\beta'\vec{z}\,\middle|\,R\alpha'\vec{u}\left(\alpha+\frac{\pi}{2}\right)\right].$$

According to the relation on the moments of a torsor:

$$\overrightarrow{\mathcal{M}_{\mathrm{B}}}\left\{{}^{0}_{2}\right\} = \overrightarrow{\mathcal{M}_{\mathrm{A}}}\left\{{}^{0}_{2}\right\} + \overrightarrow{\mathrm{BA}}\wedge\vec{s}\left\{{}^{0}_{2}\right\} = R\alpha'\vec{u}\left(\alpha+\frac{\pi}{2}\right) + r\vec{u}(\beta)\wedge\beta'\vec{z}\,,$$

$$\overrightarrow{\mathcal{M}_{\mathrm{B}}}\left\{{}^{0}_{2}\right\} = R\alpha'\vec{u}\left(\alpha+\frac{\pi}{2}\right) - r\beta'\vec{u}\left(\beta+\frac{\pi}{2}\right)$$

the moment of torsor $\left\{{}^{0}_{2}\right\}$ is known at point B, which means that the position of the instantaneous center of rotation $\mathrm{I}\left\{{}^{0}_{2}\right\}$ is given by the relation

$$\overrightarrow{\mathrm{BI}\left\{{}^{0}_{2}\right\}} = \frac{\vec{s}\left\{{}^{0}_{2}\right\}\wedge\overrightarrow{\mathcal{M}_{\mathrm{B}}}\left\{{}^{0}_{2}\right\}}{\vec{s}\left\{{}^{0}_{2}\right\}^2} = \frac{\beta'\vec{z}\wedge\left(R\alpha'\vec{u}\left(\alpha+\frac{\pi}{2}\right) - r\beta'\vec{u}\left(\beta+\frac{\pi}{2}\right)\right)}{\beta'^2},$$

$$\text{'' ''} \quad = -R\frac{\alpha'}{\beta'}\vec{u}(\alpha) + r\vec{u}(\beta) = -R\frac{\alpha'}{\beta'}\vec{u}(\alpha) + \overrightarrow{\mathrm{BA}}$$

meaning $\overrightarrow{\mathrm{AI}\left\{{}^{0}_{2}\right\}} = -R\frac{\alpha'}{\beta'}\vec{u}(\alpha)$.

– *Question 2(c)*: Determine $\left\{{}^{1}_{2}\right\}_{\mathrm{A}}$ and $I\left\{{}^{1}_{2}\right\}$

By combining motions, there is

$$\left\{{1 \atop 2}\right\} = \left\{{0 \atop 2}\right\} + \left\{{1 \atop 0}\right\} = \left\{{0 \atop 2}\right\} - \left\{{0 \atop 1}\right\},$$

$$\vec{s}\left\{{1 \atop 2}\right\} = \vec{s}\left\{{0 \atop 2}\right\} - \vec{s}\left\{{0 \atop 1}\right\} = (\beta' - \alpha')\vec{z},$$

$$\overrightarrow{\mathcal{M}_{A}}\left\{{1 \atop 2}\right\} = \overrightarrow{\mathcal{M}_{A}}\left\{{0 \atop 2}\right\} - \overrightarrow{\mathcal{M}_{A}}\left\{{0 \atop 1}\right\} = \vec{0},$$

$$\Rightarrow \quad \left\{{1 \atop 2}\right\}_{A} = \left[(\beta' - \alpha')\vec{z} \middle| \vec{0}\right].$$

Point A is shared by the two solids (1) and (2)

$$\overrightarrow{AI}\left\{{1 \atop 2}\right\} = \frac{\vec{s}\left\{{1 \atop 2}\right\} \wedge \overrightarrow{\mathcal{M}_{A}}\left\{{1 \atop 2}\right\}}{\vec{s}\left\{{1 \atop 2}\right\}^{2}} = \vec{0} \quad \Rightarrow \quad A \equiv I\left\{{1 \atop 2}\right\}.$$

Solid (2) is, therefore, linked as the pivot of axis $\left(A|\vec{z}\right)$ in relation to (1).

Question 2.(d): Determine $\left\{{0 \atop 3}\right\}_{B}$, $\overrightarrow{\mathcal{M}_{A}}\left\{{0 \atop 3}\right\}$ and $I\left\{{0 \atop 3}\right\}$.

The solid (3), which carries the slide along which solid (2) slides, undergoes a pure rotational motion relative to (0) through the pivot linkage of axis $\left(B|\vec{z}\right)$ and follows the rotation of the rod AF; as a result:

$$\vec{s}\left\{{0 \atop 3}\right\} = \beta'\vec{z} \quad \text{and} \quad \overrightarrow{\mathcal{M}_{B}}\left\{{0 \atop 3}\right\} = \vec{0},$$

$$\overrightarrow{\mathcal{M}_{A}}\left\{{0 \atop 3}\right\} = \overrightarrow{\mathcal{M}_{B}}\left\{{0 \atop 3}\right\} + \vec{s}\left\{{0 \atop 3}\right\} \wedge \overrightarrow{BA} = \beta'\vec{z} \wedge r\vec{u}(\beta) = r\beta'\vec{u}\left(\beta + \frac{\pi}{2}\right).$$

Point B belongs to solid (3); as a result we can write:

$$\overrightarrow{BI}\left\{{0 \atop 3}\right\} = \frac{\vec{s}\left\{{0 \atop 3}\right\} \wedge \overrightarrow{\mathcal{M}_{B}}\left\{{0 \atop 3}\right\}}{\vec{s}\left\{{0 \atop 3}\right\}^{2}} = \vec{0}.$$

Point B is logically the instantaneous center of rotation $I\left\{ {0 \atop 3} \right\}$ of the solid.

– *Question 2(e)*: Determine $\left\{ {2 \atop 3} \right\}_B$:

$$\left\{ {2 \atop 3} \right\} = \left\{ {0 \atop 3} \right\} - \left\{ {0 \atop 2} \right\},$$

$$\vec{s}\left\{ {2 \atop 3} \right\} = \vec{s}\left\{ {0 \atop 3} \right\} - \vec{s}\left\{ {0 \atop 2} \right\} = \beta' \vec{z} - \beta' \vec{z} = \vec{0},$$

$$\overrightarrow{\mathcal{M}_B}\left\{ {2 \atop 3} \right\} = \overrightarrow{\mathcal{M}_B}\left\{ {0 \atop 3} \right\} - \overrightarrow{\mathcal{M}_B}\left\{ {0 \atop 2} \right\} = \vec{0} - R\alpha' \vec{u}\left(\alpha + \frac{\pi}{2}\right) + r\beta' \vec{u}\left(\beta + \frac{\pi}{2}\right),$$

$$\Rightarrow \quad \left\{ {2 \atop 3} \right\}_B = \left[\vec{0} \middle| -R\alpha' \vec{u}\left(\alpha + \frac{\pi}{2}\right) + r\beta' \vec{u}\left(\beta + \frac{\pi}{2}\right) \right].$$

The motion of (3) relative to (2) is a translation of axis $\left(B|\vec{u}(\beta)\right)$; B is mobile along the rod AF, and its velocity, and the velocity of any other point on the rod, is

$$\overrightarrow{v^{(2)}}(B) = \overrightarrow{\mathcal{M}_B}\left\{ {2 \atop 3} \right\} = -R\alpha' \vec{u}\left(\alpha + \frac{\pi}{2}\right) + r\beta' \vec{u}\left(\beta + \frac{\pi}{2}\right).$$

We can also find this velocity by considering the motion of point B in the fixed centroid (2) linked to the bar:

$$\overrightarrow{v^{(2)}}(B) = \frac{d^{(2)}}{dt}\overrightarrow{AB} = \frac{d^{(2)}}{dt}\left(-r\vec{u}(\beta)\right) = -r'\vec{u}(\beta).$$

By comparing the two expressions of $\overrightarrow{v^{(2)}}(B)$ and by projecting the vector $\vec{u}\left(\alpha + \frac{\pi}{2}\right)$ onto the basis (2), this yields the differential equations that govern the motion of the mechanical group:

$$\vec{u}\left(\alpha + \frac{\pi}{2}\right) = \left[\vec{u}\left(\alpha + \frac{\pi}{2}\right) \cdot \vec{u}(\beta)\right] \vec{u}(\beta) \ldots$$

$$\ldots + \left[\vec{u}\left(\alpha + \frac{\pi}{2}\right) \cdot \vec{u}\left(\beta + \frac{\pi}{2}\right)\right] \vec{u}\left(\beta + \frac{\pi}{2}\right),$$

$$\vec{u}\left(\alpha+\frac{\pi}{2}\right)=\cos\widehat{\vec{u}\left(\alpha+\frac{\pi}{2}\right),\vec{u}(\beta)}\,\vec{u}(\beta)\ldots$$

$$\ldots+\cos\widehat{\vec{u}\left(\alpha+\frac{\pi}{2}\right),\vec{u}\left(\beta+\frac{\pi}{2}\right)}\,\vec{u}\left(\beta+\frac{\pi}{2}\right),$$

$$\widehat{\vec{u}\left(\alpha+\frac{\pi}{2}\right),\vec{u}(\beta)}=\widehat{\vec{u}\left(\alpha+\frac{\pi}{2}\right),\vec{x}}+\widehat{\vec{x},\vec{u}(\beta)}=-\left[\frac{\pi}{2}-(\beta-\alpha)\right],$$

$$\widehat{\vec{u}\left(\alpha+\frac{\pi}{2}\right),\vec{u}\left(\beta+\frac{\pi}{2}\right)}=\widehat{\vec{u}\left(\alpha+\frac{\pi}{2}\right),\vec{x}}+\widehat{\vec{x},\vec{u}\left(\beta+\frac{\pi}{2}\right)},$$

$$\text{'' ''}\qquad=-\alpha-\frac{\pi}{2}+\beta+\frac{\pi}{2}=(\beta-\alpha)$$

$$\Rightarrow\vec{u}\left(\alpha+\frac{\pi}{2}\right)=\sin(\beta-\alpha)\vec{u}(\beta)+\cos(\beta-\alpha)\vec{u}\left(\beta+\frac{\pi}{2}\right).$$

As a result, this equation:

$$-R\alpha'\vec{u}\left(\alpha+\frac{\pi}{2}\right)+r\beta'\vec{u}\left(\beta+\frac{\pi}{2}\right)=-r'\vec{u}(\beta)$$

results in the following relations

$$\begin{cases}r'=R\alpha'\sin(\beta-\alpha)\\ r\beta'=R\alpha'\cos(\beta-\alpha)\end{cases},$$

which complete the equation obtained in the first part of the problem, which was

$$\frac{\beta'}{\alpha'}=R\frac{R-a\cos\alpha}{R^2-2aR\cos\alpha+a^2}.$$

– *Question 2(f)*: Further determine the torsor $\left\{\begin{smallmatrix}1\\3\end{smallmatrix}\right\}_A$ in two manners, (a) by the combination of the motions of the three solids in relation to (0), (b) by the combination of their motions in relation to (2):

a) $\left\{\begin{smallmatrix}1\\3\end{smallmatrix}\right\}_A=\left\{\begin{smallmatrix}0\\3\end{smallmatrix}\right\}_A-\left\{\begin{smallmatrix}0\\1\end{smallmatrix}\right\}_A,$

$$\vec{s}\left\{{1 \atop 3}\right\} = \vec{s}\left\{{0 \atop 3}\right\} - \vec{s}\left\{{0 \atop 1}\right\} = (\beta' - \alpha')\vec{z}\,,$$

$$\overrightarrow{\mathcal{M}}_{\mathrm{A}}\left\{{1 \atop 3}\right\} = \overrightarrow{\mathcal{M}}_{\mathrm{A}}\left\{{0 \atop 3}\right\} - \overrightarrow{\mathcal{M}}_{\mathrm{A}}\left\{{0 \atop 1}\right\} = r\beta'\vec{u}\left(\beta + \frac{\pi}{2}\right) - R\alpha'\vec{u}\left(\alpha + \frac{\pi}{2}\right)$$

$$\Rightarrow \quad \left\{{1 \atop 3}\right\}_{\mathrm{A}} = \left[(\beta' - \alpha')\vec{z}\,\middle|\, r\beta'\vec{u}\left(\beta + \frac{\pi}{2}\right) - R\alpha'\vec{u}\left(\alpha + \frac{\pi}{2}\right)\right];$$

b) $\left\{{1 \atop 3}\right\}_{\mathrm{A}} = \left\{{2 \atop 3}\right\}_{\mathrm{A}} + \left\{{1 \atop 2}\right\}_{\mathrm{A}}$,

$$\vec{s}\left\{{1 \atop 3}\right\}_{\mathrm{A}} = \vec{s}\left\{{2 \atop 3}\right\}_{\mathrm{A}} + \vec{s}\left\{{1 \atop 2}\right\}_{\mathrm{A}} = \vec{0} + (\beta' - \alpha')\vec{z}\,,$$

$$\overrightarrow{\mathcal{M}}_{\mathrm{A}}\left\{{1 \atop 3}\right\} = \overrightarrow{\mathcal{M}}_{\mathrm{A}}\left\{{2 \atop 3}\right\} + \overrightarrow{\mathcal{M}}_{\mathrm{A}}\left\{{1 \atop 2}\right\} = \overrightarrow{\mathcal{M}}_{\mathrm{B}}\left\{{2 \atop 3}\right\} + \vec{s}\left\{{2 \atop 3}\right\} \wedge \overrightarrow{\mathrm{BA}} + \vec{0}$$

$$\text{'' ''} \quad = -R\alpha'\vec{u}\left(\alpha + \frac{\pi}{2}\right) + r\beta'\vec{u}\left(\beta + \frac{\pi}{2}\right)$$

2.7. Relative motion of two rigid solids in contact

2.7.1. *Position of the problem*

We consider the two solids (S_1) and (S_2), attached respectively to the frames $\langle 1 \rangle \equiv \langle \mathrm{O}_1 | \vec{x}_1\, \vec{y}_1\, \vec{z}_1 \rangle$ and $\langle 2 \rangle \equiv \langle \mathrm{O}_2 | \vec{x}_2\, \vec{y}_2\, \vec{z}_2 \rangle$, and for which their motion in relation to a frame $\langle \lambda \rangle$ is known, resulting in the following velocity-distributing torsors

$$\begin{cases} \left\{{\lambda \atop 1}\right\}_{\mathrm{O}_1} = \left[\overrightarrow{\omega_1^{\lambda}}\,\middle|\,\overrightarrow{v^{(\lambda)}}(\mathrm{O}_1)\right] \\ \left\{{\lambda \atop 2}\right\}_{\mathrm{O}_2} = \left[\overrightarrow{\omega_2^{\lambda}}\,\middle|\,\overrightarrow{v^{(\lambda)}}(\mathrm{O}_2)\right] \end{cases}.$$

2.7.2. *Velocity-distributing torsors*

At any point Q of the frame $\langle \lambda \rangle$, it is possible to write out the two following relations, which do not express velocities but simply moments since the point considered is neither fixed in relation to (S_1) nor in relation to (S_2).

$$\begin{cases} \overrightarrow{\mathcal{M}_Q}\{{}^{\lambda}_{1}\} = \overrightarrow{v^{(\lambda)}}(O_1) + \overrightarrow{\omega_1^{\lambda}} \wedge \overrightarrow{O_1Q} \\ \overrightarrow{\mathcal{M}_Q}\{{}^{\lambda}_{2}\} = \overrightarrow{v^{(\lambda)}}(O_2) + \overrightarrow{\omega_2^{\lambda}} \wedge \overrightarrow{O_2Q} \end{cases}.$$

Seeing as

$$\{{}^{1}_{2}\} = \{{}^{\lambda}_{2}\} + \{{}^{1}_{\lambda}\} = \{{}^{\lambda}_{2}\} - \{{}^{\lambda}_{1}\},$$

this means that

$$\begin{aligned} \{{}^{1}_{2}\}_Q &= \left[\overrightarrow{\omega_2^{1}} \,\middle|\, \overrightarrow{\mathcal{M}_Q}\{{}^{1}_{2}\} \right] \\ \text{""} &= \left[\overrightarrow{\omega_2^{\lambda}} - \overrightarrow{\omega_1^{\lambda}} \,\middle|\, \overrightarrow{v^{(\lambda)}}(O_2) + \overrightarrow{\omega_2^{\lambda}} \wedge \overrightarrow{O_2Q} - \overrightarrow{v^{(\lambda)}}(O_1) - \overrightarrow{\omega_1^{\lambda}} \wedge \overrightarrow{O_1Q} \right]. \end{aligned}$$

2.7.3. *Characterization of motions*

NOTE 2.6.– We shall now use a property of the projection of vectors that is usefully developed in Volume 2 [BOR 16a]. Let us consider the direction defined by the unit vector $\vec{u}$ and the plane to which it is orthogonal; any vector $\vec{p}$ can be projected onto it following the law

$$\vec{p} = (\vec{p}\cdot\vec{u})\vec{u} + \vec{u} \wedge (\vec{p} \wedge \vec{u}).$$

We assume that the two surfaces (Σ_1) and (Σ_2) that, respectively, limit the two solids (S_1) and (S_2) are tangential to each other at a time t and in a point M; let $\vec{n}$ be the unit vector normal to the plane that is tangential to the two surfaces in M. We can write

$$\begin{cases} \vec{s}\{{}^{1}_{2}\} = \overrightarrow{\omega_2^{1}} = (\overrightarrow{\omega_2^{1}}\cdot\vec{n})\vec{n} + \vec{n} \wedge (\overrightarrow{\omega_2^{1}} \wedge \vec{n}) \\ \overrightarrow{\mathcal{M}_M}\{{}^{1}_{2}\} = \overrightarrow{m_M} = (\overrightarrow{m_M}\cdot\vec{n})\vec{n} + \vec{n} \wedge (\overrightarrow{m_M} \wedge \vec{n}) \end{cases}.$$

This expression of the reduction elements of the velocity-distributing torsor of the motion of (S_2) relative to the (S_1), highlights the particular characteristics of the relative motion of the two solids. We can distinguish four main forms: roll, slide, pivot and penetration.

2.7.3.1. *Revolving rate and rolling rate*

The projection of the rate of rotation allows us to identify the nature of the motion; it shows how the solid (S_2) rolls over (S_1) while pivoting and expresses the following two characteristics of motion:

– $\left(\overrightarrow{\omega_1^2}\cdot\vec{n}\right)\vec{n}$, *revolving rate* around $\vec{n}$ of (S_2) in relation to (S_1);

– $\vec{n}\wedge\left(\overrightarrow{\omega_2^1}\wedge\vec{n}\right)=\overrightarrow{\omega_1^2}-\left(\overrightarrow{\omega_2^1}\cdot\vec{n}\right)\vec{n}$, *rolling rate* of ($S_2$) on ($S_1$).

2.7.3.2. *Normal penetration velocity and sliding velocity*

In the same way, by respectively projecting the moment of torsor $\left\{{}^1_2\right\}$ at M onto $\vec{n}$ and onto the normal plane to this vector, the characteristics of the local velocity at which the motion takes place are highlighted:

– $\left(\overrightarrow{m_M}\cdot\vec{n}\right)\vec{n}$, *normal penetration velocity* in the direction $\vec{n}$ of (S_2) into (S_1) at M;

– $\vec{n}\wedge\left(\overrightarrow{m_M}\wedge\vec{n}\right)=\overrightarrow{m_M}-\left(\overrightarrow{m_M}\cdot\vec{n}\right)\vec{n}$, *sliding velocity* of (S_2) on (S_1).

2.7.4. *Nature of the contact between (S_1) and (S_2)*

2.7.4.1. *General interpretation*

By interpreting these four characteristic parameters of the contact between two solids, in the case where $\overrightarrow{\omega_2^1}\cdot\vec{n}\neq 0$ and $\overrightarrow{\omega_2^1}\wedge\vec{n}\neq\vec{0}$, we can state that:

– the revolving rate $\left(\overrightarrow{\omega_1^2}\cdot\vec{n}\right)\vec{n}$, collinear to the normal unit vector to the plane of contact, expresses the relative pivot of the two solids, one in relation to the other, around the normal axis to the point of contact M;

– the rolling rate $\vec{n}\wedge\left(\overrightarrow{\omega_2^1}\wedge\vec{n}\right)$ is the projection of the relative rate of rotation $\overrightarrow{\omega_2^1}$ onto the plane of contact; it expresses the apparent rotation of the two solids in relation to each other, which is their roll one over the other at the point of contact M during their relative motion;

– the penetration velocity $\left(\overrightarrow{\mathcal{M}_{\mathrm{M}}}\left\{\begin{smallmatrix}1\\2\end{smallmatrix}\right\}\cdot\vec{n}\right)\vec{n}$, normal to the plane of contact, depending on its orientation indicates that either the two solids move away with loss of contact between them, or collide with one another. This collision must be dealt with in its own right using the appropriate equations from collision theory;

– the sliding velocity $\vec{n}\wedge\left(\overrightarrow{\mathcal{M}_{\mathrm{M}}}\left\{\begin{smallmatrix}1\\2\end{smallmatrix}\right\}\wedge\vec{n}\right)$, a projection of the moment $\overrightarrow{m_{\mathrm{M}}}$, which can be considered as a velocity, onto the plane of contact, indicates that an effect of sliding can be added to the roll taking place between the two solids. The instantaneous velocities of the point of contact on the two solids are not the same.

2.7.4.2. *Non-sliding roll*

Let us now consider the trajectories $\Gamma_1^{(t_i,t_f)}(\mathrm{M})$ and $\Gamma_2^{(t_i,t_f)}(\mathrm{M})$ of the point of contact M on the surfaces (Σ_1) and (Σ_2), which envelop the two solids; these are the locations of the points of these envelops that at a time t coincide with the point of contact:

– point M describes a motion of velocity $\overrightarrow{v^{(1)}}(\mathrm{M}|t)$ over the trajectory $\Gamma_1^{(t_i,t_f)}(\mathrm{M})$;

– point M describes a motion of velocity $\overrightarrow{v^{(2)}}(\mathrm{M}|t)$ over the trajectory $\Gamma_2^{(t_i,t_f)}(\mathrm{M})$.

These are vectors, which at the instant t considered, both belong to the tangential plane shared by the two solids at the point of contact.

According to the law of combination of motions:

$$\overrightarrow{v^{(1)}}(\mathrm{M}|t)-\overrightarrow{v^{(2)}}(\mathrm{M}|t)=\overrightarrow{\mathcal{M}_{\mathrm{M}|t}}\{\tfrac{1}{2}\},$$

at the instant t considered, $\overrightarrow{\mathcal{M}_{\mathrm{M}|t}}\{\tfrac{1}{2}\}$ is contained within the tangential plane.

If the two solids remain in contact during a time interval that is not equal to zero, the normal penetration velocity of (S_2) into (S_1), expressed by $\left(\overrightarrow{\mathcal{M}_{\mathrm{M}}}\{\tfrac{1}{2}\}\cdot\vec{n}\right)\vec{n}$, remains equal to zero over this time interval, and there is no change of the state of contact. We can deduce the following observations:

– either the sliding velocity $\vec{n}\wedge\left(\overrightarrow{m_{\mathrm{M}}}\wedge\vec{n}\right)$ is not equal to zero and this means that (S_2) slides on (S_1) with a roll and a pivot;

– or this sliding velocity is equal to zero and there is a roll without sliding. In this case (S_2) is said to roll and pivot without sliding on (S_1).

If $\vec{n}\cdot\overrightarrow{\omega_2^1}=0$, the revolving rate is equal to 0: $\vec{n}\wedge\left(\overrightarrow{\omega_2^1}\wedge\vec{n}\right)=\vec{0}$; there can only then be roll.

If $\vec{n}\wedge\left(\overrightarrow{\omega_2^1}\wedge\vec{n}\right)=\vec{0}$, there is no roll, and only the pivot can remain.

2.7.5. *Exercises*

2.7.5.1. *Exercise 19 – Relative motion of the two solids*

A roller and plate based drive (which models a bevel gear assembly) is installed around a device linked to the frame $\langle 0\rangle\equiv\left\langle \mathrm{O}|\overrightarrow{x_0}\,\overrightarrow{y_0}\,\overrightarrow{z_0}\right\rangle$. The group is made up of:

– a solid (1), linked to the frame $\langle 1\rangle \equiv \left\langle O_1 | \vec{x_1}\, \vec{y_1}\, \vec{z_1} \right\rangle$ and which places the role of the spinning plate, linked as a pivot of axis $\left(O_1 | \vec{x_0}\right)$ with the device;

– a solid (2), linked to the frame $\langle 2\rangle \equiv \left\langle O_2 | \vec{x_2}\, \vec{y_2}\, \vec{z_2} \right\rangle$ and comprising the drive roller of the plate, linked as the pivot of axis $\left(O_2 | \vec{z_0}\right)$ with the device.

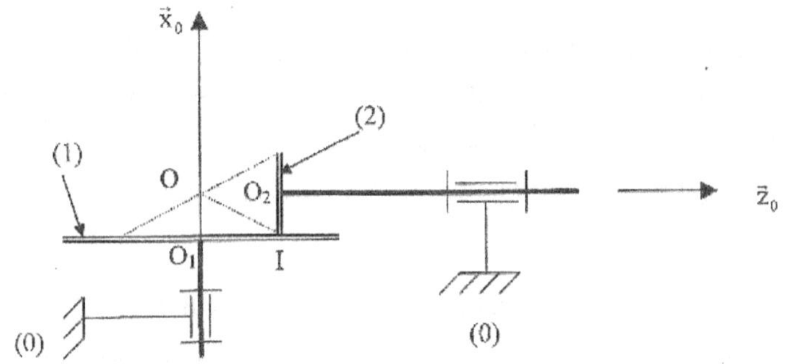

The disc represents the roller of center O_2 and radius R, integrated into the solid (2), and has its point contact at I with the plate of the solid (1).

We write:

– $\alpha = \widehat{\vec{y_0}, \vec{y_1}} = \widehat{\vec{z_0}, \vec{z_1}}$ and $\vec{x_1} \equiv \vec{x_0}$,

– $\beta = \widehat{\vec{x_0}, \vec{x_2}} = \widehat{\vec{y_0}, \vec{y_2}}$ and $\vec{z_2} \equiv \vec{z_0}$,

where α and β are two derivable functions of time t.

– *Question 1*: Using their reduction elements at the points, respectively, indicated, determine the velocity-distributing torsors:

$$\left\{ {0 \atop 1} \right\}_{/O_1} , \left\{ {0 \atop 2} \right\}_{/O_2} , \left\{ {0 \atop 2} \right\}_{/O_1} , \left\{ {1 \atop 2} \right\}_{/O_1} .$$

The solid (1) is undergoing a pure rotational motion of angle α around the axis $\left(O_1|\overrightarrow{x_0}\right)$, so:

$$\overrightarrow{\omega_1^0} = \alpha'\overrightarrow{x_0} \;,\; \overrightarrow{v^{(1)}}(O_1) = \vec{0} \;\Rightarrow\; \left\{ {}^0_1 \right\}_{/O_1} = \left[\alpha'\overrightarrow{x_0} \middle| \vec{0} \right].$$

The solid (2) is undergoing a pure rotational motion of angle β around the axis $\left(O_2|\overrightarrow{z_0}\right)$, so:

$$\overrightarrow{\omega_2^0} = \beta'\overrightarrow{z_0} \;,\; \overrightarrow{v^{(2)}}(O_2) = \vec{0} \;\Rightarrow\; \left\{ {}^0_2 \right\}_{/O_2} = \left[\beta'\overrightarrow{z_0} \middle| \vec{0} \right].$$

$$\overrightarrow{\mathcal{M}_{O_1}}\left\{ {}^0_2 \right\} = \overrightarrow{\mathcal{M}_{O_2}}\left\{ {}^0_2 \right\} + \overrightarrow{O_1O_2} \wedge \vec{s}\left\{ {}^0_2 \right\} = \left(\overrightarrow{O_1I} + \overrightarrow{IO_2} \right) \wedge \beta'\overrightarrow{z_0} \,,$$

$$\overrightarrow{\mathcal{M}_{O_1}}\left\{ {}^0_2 \right\} = \left(d\,\overrightarrow{z_0} + R\vec{x}_0 \right) \wedge \beta'\overrightarrow{z_0} = -R\beta'\overrightarrow{y_0} \,,$$

$$\Rightarrow \left\{ {}^0_2 \right\}_{/O_1} = \left[\beta'\overrightarrow{z} \middle| -R\beta'\overrightarrow{y_0} \right].$$

$$\left\{ {}^1_2 \right\} = \left\{ {}^0_2 \right\} - \left\{ {}^0_1 \right\} \;\Rightarrow\; \left\{ {}^1_2 \right\}_{/O_1} = \left[-\alpha'\overrightarrow{x_0} + \beta'\overrightarrow{z_0} \middle| -R\beta'\overrightarrow{y_0} \right].$$

– *Question 2*: Calculate $\overrightarrow{\mathcal{M}_I}\left\{ {}^1_2 \right\}$.

$$\overrightarrow{\mathcal{M}_I}\left\{ {}^1_2 \right\} = \overrightarrow{\mathcal{M}_{O_1}}\left\{ {}^1_2 \right\} + \vec{s}\left\{ {}^1_2 \right\} \wedge \overrightarrow{O_1I}$$

$$\text{" "} \quad = -R\beta'\overrightarrow{y_0} + \left(-\alpha'\overrightarrow{x_0} + \beta'\overrightarrow{z_0} \right) \wedge d\,\overrightarrow{z_0} = d\alpha'\overrightarrow{y_0} \,,$$

$$\Rightarrow \overrightarrow{\mathcal{M}_I}\left\{ {}^1_2 \right\} = (d\alpha' - R\beta')\overrightarrow{y_0} \,.$$

– *Question 3*: Assuming there is no sliding at I between (1) and (2), deduce the expression of β' as a function of α'.

If there is no sliding at I between (1) and (2), there is:

$$\overrightarrow{v^{(1)}}(I) = \overrightarrow{v^{(2)}}(I) \;\Rightarrow\; \overrightarrow{v^{(1)}}(I) - \overrightarrow{v^{(2)}}(I) = \overrightarrow{\mathcal{M}_I}\left\{ {}^1_2 \right\} = \vec{0} \,,$$

$$\Rightarrow d\alpha' = R\beta' \,.$$

– *Question 4(a)*: Determine $\overrightarrow{v^{(2)}}(\mathrm{I})$.

$$\overrightarrow{v^{(2)}}(\mathrm{I}) = \frac{d^{(2)}}{dt}\overrightarrow{\mathrm{O_2 I}} = -R\frac{d^{(2)}}{dt}\left(\overrightarrow{x_0}\right),$$

and $\frac{d^{(2)}}{dt}\overrightarrow{x_0} = \frac{d^{(0)}}{dt}\overrightarrow{x_0} + \overrightarrow{\omega_0^2} \wedge \overrightarrow{x_0} = -\beta'\overrightarrow{z_0} \wedge \overrightarrow{x_0} = -\beta'\overrightarrow{y_0}$,

$$\Rightarrow \overrightarrow{v^{(2)}}(\mathrm{I}) = R\beta'\overrightarrow{y_0}.$$

– *Question 4(b)*: Determine $\overrightarrow{v^{(1)}}(\mathrm{I})$.

$$\overrightarrow{v^{(1)}}(\mathrm{I}) = \frac{d^{(1)}}{dt}\overrightarrow{\mathrm{O_1 I}} = d\frac{d^{(1)}}{dt}\overrightarrow{z_0} = d\left(\frac{d^{(0)}}{dt}\overrightarrow{z_0} + \overrightarrow{\omega_0^1} \wedge \overrightarrow{z_0}\right) = d\alpha'\overrightarrow{y_0}.$$

– *Question 4(c)*: Show how the equation for the non-sliding at I can be obtained from these two vectors.

If there is no sliding at the point I between the two solids, the velocities of this point are equal in relation to each of the solids, and the previously established equation of non-sliding is obtained again:

$$d\alpha' = R\beta'.$$

2.7.5.2. *Exercise 20 – Solids in point contact*

In the frame $\langle\lambda\rangle \equiv \left\langle \mathrm{O}\middle|\vec{x}\,\vec{y}\,\vec{z}\right\rangle$, we consider the mechanical system made up of:

– a disk (1), linked to the frame $\langle 1\rangle \equiv \left\langle \mathrm{O_1}\middle|\overrightarrow{x_1}\,\overrightarrow{y_1}\,\overrightarrow{z_1}\right\rangle$, movable in relation to $\langle\lambda\rangle$ through a sliding-pivot linkage characterized by

$$\overrightarrow{z_1} \equiv \vec{z} \quad , \quad \widehat{\vec{x},\overrightarrow{x_1}} = \alpha_1(t) \text{ meaured on } \vec{z} \quad , \quad \overrightarrow{\mathrm{OO_1}} = z_1(t)\vec{z};$$

– a disk (2), linked to the frame $\langle 2\rangle \equiv \left\langle \mathrm{O_2}\middle|\overrightarrow{x_2}\,\overrightarrow{y_2}\,\overrightarrow{z_2}\right\rangle$, movable in relation to $\langle\lambda\rangle$ through a sliding-pivot linkage characterized by

$$\vec{z_2} \equiv \vec{z} \quad , \quad \widehat{\vec{x},\vec{x_2}} = \alpha_2(t) \text{ measured on } \vec{z} \quad , \quad \overrightarrow{OO_2} = z_2(t)\vec{z};$$

– a solid (S), linked to the frame $\langle S\rangle \equiv \langle O_S|\overrightarrow{x_S}\,\overrightarrow{y_S}\,\overrightarrow{z_S}\rangle$, made up of a circular disc of center O_S, of radius 2ℓ, of *non-negligible thickness* 2ℓ, and of a rectilinear rod of axis $(O_S|\overrightarrow{z_S})$. (S) is in motion relative to $\langle\lambda\rangle$, through a ball-and-socket link, which keeps fixed its point O'_S defined by

$$O'_S \equiv O \quad , \quad \overrightarrow{O_S O'_S} = 5\ell\overrightarrow{z_S}\ .$$

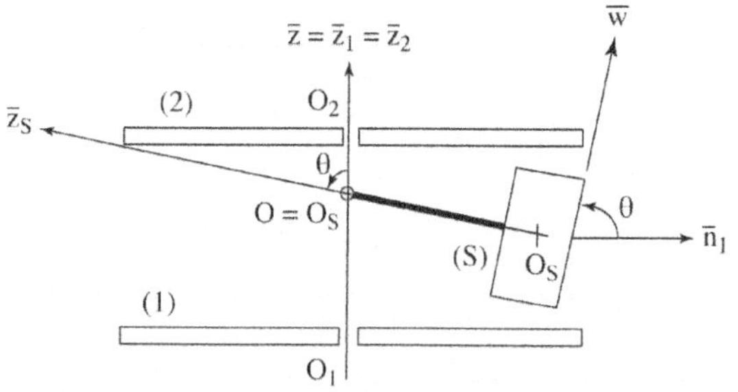

We locate the situation of the solid (S) by the three Euler angles ψ,θ,φ, which define the angular orientation of the basis $(\overrightarrow{x_S}\,\overrightarrow{y_S}\,\overrightarrow{z_S})$ in relation to the basis $(\vec{x}\,\vec{y}\,\vec{z})$ by writing:

$\vec{z} \wedge \overrightarrow{z_S} = \vec{n}\sin\theta$ where $\vec{n}$ is the nodal vector of the Eulerian orientation

$$\vec{n} = \vec{u}(\psi) = \cos\psi\vec{x} + \sin\psi\vec{y}$$

$$\vec{z} \wedge \vec{n} = \overrightarrow{n_1} = \vec{u}\left(\psi + \frac{\pi}{2}\right)$$

$$\overrightarrow{z_S} \wedge \vec{n} = \vec{w}$$

All the required vectors are expressed in the basis $(\vec{n}\,\overrightarrow{n_1}\,\vec{z})$.

– *Question 1*: Determine the reduction elements in O of the kinematic torsor $\{^{\lambda}_{S}\}$ and calculate its moment $\overrightarrow{\mathcal{M}_Q}\{^{\lambda}_{S}\}$ at the point Q defined by $\overrightarrow{OQ} = b\vec{n_1} + c\vec{z}$.

The solid (S) has a ball-and-socket linkage relative to the frame $\langle\lambda\rangle$; it is, therefore, located within it through the Euler angles ψ, θ, φ:

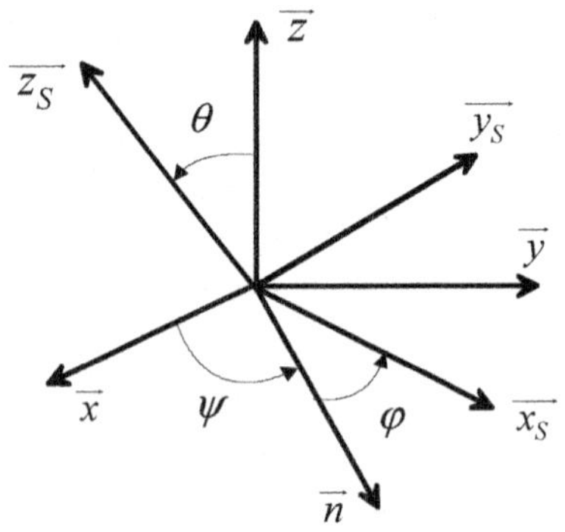

$$\overrightarrow{\omega^{\lambda}_{S}} = \psi'\vec{z} + \theta'\vec{n} + \varphi'\vec{z_S} = \psi'\vec{z} + \theta'\vec{n} + \varphi'\left(-\sin\theta\,\vec{n_1} + \cos\theta\,\vec{z}\right)$$

$$'' \;\; = \theta'\vec{n} - \varphi'\sin\theta\,\vec{n_1} + (\varphi'\cos\theta + \psi')\vec{z}$$

Point O is fixed during this motion $\Rightarrow \overrightarrow{v^{(\lambda)}}(O) = \vec{0}$.

$$\{^{\lambda}_{S}\}_O = \left[\theta'\vec{n} - \varphi'\sin\theta\,\vec{n_1} + (\varphi'\cos\theta + \psi')\vec{z}\,\middle|\,\vec{0}\right].$$

$$\overrightarrow{\mathcal{M}_Q}\{^{\lambda}_{S}\} = \overrightarrow{\mathcal{M}_O}\{^{\lambda}_{S}\} + \vec{s}\{^{\lambda}_{S}\} \wedge \overrightarrow{OQ}$$

$$'' \;\; '' \;\; = \left(\theta'\vec{n} - \varphi'\sin\theta\,\vec{n_1} + (\varphi'\cos\theta + \psi')\vec{z}\right) \wedge \left(b\vec{n_1} + c\vec{z}\right).$$

$$'' \;\; '' \;\; = -[c\varphi'\sin\theta + b(\varphi'\cos\theta + \psi')]\vec{n} - c\theta'\vec{n_1} + b\theta'\vec{z}$$

– *Question 2*: Determine the reduction elements at O_1 of the kinematic torsor $\{^{\lambda}_{1}\}$ and calculate its moment $\overrightarrow{\mathcal{M}_K}\{^{\lambda}_{1}\}$ at a point K defined by $\overrightarrow{O_1K} = B_1\vec{n_1} + C_1\vec{z}$.

The disk (1) is in rotation around the axis $\left(\mathrm{O}_1|\vec{z_1} \equiv \vec{z}\right)$ by an angle $\widehat{\vec{x},\vec{x_1}} = \alpha_1(t)$, and as a result its rate of rotation during its motion in relation to $\langle\lambda\rangle$ is

$$\overrightarrow{\omega_1^{\lambda}} = \vec{s}\left\{{}^{\lambda}_{1}\right\} = \alpha_1' \vec{z} .$$

Point O_1 has a velocity

$$\overrightarrow{v^{(\lambda)}}(\mathrm{O}_1) = \frac{d^{(\lambda)}}{dt}\overrightarrow{\mathrm{OO}_1} = z_1' \vec{z} ,$$

$$\left\{{}^{\lambda}_{1}\right\}_{\mathrm{O}_1} = \left[\alpha_1' \vec{z} \middle| z_1' \vec{z}\right].$$

$$\overrightarrow{\mathcal{M}_{\mathrm{K}}}\left\{{}^{\lambda}_{1}\right\} = \overrightarrow{\mathcal{M}_{\mathrm{O}_1}}\left\{{}^{\lambda}_{1}\right\} + \vec{s}\left\{{}^{\lambda}_{1}\right\} \wedge \overrightarrow{\mathrm{O}_1\mathrm{K}} = z_1' \vec{z} + \alpha_1' \vec{z} \wedge \left(B_1 \vec{n_1} + C_1 \vec{z}\right)$$
$$\text{'' ''} \quad = -B_1 \alpha_1' \vec{n} + z_1' \vec{z}$$

– *Question 3*: Determine the reduction elements at O_2 of the kinematic torsor $\left\{{}^{\lambda}_{2}\right\}$ and calculate its moment $\overrightarrow{\mathcal{M}_{\mathrm{H}}}\left\{{}^{\lambda}_{2}\right\}$ at a point H defined by $\overrightarrow{\mathrm{O}_2\mathrm{H}} = B_2 \vec{n_1} + C_2 \vec{z}$.

The disk (2) is in rotation around the axis $\left(\mathrm{O}_2|\vec{z_2} \equiv \vec{z}\right)$ by an angle $\widehat{\vec{x},\vec{x_2}} = \alpha_2(t)$, and as a result, at the point O_2 such that $\overrightarrow{\mathrm{OO}_2} = z_2(t)\vec{z}$, the reduction elements of torsor $\left\{{}^{\lambda}_{2}\right\}$ associated to its motion relative to $\langle\lambda\rangle$, are

$$\overrightarrow{\omega_2^{\lambda}} = \vec{s}\left\{{}^{\lambda}_{2}\right\} = \alpha_2' \vec{z} \quad ; \quad \overrightarrow{v^{(\lambda)}}(\mathrm{O}_2) = \frac{d^{(\lambda)}}{dt}\overrightarrow{\mathrm{OO}_2} = z_2' \vec{z} ,$$

meaning $\left\{{}^{\lambda}_{2}\right\}_{\mathrm{O}_2} = \left[\alpha_2' \vec{z} \middle| z_2' \vec{z}\right]$.

$$\overrightarrow{\mathcal{M}_{\mathrm{H}}}\left\{{}^{\lambda}_{1}\right\} = \overrightarrow{\mathcal{M}_{\mathrm{O}_2}}\left\{{}^{\lambda}_{2}\right\} + \vec{s}\left\{{}^{\lambda}_{2}\right\} \wedge \overrightarrow{\mathrm{O}_2\mathrm{H}} = z_2' \vec{z} + \alpha_2' \vec{z} \wedge \left(B_2 \vec{n_1} + C_2 \vec{z}\right)$$
$$\text{'' ''} \quad = -B_2 \alpha_2' \vec{n} + z_2' \vec{z}$$

– *Question 4*: Deduce $\overrightarrow{\mathcal{M}_{\mathrm{Q}}}\genfrac{\{}{\}}{0pt}{}{1}{S}$ and $\overrightarrow{\mathcal{M}_{\mathrm{Q}}}\genfrac{\{}{\}}{0pt}{}{2}{S}$.

By the combination of different motions, we obtain

$$\overrightarrow{\mathcal{M}_{\mathrm{Q}}}\genfrac{\{}{\}}{0pt}{}{1}{S} = \overrightarrow{\mathcal{M}_{\mathrm{Q}}}\genfrac{\{}{\}}{0pt}{}{\lambda}{S} - \overrightarrow{\mathcal{M}_{\mathrm{Q}}}\genfrac{\{}{\}}{0pt}{}{\lambda}{1} \quad ; \quad \overrightarrow{\mathcal{M}_{\mathrm{Q}}}\genfrac{\{}{\}}{0pt}{}{2}{S} = \overrightarrow{\mathcal{M}_{\mathrm{Q}}}\genfrac{\{}{\}}{0pt}{}{\lambda}{S} - \overrightarrow{\mathcal{M}_{\mathrm{Q}}}\genfrac{\{}{\}}{0pt}{}{\lambda}{2},$$

$$\overrightarrow{\mathcal{M}_{\mathrm{Q}}}\genfrac{\{}{\}}{0pt}{}{\lambda}{1} = \overrightarrow{\mathcal{M}_{\mathrm{O_1}}}\genfrac{\{}{\}}{0pt}{}{\lambda}{1} + \vec{s}\genfrac{\{}{\}}{0pt}{}{\lambda}{1} \wedge \overrightarrow{\mathrm{O_1Q}}.$$

This last expression is deduced from the calculation of $\overrightarrow{\mathcal{M}_{\mathrm{K}}}\genfrac{\{}{\}}{0pt}{}{\lambda}{1} = -B_1\alpha_1'\vec{n} + z_1'\vec{z}$, in writing $B_1 = b$, which gives

$$\begin{aligned}\overrightarrow{\mathcal{M}_{\mathrm{Q}}}\genfrac{\{}{\}}{0pt}{}{1}{S} &= \overrightarrow{\mathcal{M}_{\mathrm{Q}}}\genfrac{\{}{\}}{0pt}{}{\lambda}{S} - \overrightarrow{\mathcal{M}_{\mathrm{Q}}}\genfrac{\{}{\}}{0pt}{}{\lambda}{1} \\ \text{" "} \quad &= -[\,c\varphi'\sin\theta + b(\varphi'\cos\theta + \psi' - \alpha_1')\,]\vec{n} - c\theta'\vec{n_1} + (b\theta' - z_1')\vec{z}\end{aligned}.$$

In the same way, $\overrightarrow{\mathcal{M}_{\mathrm{Q}}}\genfrac{\{}{\}}{0pt}{}{2}{S}$ can be deduced from $\overrightarrow{\mathcal{M}_{\mathrm{H}}}\genfrac{\{}{\}}{0pt}{}{\lambda}{2}$ by substituting the coordinates of $\overrightarrow{\mathrm{O_2Q}}$ for those of $\overrightarrow{\mathrm{O_2H}}$, mainly b for B_2, in the following operation

$$\begin{aligned}\overrightarrow{\mathcal{M}_{\mathrm{Q}}}\genfrac{\{}{\}}{0pt}{}{2}{S} &= \overrightarrow{\mathcal{M}_{\mathrm{Q}}}\genfrac{\{}{\}}{0pt}{}{\lambda}{S} - \overrightarrow{\mathcal{M}_{\mathrm{Q}}}\genfrac{\{}{\}}{0pt}{}{\lambda}{2} = \overrightarrow{\mathcal{M}_{\mathrm{Q}}}\genfrac{\{}{\}}{0pt}{}{\lambda}{S} - \overrightarrow{\mathcal{M}_{\mathrm{O_2}}}\genfrac{\{}{\}}{0pt}{}{\lambda}{2} - \vec{s}\genfrac{\{}{\}}{0pt}{}{\lambda}{2} \wedge \overrightarrow{\mathrm{O_2Q}} \\ \text{" "} \quad &= -[\,c\varphi'\sin\theta + b(\varphi'\cos\theta + \psi' - \alpha_2')\,]\vec{n} - c\theta'\vec{n_1} + (b\theta' - z_2')\vec{z}\end{aligned}.$$

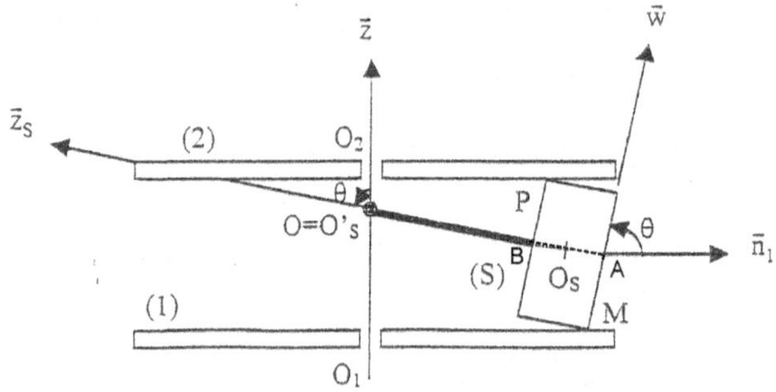

– *Question 5*: We assume in the following that (S) is in constant contact with the (1) at a point M and in constant contact with the (2) at a point P according to the figure above. Write out two relations between z_1 and θ on the one hand, and between z_2 and θ on the other hand, which reflect the constant nature of these two contacts. Calculate the two moments $\overrightarrow{\mathcal{M}_{\mathrm{M}}}\left\{\begin{smallmatrix}1\\S\end{smallmatrix}\right\}$ and $\overrightarrow{\mathcal{M}_{\mathrm{P}}}\left\{\begin{smallmatrix}2\\S\end{smallmatrix}\right\}$:

$$\begin{aligned} z_1 &= \overrightarrow{\mathrm{OM}}\cdot\vec{z} = \left(\overrightarrow{\mathrm{OO}_S}+\overrightarrow{\mathrm{O}_S\mathrm{A}}+\overrightarrow{\mathrm{AM}}\right)\cdot\vec{z} = \left(-5\ell\,\overrightarrow{z_S}-\ell\,\overrightarrow{z_S}-2\ell\,\vec{w}\right)\cdot\vec{z} \\ '' &= -6\ell\cos\theta-2\ell\sin\theta \end{aligned},$$

$$\begin{aligned} z_2 &= \overrightarrow{\mathrm{OP}}\cdot\vec{z} = \left(\overrightarrow{\mathrm{OO}_S}+\overrightarrow{\mathrm{O}_S\mathrm{B}}+\overrightarrow{\mathrm{BP}}\right)\cdot\vec{z} = \left(-5\ell\,\overrightarrow{z_S}+\ell\,\overrightarrow{z_S}+2\ell\,\vec{w}\right)\cdot\vec{z} \\ '' &= -4\ell\cos\theta+2\ell\sin\theta \end{aligned}.$$

For the respective calculations of the moments at M and P of the torsors $\left\{\begin{smallmatrix}1\\S\end{smallmatrix}\right\}$ and $\left\{\begin{smallmatrix}2\\S\end{smallmatrix}\right\}$, we use the previous calculation for the respective moments of torsors at Q:

$$\begin{aligned} \overrightarrow{\mathrm{OM}} &= \left(\overrightarrow{\mathrm{OM}}\cdot\vec{n}\right)\vec{n}+\left(\overrightarrow{\mathrm{OM}}\cdot\overrightarrow{n_1}\right)\overrightarrow{n_1}+\left(\overrightarrow{\mathrm{OM}}\cdot\vec{z}\right)\vec{z} \\ '''' &= 2\ell(3\sin\theta-\cos\theta)\overrightarrow{n_1}-2\ell(3\cos\theta+\sin\theta)\vec{z} \end{aligned},$$

$$\begin{aligned} \overrightarrow{\mathrm{OP}} &= \left(\overrightarrow{\mathrm{OP}}\cdot\vec{n}\right)\vec{n}+\left(\overrightarrow{\mathrm{OP}}\cdot\overrightarrow{n_1}\right)\overrightarrow{n_1}+\left(\overrightarrow{\mathrm{OP}}\cdot\vec{z}\right)\vec{z} \\ '''' &= 2\ell(2\sin\theta+\cos\theta)\overrightarrow{n_1}-2\ell(2\cos\theta-\sin\theta)\vec{z} \end{aligned},$$

$$\begin{aligned} \overrightarrow{\mathcal{M}_{\mathrm{M}}}\left\{\begin{smallmatrix}1\\S\end{smallmatrix}\right\} &= 2\ell[\varphi'-(3\sin\theta-\cos\theta)(\psi'-\alpha_1')]\vec{n}\ \ldots \\ &\ldots\ +2\ell(3\cos\theta+\sin\theta)\theta'\overrightarrow{n_1}+[2\ell(3\sin\theta-\cos\theta)\theta'-z_1']\vec{z} \end{aligned},$$

$$\begin{aligned} \overrightarrow{\mathcal{M}_{\mathrm{P}}}\left\{\begin{smallmatrix}1\\S\end{smallmatrix}\right\} &= -2\ell[\varphi'+(2\sin\theta+\cos\theta)(\psi'-\alpha_2')]\vec{n}\ \ldots \\ &\ldots\ +2\ell(2\cos\theta-\sin\theta)\theta'\overrightarrow{n_1}+[2\ell(2\sin\theta+\cos\theta)\theta'-z_2']\vec{z} \end{aligned}.$$

Seeing as the contact is constant between (S) and the two disks (1) and (2), the component along $\vec{z}$ of these two moments, called the penetration velocity, is equal to zero, which means

$$\overrightarrow{\mathcal{M}_{\mathrm{M}}}\left\{{}^{1}_{S}\right\}\cdot\vec{z}=\vec{0} \quad ; \quad \overrightarrow{\mathcal{M}_{\mathrm{P}}}\left\{{}^{2}_{S}\right\}\cdot\vec{z}=\vec{0},$$

$$\Rightarrow \quad \begin{cases} z_1' = 2\ell(3\sin\theta-\cos\theta)\theta' \\ z_2' = 2\ell(2\sin\theta+\cos\theta)\theta' \end{cases}$$

In the following questions we assume that (S) rolls and pivots without sliding on (1) and on (2).

– *Question 6(a)*: Write the conditions that express an absence of sliding at M and at P.

The relative shifting of the solid (S) on the disks (1) and (2), reflected in the two torsors $\left\{{}^{1}_{S}\right\}$ and $\left\{{}^{2}_{S}\right\}$, takes place without sliding at the points of contact when the component on the plane of contact of their moments at the corresponding point is equal to zero. The normal vector to the planes of contact at M and at P being the vector $\vec{z}$, the plane of contact parallel to $\Pi\left(\vec{n},\vec{n_1}\right)$ in both cases, we can write

$$\begin{cases} \overrightarrow{\mathcal{M}_{\mathrm{M}}}\left\{{}^{1}_{S}\right\}\cdot\vec{n}=\vec{0} \\ \overrightarrow{\mathcal{M}_{\mathrm{M}}}\left\{{}^{1}_{S}\right\}\cdot\vec{n_1}=\vec{0} \end{cases} ; \quad \begin{cases} \overrightarrow{\mathcal{M}_{\mathrm{P}}}\left\{{}^{2}_{S}\right\}\cdot\vec{n}=\vec{0} \\ \overrightarrow{\mathcal{M}_{\mathrm{P}}}\left\{{}^{2}_{S}\right\}\cdot\vec{n_1}=\vec{0} \end{cases}.$$

This provides us with a system of four scalar equations

$$\begin{cases} \varphi' - (3\sin\theta-\cos\theta)(\psi'-\alpha_1') = 0 \\ (3\cos\theta+\sin\theta)\theta' = 0 \\ \varphi' + (2\sin\theta+\cos\theta)(\psi'-\alpha_2') = 0 \\ (2\cos\theta-\sin\theta)\theta' = 0 \end{cases}.$$

– *Question 6(b)*: Deduce that θ remains constant and express ψ' and φ' as functions of α_1' and d_2.

A first result given by two of the equations above is

$$\theta' = 0 \quad \Rightarrow \quad \theta = \theta_0 \neq 0\ .$$

There is, therefore, a privileged incline θ_0 of the solid (S) for a drive without sliding of the two disks.

The two other equations above give us (with $\tan\theta_0 \neq 0$)

$$\begin{cases} \psi' = \dfrac{3\alpha_1' + 2\alpha_2'}{5} + \dfrac{\alpha_2' - \alpha_1'}{5\tan\theta_0} \\ \varphi' = (3\sin\theta_0 - \cos\theta_0)\left(2 + \dfrac{1}{\tan\theta_0}\right)\dfrac{\alpha_2' - \alpha_1'}{5} \end{cases}.$$

– *Question 6(c)*: In this case, determine the reduction elements at O_2 of the kinematic torsor $\left\{{}^1_2\right\}$.

$$\left\{{}^1_2\right\} = \left\{{}^1_S\right\} - \left\{{}^2_S\right\} = \left\{{}^\lambda_2\right\} - \left\{{}^\lambda_1\right\},$$

$$\vec{s}\left\{{}^1_2\right\} = \vec{s}\left\{{}^\lambda_2\right\} - \vec{s}\left\{{}^\lambda_1\right\} = (\alpha_2' - \alpha_1')\vec{z},$$

$$\overrightarrow{\mathcal{M}_{O_2}}\left\{{}^1_2\right\} = \overrightarrow{\mathcal{M}_{O_2}}\left\{{}^1_S\right\} - \overrightarrow{\mathcal{M}_{O_2}}\left\{{}^2_S\right\}.$$

By substituting the coordinates of the point $O_2\left(\overrightarrow{OO_2} = z_2\vec{z}\right)$ for those of Q in the expressions of $\overrightarrow{\mathcal{M}_Q}\left\{{}^1_S\right\}, \overrightarrow{\mathcal{M}_Q}\left\{{}^2_S\right\}$, which means writing $b = 0, c = z_2$, we obtain

$$\overrightarrow{\mathcal{M}_{O_2}}\left\{{}^1_2\right\} = (z_2' - z_1')\vec{z} = \vec{0},$$

since $\theta' = 0$ and, as a result, z_1 and z_2 are constant. In this configuration, the two disks stay at an equal distance one from the other and undergo a simple rotation around the axis $\left(O|\vec{z}\right)$.

2.8. Other exercises

2.8.1. *Exercise 21 – Motion with fixed plane*

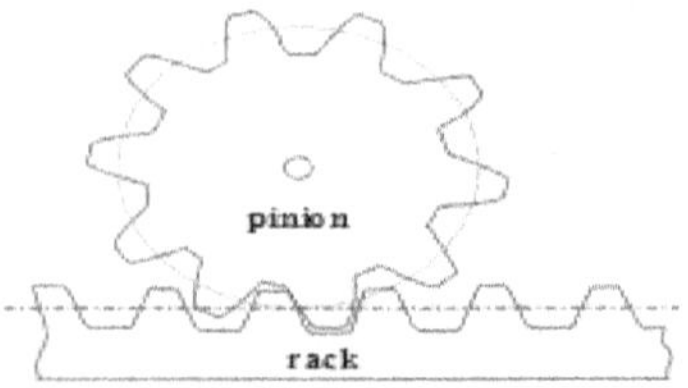

The kinematics of a group made of a rack (mobile part) and a rack (fixed part) can be modeled as a disc (*S*), linked to the frame $\langle S\rangle \equiv \langle O_S|\overrightarrow{x_S}\,\overrightarrow{y_S}\,\overrightarrow{z_S}\rangle$, which rolls and slides along a line in its plane, fixed in the reference frame $\langle \lambda\rangle \equiv \langle O_\lambda|\overrightarrow{x_\lambda}\,\overrightarrow{y_\lambda}\,\overrightarrow{z_\lambda}\rangle$, according to the following diagram:

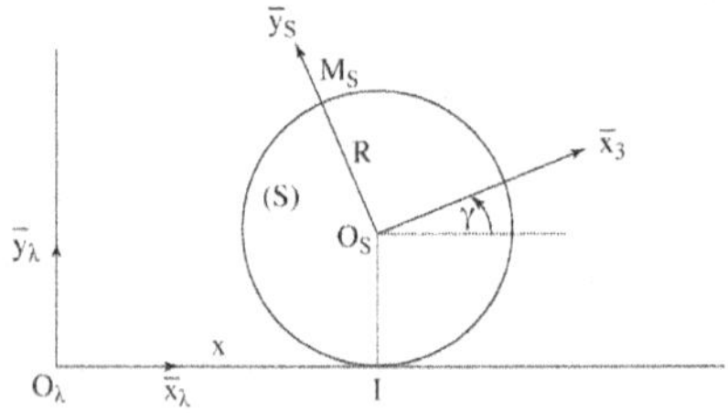

The situation of (S) in $\langle \lambda\rangle$ then depends on two parameters $x(t)$ and $\gamma(t)$ defined by

$$\overrightarrow{O_\lambda O_S} = x\overrightarrow{x_\lambda} + R\overrightarrow{y_\lambda} \quad \text{and} \quad \gamma = \widehat{\overrightarrow{x_\lambda},\overrightarrow{x_S}} = \widehat{\overrightarrow{y_\lambda},\overrightarrow{y_S}}$$

measured on $\overrightarrow{z_S} = \overrightarrow{z_\lambda} = \vec{z}$.

– *Question 1*: Determine the kinematic torsor $\left\{{}^{\lambda}_{S}\right\}_{O_S}$ and its scalar invariant I.

$$\left\{{}^{\lambda}_{S}\right\}_{O_S} = \left[\overrightarrow{\omega_S^\lambda}\,\middle|\,\overrightarrow{v^{(\lambda)}}(O_S)\right].$$

- Determination of $\vec{s}\left\{{}^{\lambda}_{S}\right\} = \overrightarrow{\omega_S^\lambda}$.

To determine the rate of rotation $\overrightarrow{\omega_S^\lambda}$, seeing as the motion is in a fixed plane, we can consider any rotation of the solid takes place around the normal axis to this plane, $\Pi\left(\overrightarrow{x_\lambda},\overrightarrow{y_\lambda}\right)$, meaning it takes place around the axis $\overrightarrow{z_\lambda}$. Seeing as the angular parameter of this motion is γ, we can write

$$\overrightarrow{\omega_S^\lambda} = \gamma' \overrightarrow{z_\lambda} .$$

To establish this rate of rotation, we can also start with the results presented in Volume 2 [BOR 16a] on the derivation of the vectorial functions, which is recalled below.

During the temporal motion of the basis $(E) \equiv \left(\vec{X}\,\vec{Y}\,\vec{Z}\right)$ in relation to the base (e), the rate of rotation $\overrightarrow{\omega_E^e}$ is written

$$\overrightarrow{\omega_E^e} = {}_t\vec{\delta}(e,E) = {}^i\vec{\delta}(e,E)\,q_i' \quad \text{with } {}^i\vec{\delta}(e,E) = {}^i\delta_1\vec{X} + {}^i\delta_2\vec{Y} + {}^i\delta_3\vec{Z} ,$$

where
$$\begin{cases} {}^i\delta_1(e,E) = \vec{Z}\cdot\dfrac{\partial^{(e)}\vec{Y}}{\partial q_i} = -\vec{Y}\cdot\dfrac{\partial^{(e)}\vec{Z}}{\partial q_i} \\ {}^i\delta_2(e,E) = \vec{X}\cdot\dfrac{\partial^{(e)}\vec{Z}}{\partial q_i} = -\vec{Z}\cdot\dfrac{\partial^{(e)}\vec{X}}{\partial q_i} \\ {}^i\delta_3(e,E) = \vec{Y}\cdot\dfrac{\partial^{(e)}\vec{X}}{\partial q_i} = -\vec{X}\cdot\dfrac{\partial^{(e)}\vec{Y}}{\partial q_i} \end{cases}.$$

The only angular parameter that intervenes here is γ

$$\overrightarrow{\omega_S^\lambda} = \overrightarrow{{}^\gamma\delta(\lambda,S)}\,\gamma' = {}^\gamma\delta_1\overrightarrow{x_S} + {}^\gamma\delta_2\overrightarrow{y_S} + {}^\gamma\delta_3\overrightarrow{z_S} ,$$

with the basis change table

$p(\lambda,S) \Leftrightarrow$

	$\overrightarrow{x_S}$	$\overrightarrow{y_S}$	$\overrightarrow{z_S}$
$\overrightarrow{x_\lambda}$	$\cos\gamma$	$-\sin\gamma$	0
$\overrightarrow{y_\lambda}$	$\sin\gamma$	$\cos\gamma$	0
$\overrightarrow{z_\lambda}$	0	0	1

$$\Rightarrow \quad {}^\gamma\delta_1 = 0, {}^\gamma\delta_2 = 0 , {}^\gamma\delta_3 = 1 .$$

We again find $\overrightarrow{\omega_S^\lambda} = \gamma' \overrightarrow{z_S} = \gamma' \overrightarrow{z_\lambda}$.

- Determination of $\overrightarrow{\mathcal{M}_{O_S}}\{{}^\lambda_S\} = \overrightarrow{v^{(\lambda)}}(O_S)$.

$$\overrightarrow{v^{(\lambda)}}(O_S) = \frac{d^{(\lambda)}}{dt}\overrightarrow{O_\lambda O_S} = \frac{d^{(\lambda)}}{dt}\left(x\overrightarrow{x_\lambda} + R\overrightarrow{y_\lambda}\right) = x'\overrightarrow{x_\lambda}.$$

$$\Rightarrow \quad \{{}^\lambda_S\}_{O_S} = \left[\gamma'\overrightarrow{z_\lambda} \middle| x'\overrightarrow{x_\lambda}\right], \text{ with } I\{{}^\lambda_S\} = \gamma'\overrightarrow{z_\lambda}\cdot x'\overrightarrow{x_\lambda} = 0.$$

The torsor $\{{}^\lambda_S\}$ is a sliding torsor that can be reduced to as single vector carried by the central axis of this torsor.

– *Question 2*: We consider the point M_S on the circumference of (S) defined by $\overrightarrow{O_S M_S} = R\overrightarrow{y_S}$. Determine $\overrightarrow{v^{(\lambda)}}(M_S)$

$$O_S \text{ and } M_S \in (S) \quad \Rightarrow \quad \overrightarrow{v^{(\lambda)}}(M_S) = \overrightarrow{v^{(\lambda)}}(O_S) + \overrightarrow{\omega_S^\lambda} \wedge \overrightarrow{O_S M_S},$$

$$\begin{aligned} \overrightarrow{v^{(\lambda)}}(M_S) &= x'\overrightarrow{x_\lambda} + \gamma'\overrightarrow{z_S} \wedge R\overrightarrow{y_S} = x'\overrightarrow{x_\lambda} - R\gamma'\overrightarrow{x_S} \\ \text{'' ''} \quad &= x'\overrightarrow{x_\lambda} - R\gamma'\left(\cos\gamma\overrightarrow{x_\lambda} + \sin\gamma\overrightarrow{y_\lambda}\right) \end{aligned},$$

$$\Rightarrow \quad \overrightarrow{v^{(\lambda)}}(M_S) = (x' - R\gamma'\cos\gamma)\overrightarrow{x_\lambda} - R\gamma'\sin\gamma\overrightarrow{y_\lambda}.$$

We can also write

$$\begin{aligned} \overrightarrow{v^{(\lambda)}}(M_S) &= \frac{d^{(\lambda)}}{dt}\overrightarrow{O_\lambda M_S} = \frac{d^{(\lambda)}}{dt}\left(\overrightarrow{O_\lambda O_S} + \overrightarrow{O_S M_S}\right) \\ \text{'' ''} \quad &= \overrightarrow{v^{(\lambda)}}(O_S) + \frac{d^{(\lambda)}}{dt}\overrightarrow{O_S M_S} \end{aligned},$$

with $\frac{d^{(\lambda)}}{dt}\overrightarrow{O_S M_S} - \frac{d^{(S)}}{dt}\overrightarrow{O_S M_S} = \overrightarrow{\omega_S^\lambda} \wedge \overrightarrow{O_S M_S} = -R\gamma'\overrightarrow{x_S}$, as $\frac{d^{(S)}}{dt}\overrightarrow{O_S M_S} = \vec{0}$.

We again find the expression of $\overrightarrow{v^{(\lambda)}}(M_S)$.

– *Question 3*: $\overrightarrow{O_SQ} = X\overrightarrow{x_\lambda} + Y\overrightarrow{y_\lambda} + Z\overrightarrow{z_\lambda}$ is given. Determine the group of points Q such that $\overrightarrow{\mathcal{M}_Q}\{{}^\lambda_S\} = \vec{0}$ (central axis of the torsor)

$$\overrightarrow{\mathcal{M}_Q}\{{}^\lambda_S\} = \overrightarrow{\mathcal{M}_{O_S}}\{{}^\lambda_S\} + \vec{s}\{{}^\lambda_S\} \wedge \overrightarrow{O_SQ}$$

$$'' \quad '' \quad = x'\overrightarrow{x_\lambda} + \gamma'\overrightarrow{z_\lambda} \wedge \left(X\overrightarrow{x_\lambda} + Y\overrightarrow{y_\lambda} + Z\overrightarrow{z_\lambda}\right),$$

$$\overrightarrow{\mathcal{M}_Q}\{{}^\lambda_S\} = (x' - Y\gamma')\overrightarrow{x_\lambda} + X\gamma'\overrightarrow{y_\lambda} = \vec{0} \quad , \quad \forall x' \text{ et } \gamma',$$

$$\Rightarrow \quad X = 0 \quad , \quad Y = \frac{x'}{\gamma'} \quad \text{and} \quad \overrightarrow{O_SQ} = \frac{x'}{\gamma'}\overrightarrow{y_\lambda} + Z\overrightarrow{z_\lambda}.$$

The central axis (Δ) of the torsor is made up of all points Q such that $\overrightarrow{O_SQ} = \frac{x'}{\gamma'}\overrightarrow{y_\lambda} + Z\overrightarrow{z_\lambda}$. This axis is collinear to $\overrightarrow{z_\lambda}$ and cuts through the plane $\Pi\left(\overrightarrow{x_\lambda}, \overrightarrow{y_\lambda}\right)$, whose level Z is equal to zero, to the point Q_0, the instantaneous center of rotation of the sliding-torsor $\{{}^\lambda_S\}$, located by

$$\overrightarrow{O_SQ_0} = \frac{x'}{\gamma'}\overrightarrow{y_\lambda}.$$

– *Question 4*: Determine the condition of non-sliding roll of (S) in the plane $\Pi\left(O_\lambda|\overrightarrow{x_\lambda}, \overrightarrow{y_\lambda}\right)$ as well as the instantaneous center of rotation I of this fixed plane motion.

Let I be the point of contact of the disc (S) with the axis $\left(O_\lambda|\overrightarrow{x_\lambda}\right)$. The position of this point on the disc and on the axis is variable when the disc rolls; the disc rolls without slipping along the axis if, during the motion, point I moves by the same length, in other words if its motion velocity is the same; this is represented by the equation

$$\overrightarrow{v^{(\lambda)}}(I) = \overrightarrow{v^{(S)}}(I) \quad \Rightarrow \quad \frac{d^{(\lambda)}}{dt}\overrightarrow{O_\lambda I} = \frac{d^{(\lambda)}}{dt}\left(\overrightarrow{O_\lambda O_S} + \overrightarrow{O_S I}\right) = \frac{d^{(S)}}{dt}\overrightarrow{O_S I},$$

meaning $\overrightarrow{v^{(\lambda)}}(\mathrm{O}_S)+\frac{d^{(\lambda)}}{dt}\overrightarrow{\mathrm{O}_s\mathrm{I}}-\frac{d^{(S)}}{dt}\overrightarrow{\mathrm{O}_s\mathrm{I}}=\overrightarrow{v^{(\lambda)}}(\mathrm{O}_S)+\overrightarrow{\omega_S^{\lambda}}\wedge\overrightarrow{\mathrm{O}_s\mathrm{I}}=\vec{0}$.

The condition of a non-sliding roll at I is seen in the equation

$$\overrightarrow{\mathcal{M}_{\mathrm{O}_S}}\left\{{}^{\lambda}_{S}\right\}+\vec{s}\left\{{}^{\lambda}_{S}\right\}\wedge\overrightarrow{\mathrm{O}_S\mathrm{I}}=\overrightarrow{\mathcal{M}_I}\left\{{}^{\lambda}_{S}\right\}=\vec{0},$$

which means that the point I is then located on the central axis (Δ) of the torsor $\left\{{}^{\lambda}_{S}\right\}$:

$$\overrightarrow{\mathcal{M}_I}\left\{{}^{\lambda}_{S}\right\}=x'+\gamma'\overrightarrow{z_\lambda}\wedge-R\overrightarrow{y_\lambda}=(x'+R\gamma')\overrightarrow{x_\lambda}=\vec{0}\cdot$$

The condition of a non-sliding roll at I is written

$$x'+R\gamma'=0\,.$$

With this condition, the point of contact becomes the instantaneous center of rotation of the torsor $\left\{{}^{\lambda}_{S}\right\}$.

2.8.2. *Exercise 22 – Combination of motions*

In the frame $\langle 0\rangle\equiv\left\langle \mathrm{A}\middle|\overrightarrow{x_0}\,\overrightarrow{y_0}\,\overrightarrow{z_0}\right\rangle$, which is linked to the device, we consider the mechanical device whose goal is to transform:

– the rotational motion around a fixed axis of the input shaft (1);

– into a translational motion of the exit bar (3), parallel to the input shaft.

The input shaft (1), linked to two frames:

$$\langle 1\rangle\equiv\left\langle \mathrm{B}\middle|\overrightarrow{x_1}\,\overrightarrow{y_1}\,\overrightarrow{z_1}\right\rangle\quad;\quad\langle 1'\rangle\equiv\left\langle \mathrm{B}\middle|\overrightarrow{x_1}\,\overrightarrow{v_1}\,\overrightarrow{w_1}\right\rangle,$$

– articulated with the device (0) through a pivot (pivot linkage) of axis $\left(\mathrm{A}\middle|\overrightarrow{y_0}\right)\equiv\left(\mathrm{B}\middle|\overrightarrow{y_1}\right)$;

– linked by the pivot (pivot linkage) of axis $\left(B|\overrightarrow{v_1}\right)$ to the intermediate fork (2) associated to the frame $\left\langle B|\overrightarrow{x_2}\,\overrightarrow{y_2}\,\overrightarrow{z_2}\right\rangle$.

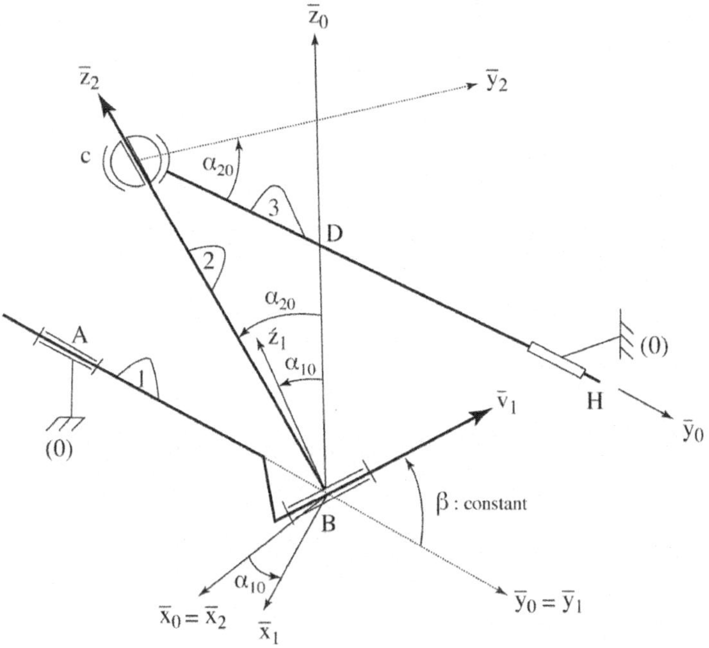

We give,

$$\alpha_{10}(t) = \widehat{\overrightarrow{z_0},\overrightarrow{z_1}} = \widehat{\overrightarrow{x_0},\overrightarrow{x_1}} \qquad \text{measured on } \overrightarrow{y_0} = \overrightarrow{y_1}$$

$$\beta = \widehat{\overrightarrow{y_1},\overrightarrow{v_1}} = \widehat{\overrightarrow{z_1},\overrightarrow{w_1}} \qquad \text{measured on } \overrightarrow{x_1},$$

where β is a given angle, independent of t, contained between 0 and $\frac{\pi}{4}$.

Furthermore, we have

$$\overrightarrow{v_1}\cdot\overrightarrow{z_2} = 0\,.$$

The exit bar CH, the solid (3), linked to the frame $\left\langle C|\overrightarrow{x_0}\,\overrightarrow{y_0}\,\overrightarrow{z_0}\right\rangle$, is in translation on the axis $\left(D|\overrightarrow{y_0}\right)$ in relation to the device (0), attached here at $\left\langle D|\overrightarrow{x_0}\,\overrightarrow{y_0}\,\overrightarrow{z_0}\right\rangle$, and guided by a ball-and-socket (spherical linkage) of center C.

The intermediate fork (2) activates the ball-and-socket through a pivoting and sliding device of axis $\left(C|\overrightarrow{z_2}\right)$.

b and c being two given lengths, independent of time, we can write:

- $\overrightarrow{AB} = b\overrightarrow{y_0}$;

- $\overrightarrow{BD} = c\overrightarrow{z_0}$;

- $\overrightarrow{BC} = z(t)\overrightarrow{z_2}$, which defines the translation of (2) in relation to (3);

- $\overrightarrow{CD} = y(t)\overrightarrow{y_0}$, which defines the translation of (3) in relation to (0).

The orientation of the intermediate fork (2) in relation to the device (0) is defined by

$$\alpha_{20}(t) = \widehat{\overrightarrow{y_0},\overrightarrow{y_2}} = \widehat{\overrightarrow{z_0},\overrightarrow{z_{21}}} \qquad \text{measured on } \overrightarrow{x_0} = \overrightarrow{x_2}.$$

– *Question 1*: Make the three planar figures representing the rotations of angle $\alpha_{10}, \beta, \alpha_{20}$.

The input shaft (1), linked to the frame $\langle 1\rangle \equiv \left\langle B|\overrightarrow{x_1}\,\overrightarrow{y_1}\,\overrightarrow{z_1}\right\rangle$, is in rotation through a pivot link around the axis $\left(A|\overrightarrow{y_0}\right) \equiv \left(B|\overrightarrow{y_1}\right)$. There is, therefore, a first rotation of angle $\alpha_{10}(t)$ relative to the device (0) along that axis (Figure 1).

Rod BC of the fork (2), directed along the axis $\left(B|\overrightarrow{z_2}\right)$ with $\overrightarrow{v_1}\cdot\overrightarrow{z_2}=0$, is perpendicular to a plane $\Pi\left(B|\overrightarrow{v_1},\overrightarrow{w_1}\right)$ with an incline of constant angle $\beta=\widehat{\overrightarrow{y_1},\overrightarrow{v_1}}$ relative to plane $\Pi\left(B|\overrightarrow{x_1},\overrightarrow{y_1}\right)$ through a rotation of axis $\left(B|\overrightarrow{x_1}\right)$ (Figure 2).

Through a ball-and-socket at C in which the fork BC slides and pivots, the rotational motion applied at A to the shaft AB is transmitted to the shaft CH (3). The support device at D and the guiding device at H transform the motion of this solid (3) into a translation of axis $\left(C|\overrightarrow{y_0}\right)$.

The sliding of the ball-and-socket on the rod of the fork takes place along the axis $\left(C|\overrightarrow{z_2}\right)$ and its rotation, of angle $\alpha_{20}(t)$, takes place in the plane $\Pi\left(C|\overrightarrow{y_0},\overrightarrow{z_0}\right)$ orthogonally to the axis $\left(C|\overrightarrow{x_0}\right)$ (Figure 3).

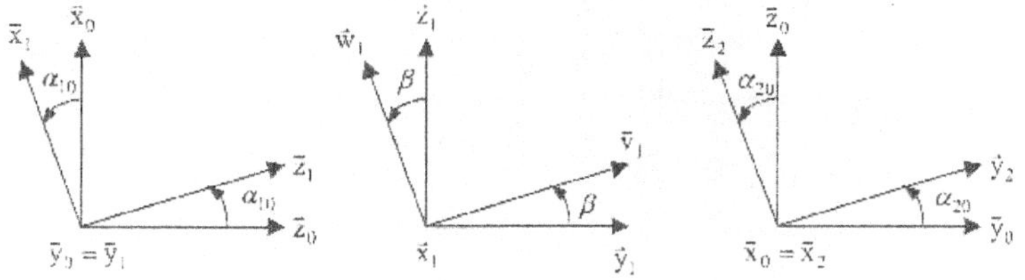

figure 1 figure 2 figure 3

– *Question 2*: Project the following relation of geometrical locking onto the basis (0) and deduce y and z as functions of α_{20}

$$\overrightarrow{BC}+\overrightarrow{CD}+\overrightarrow{DB}=\vec{0}.$$

This relation is written

$$z(t)\overrightarrow{z_2}+y(t)\overrightarrow{y_0}-c\,\overrightarrow{z_0}=\vec{0}$$

meaning $z(t)\left[-\sin\alpha_{20}\,\overrightarrow{y_0}+\cos\alpha_{20}\,\overrightarrow{z_0}\right]+y(t)\overrightarrow{y_0}-c\,\overrightarrow{z_0}=\vec{0}$, hence, the two scalar expressions representing this relation of locking

$$\begin{cases} y - z\sin\alpha_{20} = 0 \\ z\cos\alpha_{20} - c = 0 \end{cases} \quad \Rightarrow \quad \begin{cases} z(t) = \dfrac{c}{\cos\alpha_{20}(t)} \\ y(t) = c\tan\alpha_{20}(t) \end{cases}.$$

– *Question 3*: Explicit the device relation $\overrightarrow{v_1}\cdot\overrightarrow{z_2} = 0$ and express $tan\alpha_{20}$ as a function of α_{10} (input parameter)

$$\begin{aligned} \overrightarrow{v_1}\cdot\overrightarrow{z_2} &= \left(\cos\beta\,\overrightarrow{y_1} + \sin\beta\,\overrightarrow{z_1}\right)\cdot\left(-\sin\alpha_{20}\,\overrightarrow{y_0} + \cos\alpha_{20}\,\overrightarrow{z_0}\right) \\ \text{'' ''} &= \left[\cos\beta\,\overrightarrow{y_0} + \sin\beta\left(\sin\alpha_{10}\,\overrightarrow{x_0} + \cos\alpha_{10}\,\overrightarrow{z_0}\right)\right]\ldots \\ &\qquad \ldots\cdot\left(-\sin\alpha_{20}\,\overrightarrow{y_0} + \cos\alpha_{20}\,\overrightarrow{z_0}\right) = 0 \end{aligned},$$

$$\overrightarrow{v_1}\cdot\overrightarrow{z_2} = -\cos\beta\sin\alpha_{20} + \sin\beta\cos\alpha_{10}\cos\alpha_{20} = 0,$$

$$\Rightarrow \quad \tan\alpha_{20} = \tan\beta\cos\alpha_{10}.$$

– *Question 4*: For $\beta = \dfrac{\pi}{4}$ and $\alpha_{10} \in \mathbb{R}$, determine the interval of the variations of α_{20}, y, z and the expression of α_{20}', y', z' as functions of α_{10} and α_{10}'

$$\begin{cases} y = c\tan\alpha_{20} = c\tan\beta\cos\alpha_{10} \\ z = \dfrac{c}{\cos\alpha_{20}} = c\sqrt{\dfrac{1}{\cos^2\alpha_{20}}} = c\sqrt{\dfrac{\sin^2\alpha_{20} + \cos^2\alpha_{20}}{\cos^2\alpha_{20}}} = c\sqrt{1+\tan^2\alpha_{20}} \\ \text{''} = c\sqrt{1+\tan^2\beta\cos^2\alpha_{10}} \end{cases}.$$

$$\text{For } \beta = \frac{\pi}{4},\ \tan\beta = 1 \quad \Rightarrow \quad \begin{cases} \tan\alpha_{20} = \cos\alpha_{10} \\ y = c\cos\alpha_{10} \\ z = c\sqrt{1+\cos^2\alpha_{10}} \end{cases}.$$

If $\alpha_{10} \in \mathbb{R}$, $\cos\alpha_{10} \in [-1,1]$ and the three variables α_{20}, y, z exist in the intervals

$$\begin{cases} \alpha_{20} \in [-\pi/4, \pi/4] \\ y \in [-c, c] \\ z \in \left[c, c\sqrt{2}\right] \end{cases}.$$

By deriving in relation to time the three expressions above between the parameters of the system, we obtain

$$\begin{cases} \alpha'_{20} = \dfrac{-\sin\alpha_{10}}{1+\cos^2\alpha_{10}}\,\alpha'_{10} \\ y' = -c\sin\alpha_{10}\,\alpha'_{10} \\ z' = -c\dfrac{\sin\alpha_{10}\cos\alpha_{10}}{\sqrt{1+\cos^2\alpha_{10}}}\,\alpha'_{10} \end{cases}.$$

– *Question 5*: Using their components on the basis $\left(\overrightarrow{x_0}\,\overrightarrow{y_0}\,\overrightarrow{z_0}\right)$, determine the reduction elements at B of the following velocity-distributing torsors

$$\left\{ {}^{0}_{1} \right\}\ ;\ \left\{ {}^{0}_{2} \right\}\ ;\ \left\{ {}^{1}_{2} \right\}\ ;\ \left\{ {}^{0}_{3} \right\}\ ;\ \left\{ {}^{2}_{3} \right\}.$$

– $\vec{s}\left\{ {}^{0}_{1} \right\} = \overrightarrow{\omega_1^0} = \alpha'_{10}\,\overrightarrow{y_0}$ and B being fixed on the axis $\left(\mathrm{A}|\overrightarrow{y_0}\right)$,

$$\overrightarrow{\mathcal{M}_{\mathrm{B}}}\left\{ {}^{0}_{1} \right\} = \overrightarrow{v^{(0)}}(\mathrm{B}) = \vec{0},$$

$\Rightarrow$ $\left\{ {}^{0}_{1} \right\}_{\mathrm{B}} = \left[\alpha'_{10}\,\overrightarrow{y_0}\middle|\vec{0}\right]$, sliding-torsor of axis $\left(\mathrm{B}|\overrightarrow{y_0}\right)$;

– $\vec{s}\left\{ {}^{0}_{2} \right\} = \overrightarrow{\omega_2^0} = \alpha'_{20}\,\overrightarrow{x_0}$, and B also being a fixed point of the solid (2),

$$\overrightarrow{\mathcal{M}_{\mathrm{B}}}\left\{ {}^{0}_{2} \right\} = \overrightarrow{v^{(0)}}(\mathrm{B}) = \vec{0},$$

$\Rightarrow \quad \left\{ {}^{0}_{2} \right\}_{B} = \left[\alpha'_{20}\, \overrightarrow{x_0} \middle| \vec{0} \right]$, sliding-torsor of axis $\left(B | \overrightarrow{x_0} \right)$ orthogonal to that of $\left\{ {}^{0}_{1} \right\}$.

– $\left\{ {}^{1}_{2} \right\} = \left\{ {}^{0}_{2} \right\} - \left\{ {}^{0}_{1} \right\} \quad \Rightarrow \quad \left\{ {}^{1}_{2} \right\}_{B} = \left[\alpha'_{20}\, \overrightarrow{x_0} - \alpha'_{10}\, \overrightarrow{y_0} \middle| \vec{0} \right]$, sliding-torsor of axis $\left(B | \dfrac{\sin \alpha_{10}}{1 + \cos^2 \alpha_{10}} \overrightarrow{x_0} + \overrightarrow{y_0} \right)$;

– $\vec{s}\left\{ {}^{0}_{3} \right\} = \vec{0}$ as the exit bar is in translation in relation to the device (0).

The translation of the bar CH is measured in relation to the support at D which is fixed on the device, with

$$\overrightarrow{DC} = -y\, \overrightarrow{y_0} \quad \Rightarrow \quad \overrightarrow{v^{(0)}}(C) = \frac{d^{(0)}}{dt} \overrightarrow{DC} = -y'\, \overrightarrow{y_0} = \overrightarrow{\mathcal{M}_C}\left\{ {}^{0}_{3} \right\},$$

$$\overrightarrow{\mathcal{M}_B}\left\{ {}^{0}_{3} \right\} = \overrightarrow{\mathcal{M}_C}\left\{ {}^{0}_{3} \right\} + \overrightarrow{BC} \wedge \vec{s}\left\{ {}^{0}_{3} \right\} = -y'\, \overrightarrow{y_0} + z\, \overrightarrow{z_2} \wedge \vec{0} = -y'\, \overrightarrow{y_0},$$

$$\Rightarrow \quad \left\{ {}^{0}_{3} \right\}_{B} = \left[\vec{0} \middle| -y'\, \overrightarrow{y_0} \right], \text{ which is a couple-torsor;}$$

– $\left\{ {}^{2}_{3} \right\} = \left\{ {}^{0}_{3} \right\} - \left\{ {}^{0}_{2} \right\} \quad \Rightarrow \quad \left\{ {}^{2}_{3} \right\}_{B} = \left[-\alpha'_{20}\, \overrightarrow{x_0} \middle| -y'\, \overrightarrow{y_0} \right]$.

– *Question 6(a)*: Using their components on the basis $\left(\overrightarrow{x_2}\, \overrightarrow{y_2}\, \overrightarrow{z_2} \right)$, determine the vectors $\overrightarrow{\mathcal{M}_C}\left\{ {}^{0}_{2} \right\}$ and $\overrightarrow{\mathcal{M}_C}\left\{ {}^{2}_{3} \right\}$

$$\overrightarrow{\mathcal{M}_C}\left\{ {}^{0}_{2} \right\} = \overrightarrow{\mathcal{M}_B}\left\{ {}^{0}_{2} \right\} + \vec{s}\left\{ {}^{0}_{2} \right\} \wedge \overrightarrow{BC} = \vec{0} + \alpha'_{20}\, \overrightarrow{x_0} \wedge z\, \overrightarrow{z_2} = -z\alpha'_{20}\, \overrightarrow{y_2},$$

$$\overrightarrow{\mathcal{M}_C}\left\{ {}^{2}_{3} \right\} = \overrightarrow{\mathcal{M}_B}\left\{ {}^{2}_{3} \right\} + \vec{s}\left\{ {}^{2}_{3} \right\} \wedge \overrightarrow{BC} = -y'\, \overrightarrow{y_0} - \alpha'_{20}\, \overrightarrow{x_0} \wedge z\, \overrightarrow{z_2}$$

$$\text{'' ''} \quad = -y'\, \overrightarrow{y_0} + z\alpha'_{20}\, \overrightarrow{y_2}$$

– *Question 6(b)*: Deduce $\overrightarrow{v^{(2)}}(C)$ as a function of α_{20} and α_{20}', and then as a function of *z*, considering the results obtained in the first part.

As the point C is linked to the exit bar CH, and in other words to the solid (3), $\overrightarrow{\mathcal{M}_C\left\{{}^2_3\right\}}$ is the velocity of this point in relation to the basis (2), so $\overrightarrow{v^{(2)}}(C)$.

$$\overrightarrow{v^{(2)}}(C) = \overrightarrow{\mathcal{M}_C}\left\{{}^2_3\right\} = -y'\left(\cos\alpha_{20}\,\overrightarrow{y_2} - \sin\alpha_{20}\,\overrightarrow{z_2}\right) + z\alpha'_{20}\,\overrightarrow{y_2}$$
$$\text{'' ''} \quad = \left(z\alpha'_{20} - y'\cos\alpha_{20}\right)\overrightarrow{y_2} + y'\sin\alpha_{20}\,\overrightarrow{z_2}.$$

According to the locking relation established at the beginning of the problem

$$\left\{\begin{array}{lll} y = c\tan\alpha_{20} & \Rightarrow & y' = \dfrac{c}{\cos^2\alpha_{20}}\alpha'_{20} \\ c = z\cos\alpha_{20} & \Rightarrow & z' = \dfrac{c\sin\alpha_{20}}{\cos^2\alpha_{20}}\alpha'_{20} \end{array}\right\} \Rightarrow \quad y'\cos\alpha_{20} - z\alpha'_{20} = 0,$$

$$\Rightarrow \quad \overrightarrow{v^{(2)}}(C) = y'\sin\alpha_{20}\,\overrightarrow{z_2} = \frac{c\sin\alpha_{20}}{\cos^2\alpha_{20}}\alpha'_{20}\,\overrightarrow{z_2}.$$

However, the point C is also located on the fork (2); we can therefore write

$$\overrightarrow{v^{(2)}}(C) = \frac{d^{(2)}}{dt}\overrightarrow{BC} = \frac{d^{(2)}}{dt}z\,\overrightarrow{z_2} = z'\,\overrightarrow{z_2},$$

which verifies the previous relation.

– *Question 6(c)*: Determine the central axis $\Delta\left\{{}^2_3\right\}$. Let Q_0 be the point of intersection of this axis with the plane perpendicular to $\overrightarrow{x_0} = \overrightarrow{x_2}$ and which contains points B and C. Determine the vector $\overrightarrow{CQ_0}$ using its components in the basis (2), the vector $\overrightarrow{BQ_0}$ using its components in the basis (0). Make a figure.

The situation of the central rotation axis $\left(\Delta_3^2\right)$ of the torsor $\left\{ {}^2_3 \right\}$ can be determined from any point of the affine space Q of the central axis, particularly from points C and B. Thus,

$$\forall \mathrm{Q} \in \left(\Delta_3^2\right), \ \overrightarrow{\mathrm{CQ}} = \frac{\vec{s}\left\{ {}^2_3 \right\} \wedge \overrightarrow{\mathcal{M}_{\mathrm{C}}}\left\{ {}^2_3 \right\}}{\vec{s}\left\{ {}^2_3 \right\}^2} + \lambda \vec{s}\left\{ {}^2_3 \right\} = \frac{-\alpha'_{20}\,\overrightarrow{x_0} \wedge z'\,\overrightarrow{z_2}}{\alpha_{20}'^{\,2}} + \ell\,\overrightarrow{x_0} \,,$$

$$\forall \mathrm{Q} \in \left(\Delta_3^2\right), \ \overrightarrow{\mathrm{BQ}} = \frac{\vec{s}\left\{ {}^2_3 \right\} \wedge \overrightarrow{\mathcal{M}_{\mathrm{B}}}\left\{ {}^2_3 \right\}}{\vec{s}\left\{ {}^2_3 \right\}^2} + \lambda \vec{s}\left\{ {}^2_3 \right\} = \frac{-\alpha'_{20}\,\overrightarrow{x_0} \wedge -y'\,\overrightarrow{y_0}}{\alpha_{20}'^{\,2}} + \ell\,\overrightarrow{x_0} \,.$$

Its point of intersection Q_0 with the orthogonal plane $\overrightarrow{x_0} \equiv \overrightarrow{x_2}$ containing B and C, and therefore for $\ell = 0$, is located by

$$\overrightarrow{\mathrm{CQ}_0} = \frac{-\alpha'_{20}\,\overrightarrow{x_0} \wedge z'\,\overrightarrow{z_2}}{\alpha_{20}'^{\,2}} = \frac{z'}{\alpha'_{20}}\overrightarrow{y_2} = \frac{c \sin\alpha_{20}}{\cos^2\alpha_{20}}\overrightarrow{y_2} \,,$$

or by

$$\overrightarrow{\mathrm{BQ}_0} = \frac{-\alpha'_{20}\,\overrightarrow{x_0} \wedge -y'\,\overrightarrow{y_0}}{\alpha_{20}'^{\,2}} = \frac{y'}{\alpha'_{20}}\overrightarrow{z_0} = \frac{c}{\cos^2\alpha_{20}}\overrightarrow{z_0} \,.$$

We find that last result again by writing

$$\overrightarrow{\mathrm{BQ}_0} = \overrightarrow{\mathrm{BC}} + \overrightarrow{\mathrm{CQ}_0} \,,$$

$$\begin{aligned}\overrightarrow{\mathrm{BQ}_0} &= \frac{c}{\cos\alpha_{20}}\left(-\sin\alpha_{20}\,\overrightarrow{y_0} + \cos\alpha_{20}\,\overrightarrow{z_0}\right) + \frac{c\sin\alpha_{20}}{\cos^2\alpha_{20}}\left(\cos\alpha_{20}\,\overrightarrow{y_0} + \sin\alpha_{20}\,\overrightarrow{z_0}\right) \\ \text{'' ''} &= \frac{c}{\cos^2\alpha_{20}}\overrightarrow{z_0}\end{aligned} \,,$$

which shows that the central axis is unique and is not dependent on the point from which it is located.

To create a graphic representation of the situation of the central axis, let us represent the plane $\Pi\left(C|\overrightarrow{y_2},\overrightarrow{z_2}\right) \equiv \Pi\left(B|\overrightarrow{y_0},\overrightarrow{z_0}\right)$ where the motion is taking place.

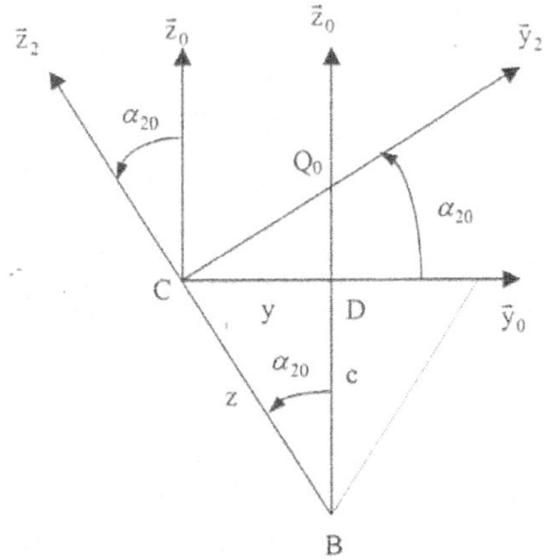

In the right-angled triangle BCD

$$\begin{cases} \cos\alpha_{20} = \dfrac{c}{z} \\ \tan\alpha_{20} = \dfrac{y}{c} \end{cases} \Rightarrow \begin{cases} z = \dfrac{c}{\cos\alpha_{20}} \\ y = c\tan\alpha_{20} \end{cases},$$

In the right-angled triangle BCQ_0

$$\begin{cases} \tan\alpha_{20} = \dfrac{CQ_0}{z} \\ \cos\alpha_{20} = \dfrac{z}{BQ_0} \end{cases} \Rightarrow \begin{cases} \overrightarrow{CQ_0} = z\tan\alpha_{20}\,\overrightarrow{y_2} = \dfrac{c\tan\alpha_{20}}{\cos\alpha_{20}}\overrightarrow{y_2} = \dfrac{c\sin\alpha_{20}}{\cos^2\alpha_{20}}\overrightarrow{y_2} \\ \overrightarrow{BQ_0} = \dfrac{z}{\cos\alpha_{20}}\overrightarrow{z_0} = \dfrac{c}{\cos^2\alpha_{20}}\overrightarrow{z_0} \end{cases}.$$

2.8.3. *Exercise 23 – Kinematics of contact in a system*

Let us consider a device (0) attached to a direct orthonormal frame $\langle 0\rangle \equiv \left\langle O|\overrightarrow{x_0}\,\overrightarrow{y_0}\,\overrightarrow{z_0}\right\rangle$. The mechanical system considered is represented in the figure below, made with $\omega < 0$.

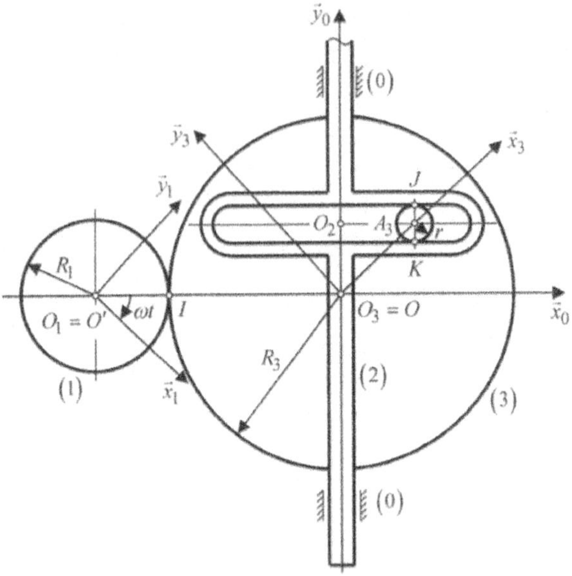

The wheel (1), of radius R_1 and which is linked to a direct orthonormal frame $\langle 1 \rangle \equiv \left\langle O_1 | \vec{x_1}\ \vec{y_1}\ \vec{z_1} \right\rangle$, has a rotational (pivot) link of axis $\left(O' | \vec{z_0} \right)$ with the device (0). There is:

$$O_1 \equiv O' \quad , \quad \vec{z_1} \equiv \vec{z_0} \equiv \vec{z} \quad , \quad \widehat{\vec{x_0}, \vec{x_1}} = \omega t \text{ with } \omega \text{ independent of time } t.$$

The wheel (3), of radius R_3 and linked to the frame $\langle 3 \rangle \equiv \left\langle O_3 | \vec{x_3}\ \vec{y_3}\ \vec{z_3} \right\rangle$, has a rotational link of axis $\left(O | \vec{z_0} \right)$.

Wheel (1) is a driving wheel and carries wheel (3).

The ram (2) is in a rectilinear translation along the axis $\left(O | \vec{y_0} \right)$ relative to the device (0).

The wheel (3) is equipped with a crankpin, of radius r, which moves in a slide that is interdependent with the ram. The position of point A_3, the center of the crankpin, is given by $\overrightarrow{O_3 A_3} = e \vec{x_3}$.

At the instant $t = 0$, there is:

$$\overrightarrow{x_1} = \overrightarrow{x_3} = \overrightarrow{x_0} \quad \text{and} \quad O_2 = O_3 = O \, .$$

– *Question 1(a)*: Express the non-sliding at the point of contact I of the wheels (1) and (3); and deduce the rotation rate vector of the wheel (3) relative to the device (0) as a function of ω, R_1 and R_3 ω.

During the non-sliding rotation of the wheels (3) and (1) one on the other, the velocities of evolution of their point of contact I on each of them must be the same, meaning

$$\begin{aligned} &\overrightarrow{v^{(1)}}(I) = \overrightarrow{v^{(3)}}(I) \\ &\Rightarrow \overrightarrow{v^{(1)}}(I) - \overrightarrow{v^{(3)}}(I) = \overrightarrow{\mathcal{M}_I}\left\{ {}^{1}_{3} \right\} = \overrightarrow{\mathcal{M}_I}\left\{ {}^{0}_{3} \right\} - \overrightarrow{\mathcal{M}_I}\left\{ {}^{0}_{1} \right\} = \vec{0}\,, \end{aligned}$$

with $\left\{ {}^{0}_{1} \right\}_{/O_1} = \left[\omega \vec{z} \middle| \vec{0} \right]$, $\left\{ {}^{0}_{3} \right\}_{/O_3} = \left[\omega_3^0 \middle| \vec{0} \right]$, hence

$$- \quad \overrightarrow{\mathcal{M}_I}\left\{ {}^{0}_{1} \right\} = \overrightarrow{\mathcal{M}_{O_1}}\left\{ {}^{0}_{1} \right\} + \vec{s}\left\{ {}^{0}_{1} \right\} \wedge \overrightarrow{O_1 I} = \omega \vec{z} \wedge R_1 \overrightarrow{x_0} = \omega R_1 \overrightarrow{y_0}\,,$$

$$- \quad \overrightarrow{\mathcal{M}_I}\left\{ {}^{0}_{3} \right\} = \overrightarrow{\mathcal{M}_{O_3}}\left\{ {}^{0}_{3} \right\} + \vec{s}\left\{ {}^{0}_{3} \right\} \wedge \overrightarrow{O_3 I} = \omega_3^0 \vec{z} \wedge -R_3 \overrightarrow{x_0} = -\omega_3^0 R_3 \overrightarrow{y_0}\,.$$

We obtain the condition of roll without slide between the wheels (1) and (3)

$$-R_3 \omega_3^0 - R_1 \omega = 0 \text{ meaning } \omega_3^0 = -\frac{R_1}{R_3}\omega\,,$$

The rotation rate of the wheel (3) in relation to the device (0) is:

$$\overrightarrow{\omega_3^0} = -\frac{R_1}{R_3}\omega \vec{z}\,.$$

– *Question 1(b)*: Determine the moments of the velocity-distributing torsor $\left\{ {}^{0}_{3} \right\}$ at the points A_3, J and K, points of contact of the crankpin with the ram.

The velocity-distributing torsor of the wheel (3) in relation to the device is

$$\left\{ {}^0_3 \right\}_{/O_3} = \left[-\frac{R_1}{R_3}\omega\vec{z} \middle| \vec{0} \right].$$

Therefore, there is

$$- \overrightarrow{\mathcal{M}_{A_3}}\left\{ {}^0_3 \right\} = \overrightarrow{v^{(0)}(A_3)} = -\frac{R_1}{R_3}\omega\vec{z} \wedge e\overrightarrow{x_3} = -e\frac{R_1}{R_3}\omega\overrightarrow{y_3},$$

$$- \overrightarrow{\mathcal{M}_J}\left\{ {}^0_3 \right\} = -\frac{R_1}{R_3}\omega\vec{z} \wedge \overrightarrow{O_3J} = -\frac{R_1}{R_3}\omega\vec{z} \wedge \left(e\overrightarrow{x_3} + r\overrightarrow{y_0} \right)$$

$$\text{" "} \quad = -e\frac{R_1}{R_3}\omega\overrightarrow{y_3} + r\frac{R_1}{R_3}\omega\overrightarrow{x_0},$$

$$- \overrightarrow{\mathcal{M}_K}\left\{ {}^0_3 \right\} = -\frac{R_1}{R_3}\omega\vec{z} \wedge \overrightarrow{O_3K} = -\frac{R_1}{R_3}\omega\vec{z} \wedge \left(e\overrightarrow{x_3} - r\overrightarrow{y_0} \right)$$

$$\text{" "} \quad = -e\frac{R_1}{R_3}\omega\overrightarrow{y_3} - r\frac{R_1}{R_3}\omega\overrightarrow{x_0}.$$

– *Question 1(c)*: Determine the torsor $\left\{ {}^0_2 \right\}$.

The ram (2) undergoes a motion of rectilinear translation in relation to the device (0) with a velocity $\overrightarrow{v^{(0)}(O_2)} = v\overrightarrow{y_0}$. Its rate of rotation is equal to zero; it is therefore a couple-torsor whose moment at any point is $v\overrightarrow{y_0}$.

$$\left\{ {}^0_2 \right\} = \left[\vec{0} \middle| v\overrightarrow{y_0} \right].$$

However, $\overrightarrow{OO_2} = \left(\overrightarrow{OA_3} \cdot \overrightarrow{y_0} \right)\overrightarrow{y_0} = \left(e\overrightarrow{x_3} \cdot \overrightarrow{y_0} \right)\overrightarrow{y_0}$.

The rotation term ω_3^0 of the basis (3) in relation to (0) being constant for a given ω, the angle $\widehat{\overrightarrow{x_0}, \overrightarrow{x_3}}$ is a linear function of time, equal to $-\frac{R_1}{R_3}\omega t$ to the nearest constant, which here can be taken as equal to zero. As a result,

$$\overrightarrow{x_3}\cdot\overrightarrow{y_0}=\sin\left(-\frac{R_1}{R_3}\omega t\right)=-\sin\left(\frac{R_1}{R_3}\omega t\right),$$

$$\overrightarrow{v^{(0)}}(O_2)=\frac{d^{(0)}}{dt}\overrightarrow{OO_2}=-e\frac{R_1}{R_3}\omega\cos\left(\frac{R_1}{R_3}\omega t\right)\overrightarrow{y_0},$$

$$\Rightarrow\left\{{}^{0}_{2}\right\}=\left[\vec{0}\left|-e\frac{R_1}{R_3}\omega\cos\left(\frac{R_1}{R_3}\omega t\right)\overrightarrow{y_0}\right.\right].$$

– *Question 2(a)*: Determine the reduction elements of the torsor $\left\{{}^{2}_{3}\right\}$ at *O*. Characterize this torsor and determine its central axis (Δ) and its intersection Q_0 with the plane $\Pi\left(O|\overrightarrow{x_0},\overrightarrow{y_0}\right)$.

$$\left\{{}^{2}_{3}\right\}=\left\{{}^{0}_{3}\right\}-\left\{{}^{0}_{2}\right\},$$

$$\vec{s}\left\{{}^{2}_{3}\right\}=\vec{s}\left\{{}^{0}_{3}\right\}-\vec{s}\left\{{}^{0}_{2}\right\}=\vec{s}\left\{{}^{0}_{3}\right\}=-\frac{R_1}{R_3}\omega\vec{z},$$

$$\overrightarrow{\mathcal{M}_O}\left\{{}^{2}_{3}\right\}=\overrightarrow{\mathcal{M}_O}\left\{{}^{0}_{3}\right\}-\overrightarrow{\mathcal{M}_O}\left\{{}^{0}_{2}\right\}=-\overrightarrow{\mathcal{M}_O}\left\{{}^{0}_{2}\right\}=e\frac{R_1}{R_3}\omega\cos\left(\frac{R_1}{R_3}\omega t\right)\overrightarrow{y_0}.$$

Its scalar invariant $\mathrm{I}\left\{{}^{2}_{3}\right\}=\vec{s}\left\{{}^{2}_{3}\right\}\cdot\overrightarrow{\mathcal{M}_O}\left\{{}^{2}_{3}\right\}$ being equal to zero, it is a sliding-torsor, representing an instantaneous rotational motion around its central axis and whose rate of rotation is carried by this axis. This central axis (Δ), collinear to $\vec{s}\left\{{}^{2}_{3}\right\}$, is the line $\left(Q_0|\overrightarrow{z_0}\right)$ defined as the location of the points Q such that $\overrightarrow{OQ}=\overrightarrow{OQ_0}+\lambda\vec{s}$ where

$$\overrightarrow{OQ_0}=\frac{\vec{s}\left\{{}^{2}_{3}\right\}\wedge\overrightarrow{\mathcal{M}_O}\left\{{}^{2}_{3}\right\}}{\vec{s}^{\,2}\left\{{}^{2}_{3}\right\}}=e\cos\left(\frac{R_1}{R_3}\omega t\right)\overrightarrow{x_0}.$$

We remark that the point Q_0 is the orthogonal projection of point A_3 onto the axis $(O_3|\vec{x_0})$; it is located on the circle of diameter O_3A_3.

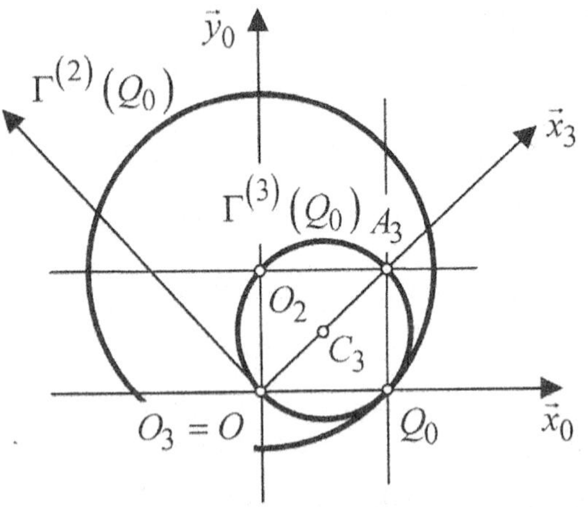

– *Question 2(b)*: Determine the moments of this torsor at the points A_3, J and K on the basis $(\vec{x_0}\ \vec{y_0}\ \vec{z_0})$. Give an interpretation for each of the vectors

– $\overrightarrow{\mathcal{M}_{A_3}}\left\{\begin{smallmatrix}2\\3\end{smallmatrix}\right\}$, $\overrightarrow{\mathcal{M}_{J}}\left\{\begin{smallmatrix}2\\3\end{smallmatrix}\right\}$, $\overrightarrow{\mathcal{M}_{K}}\left\{\begin{smallmatrix}2\\3\end{smallmatrix}\right\}$, $\overrightarrow{\mathcal{M}_{O_3}}\left\{\begin{smallmatrix}2\\3\end{smallmatrix}\right\}$.

– $\overrightarrow{\mathcal{M}_{A_3}}\left\{\begin{smallmatrix}2\\3\end{smallmatrix}\right\} = \overrightarrow{\mathcal{M}_{O}}\left\{\begin{smallmatrix}2\\3\end{smallmatrix}\right\} + \vec{s}\left\{\begin{smallmatrix}2\\3\end{smallmatrix}\right\} \wedge \overrightarrow{OA_3} = -e\frac{R_1}{R_3}\omega\sin\left(\frac{R_1}{R_3}\omega t\right)\vec{x_0}$,

– $\overrightarrow{\mathcal{M}_{J}}\left\{\begin{smallmatrix}2\\3\end{smallmatrix}\right\} = \overrightarrow{\mathcal{M}_{A_3}}\left\{\begin{smallmatrix}2\\3\end{smallmatrix}\right\} + \vec{s}\left\{\begin{smallmatrix}2\\3\end{smallmatrix}\right\} \wedge \overrightarrow{A_3J} = -\frac{R_1}{R_3}\omega\left[e\sin\left(\frac{R_1}{R_3}\omega t\right) - r\right]\vec{x_0}$,

– $\overrightarrow{\mathcal{M}_{K}}\left\{\begin{smallmatrix}2\\3\end{smallmatrix}\right\} = \overrightarrow{\mathcal{M}_{A_3}}\left\{\begin{smallmatrix}2\\3\end{smallmatrix}\right\} + \vec{s}\left\{\begin{smallmatrix}2\\3\end{smallmatrix}\right\} \wedge \overrightarrow{A_3K} = -\frac{R_1}{R_3}\omega\left[e\sin\left(\frac{R_1}{R_3}\omega t\right) + r\right]\vec{x_0}$,

– $\overrightarrow{\mathcal{M}_{O_3}}\left\{\begin{smallmatrix}2\\3\end{smallmatrix}\right\} = \overrightarrow{\mathcal{M}_{O}}\left\{\begin{smallmatrix}2\\3\end{smallmatrix}\right\} = e\frac{R_1}{R_3}\omega\cos\left(\frac{R_1}{R_3}\omega t\right)\vec{y_0}$, and, where

– $\overrightarrow{\mathcal{M}_{A_3}}\left\{\begin{smallmatrix}2\\3\end{smallmatrix}\right\} = \overrightarrow{v^{(2)}}(A_3)$ and $\overrightarrow{\mathcal{M}_{O_3}}\left\{\begin{smallmatrix}2\\3\end{smallmatrix}\right\} = \overrightarrow{v^{(2)}}(O_3)$ are the respective velocities of the points A_3 and O_3 relative to the ram (2);

– $\overrightarrow{\mathcal{M}_J}\left\{\begin{smallmatrix}2\\3\end{smallmatrix}\right\} = \overrightarrow{v^{(2)}}(J) - \overrightarrow{v^{(3)}}(J)$ and $\overrightarrow{\mathcal{M}_K}\left\{\begin{smallmatrix}2\\3\end{smallmatrix}\right\} = \overrightarrow{v^{(2)}}(K) - \overrightarrow{v^{(3)}}(K)$ are the respective slide velocities of the crankpin at J and at K in relation to the ram.

3

Kinetics of Solid Bodies

A solid body is a material system characterized by a mass and an inertia resulting from geometry, volume and mass, and, if in motion, by velocity in each of its points, therefore, its kinematics. This motion is induced by the forces exerted on the body by its external environment; its study involves significant elaboration, which is aimed at identifying and expressing all the components that describe it, and can be used for modeling it.

We have already looked into how a solid body is located, and how the field of velocities and accelerations characterizing its motion can be determined. This chapter focuses on expressing the characteristics and properties of this body which will come into play during its motion. Once this stage is completed, we shall have the elements that are required for expressing the dynamics of this body.

3.1. The mass of a continuous mechanical set ($\mathscr{D}$)

Whether it is in equilibrium or in motion, the mass of a body is one of its essential characteristics, as it naturally contributes to its energy. The manner in which a body reaches equilibrium or moves is predominantly influenced by its energy distribution and, consequently, by its inertia characteristics. This is the point where the approach of the kinetics of the solid body starts from.

3.1.1. *The notion of measure on a continuous mechanical set*

Let $(\mathscr{D})$ be a continuous mechanical set; a scalar measure μ is defined on it and an elementary measured $\mu(\mathrm{M})$, positive by definition, is associated to an arbitrary material point M of $(\mathscr{D})$.

Because of this association, the global measure of the set $(\mathscr{D})$ relative to μ is analyzed as the quantity $\mu(\mathscr{D})$ defined by the integral

$$\mu(\mathscr{D}) = \int_{\mathrm{M}\in(\mathscr{D})} \mathrm{d}\mu(\mathrm{M}),$$

where $\mu(\mathscr{D})$ is generally a triple volume integral.

3.1.2. *The volume and the mass of a continuous mechanical set*

We define the elementary measure $\mathrm{d}v(\mathrm{M})$, $\forall \mathrm{M} \in (\mathscr{D})$, and the global quantity *volume* given by the integral

$$\mathscr{V}(\mathscr{D}) = \int_{\mathrm{M}\in(\mathscr{D})} \mathrm{d}v(\mathrm{M}).$$

Similarly, we define the elementary measure $\mathrm{d}m(\mathrm{M})$, $\forall \mathrm{M} \in (\mathscr{D})$, and the *mass* of the set $(\mathscr{D})$ given by

$$m(\mathscr{D}) = \int_{\mathrm{M}\in(\mathscr{D})} \mathrm{d}m(\mathrm{M}).$$

These two definitions lead to a third local quantity $\rho(\mathrm{M})$, the *density* of the material point M which is defined by

$$\mathrm{d}\mu(\mathrm{M}) = \rho(\mathrm{M}) \times \mathrm{d}v(\mathrm{M}).$$

3.2. Center of the measure of μ on $(\mathscr{D})$

3.2.1. *Definition*

The center of the measure of μ on $(\mathscr{D})$ is the point H defined by the integral

$$\int_{M\in(\mathscr{D})} \overrightarrow{HM}\, d\mu(M) = \vec{0}.$$

In a system of reference $\langle\lambda\rangle$, with the origin O_λ, the following relation can be written relative to this origin

$$\int_{M\in(\mathscr{D})} \left(\overrightarrow{O_\lambda M} - \overrightarrow{O_\lambda H}\right) d\mu(M) = \vec{0},$$

$$\overrightarrow{O_\lambda H} \int_{M\in(\mathscr{D})} d\mu(M) = \overrightarrow{O_\lambda H}\, \mu(\mathscr{D}) = \int_{M\in(\mathscr{D})} \overrightarrow{O_\lambda M}\, d\mu(M),$$

an expression which leads to that of the position vector of the center H of the measure of μ on $(\mathscr{D})$ in the system of reference $\langle\lambda\rangle$

$$\overrightarrow{O_\lambda H} = \frac{\int_{M\in(\mathscr{D})} \overrightarrow{O_\lambda M}\, d\mu(M)}{\int_{M\in(\mathscr{D})} d\mu(M)} = \frac{\int_{M\in(\mathscr{D})} \overrightarrow{O_\lambda M}\, d\mu(M)}{\mu(\mathscr{D})}.$$

3.2.2. *Uniqueness of the center of measure*

The center of measure H does not depend on the system of reference relative to which it was analyzed; it is intrinsically attached to the mechanical set $(\mathscr{D})$ and is one of its major kinetic characteristics. We shall now prove that it is also independent of the origin taken in the system of reference; whatever it is, the above expressions remain valid.

Let O'_λ be another origin in the previous system of reference $\langle\lambda\rangle$ and let us suppose that we have obtained a point H' that satisfies the below relation:

$$\overrightarrow{O'_\lambda H'} = \frac{\int_{M\in(\mathcal{D})} \overrightarrow{O'_\lambda M}\, d\mu(M)}{\mu(\mathcal{D})} = \frac{\int_{M\in(\mathcal{D})} \left(\overrightarrow{O'_\lambda O_\lambda} + \overrightarrow{O_\lambda M}\right) d\mu(M)}{\mu(\mathcal{D})}$$

$$\Rightarrow \quad \overrightarrow{O'_\lambda H'} = \overrightarrow{O'_\lambda O_\lambda} + \overrightarrow{O_\lambda H} = \overrightarrow{O'_\lambda H},$$

which shows that the points H and H' coincide.

The center of measure is called the *center of volume* when the measure considered is the volume

$$\int_{M\in(\mathcal{D})} \overrightarrow{HM}\, dv(M) = \vec{0} \quad \Rightarrow \quad \overrightarrow{O_\lambda H} = \frac{\int_{M\in(\mathcal{D})} \overrightarrow{O_\lambda M}\, dv(M)}{\mathcal{V}(\mathcal{D})}.$$

The point G, defined by the relation below, is called the *center of the mass* (or *center of inertia*):

$$\int_{M\in(\mathcal{D})} \overrightarrow{GM}\, dm(M) = \vec{0} \quad \Rightarrow \quad \overrightarrow{O_\lambda G} = \frac{\int_{M\in(\mathcal{D})} \overrightarrow{O_\lambda M}\, dm(M)}{m(\mathcal{D})}.$$

NOTE 3.1.– if the density does not depend on M, which means that the mechanical set $(\mathcal{D})$ is homogeneous, the center H of the volume coincides with the center G of the mass.

3.2.3. *Center of measure of two disjoint sets*

Let us consider a measure μ applied to two disjoint mechanical sets $(\mathcal{D}_1)$ and $(\mathcal{D}_2)$, that is $\mu(\mathcal{D}_1)$ and $\mu(\mathcal{D}_2)$, respectively, and the corresponding centers of measure H_1 and H_2, respectively. Let H be the center of measure of the mechanical set $(\mathcal{D}) \equiv (\mathcal{D}_1) \cup (\mathcal{D}_2)$; since the two mechanical sets $(\mathcal{D}_1)$ and $(\mathcal{D}_2)$ are disjoint, we have:

$$\int_{M\in(\mathcal{D})} d\mu(M) = \int_{M\in(\mathcal{D}_1)} d\mu(M) + \int_{M\in(\mathcal{D}_2)} d\mu(M)$$

$$\Rightarrow \quad \mu(\mathcal{D}) = \mu(\mathcal{D}_1) + \mu(\mathcal{D}_2)$$

$$\overrightarrow{O_\lambda H} = \frac{\int_{M\in(\mathscr{D})} \overrightarrow{O_\lambda M}\, d\mu(M)}{\mu(\mathscr{D})}$$

$$'' \quad = \frac{\int_{M\in(\mathscr{D}_1)} \overrightarrow{O_\lambda M}\, d\mu(M)}{\mu(\mathscr{D}_1)} \frac{\mu(\mathscr{D}_1)}{\mu(\mathscr{D})} + \frac{\int_{M\in(\mathscr{D}_2)} \overrightarrow{O_\lambda M}\, d\mu(M)}{\mu(\mathscr{D}_2)} \frac{\mu(\mathscr{D}_2)}{\mu(\mathscr{D})}$$

$$'' \quad = \overrightarrow{O_\lambda H_1} \frac{\mu(\mathscr{D}_1)}{\mu(\mathscr{D})} + \overrightarrow{O_\lambda H_2} \frac{\mu(\mathscr{D}_2)}{\mu(\mathscr{D})}$$

Hence, we obtain the relation

$$\overrightarrow{O_\lambda H}\, \mu(\mathscr{D}) = \overrightarrow{O_\lambda H_1}\, \mu(\mathscr{D}_1) + \overrightarrow{O_\lambda H_2}\, \mu(\mathscr{D}_2),$$

that can also be written as

$$\overrightarrow{HH_1}\, \mu(\mathscr{D}_1) + \overrightarrow{HH_2}\, \mu(\mathscr{D}_2) = \vec{0}.$$

These three relations are applicable to a finite number of disjoint mechanical sets.

3.2.4. *Coordinates of the center of measure in a system of reference* $\langle\lambda\rangle$

Let us consider the system of reference $\langle\lambda\rangle = \left\langle O_\lambda \middle| \overrightarrow{x_\lambda}\ \overrightarrow{y_\lambda}\ \overrightarrow{z_\lambda} \right\rangle$ and the current material point M of the mechanical set $(\mathscr{D})$, whose position vector is

$$\overrightarrow{O_\lambda M} = x(M)\overrightarrow{x_\lambda} + y(M)\overrightarrow{y_\lambda} + z(M)\overrightarrow{z_\lambda},$$

and the center of measure H relative to the measure μ applied to $(\mathscr{D})$, whose position vector is

$$\overrightarrow{O_\lambda H} = x(H)\overrightarrow{x_\lambda} + y(H)\overrightarrow{y_\lambda} + z(H)\overrightarrow{z_\lambda}.$$

According to the definition of the center of measure and to the linearity of the 'integral' operator we have

$$\begin{cases} x(\mathrm{H}) = \dfrac{\int_{\mathrm{M}\in(\mathscr{D})} x(\mathrm{M})\mathrm{d}\mu(\mathrm{M})}{\mu(\mathscr{D})} \\ y(\mathrm{H}) = \dfrac{\int_{\mathrm{M}\in(\mathscr{D})} y(\mathrm{M})\mathrm{d}\mu(\mathrm{M})}{\mu(\mathscr{D})} \\ z(\mathrm{H}) = \dfrac{\int_{\mathrm{M}\in(\mathscr{D})} z(\mathrm{M})\mathrm{d}\mu(\mathrm{M})}{\mu(\mathscr{D})} \end{cases}.$$

In the system of reference $\left\langle \mathrm{H} | \overrightarrow{x_\lambda}\ \overrightarrow{y_\lambda}\ \overrightarrow{z_\lambda} \right\rangle$, the position vector of the current point is expressed as follows:

$$\overrightarrow{\mathrm{HM}} = X(\mathrm{M})\overrightarrow{x_\lambda} + Y(\mathrm{M})\overrightarrow{y_\lambda} + Z(\mathrm{M})\overrightarrow{z_\lambda}.$$

Therefore,

$$\overrightarrow{\mathrm{O}_\lambda\mathrm{M}} = \overrightarrow{\mathrm{O}_\lambda\mathrm{H}} + \overrightarrow{\mathrm{HM}} = \overrightarrow{\mathrm{O}_\lambda\mathrm{H}} + X(\mathrm{M})\overrightarrow{x_\lambda} + Y(\mathrm{M})\overrightarrow{y_\lambda} + Z(\mathrm{M})\overrightarrow{z_\lambda}$$

$$\Rightarrow \begin{cases} x(\mathrm{M}) = x(\mathrm{H}) + X(\mathrm{M}) \\ y(\mathrm{M}) = y(\mathrm{H}) + Y(\mathrm{M}) \\ z(\mathrm{M}) = z(\mathrm{H}) + Z(\mathrm{M}) \end{cases}.$$

The definition of the center of measure can also be expressed by these three relations:

$$\begin{cases} \int_{\mathrm{M}\in(\mathscr{D})} X(\mathrm{M})\mathrm{d}\mu(\mathrm{M}) = 0 \\ \int_{\mathrm{M}\in(\mathscr{D})} Y(\mathrm{M})\mathrm{d}\mu(\mathrm{M}) = 0 \\ \int_{\mathrm{M}\in(\mathscr{D})} Z(\mathrm{M})\mathrm{d}\mu(\mathrm{M}) = 0. \end{cases}$$

3.3. Interpretation of the notion of center of measure

It is worth reminding that, in the sense of Riemannian integration, the average value on an interval $[a,b]$ of a function $f(x)$ defined on this interval is given by

$$\underline{f}^{(a,b)} = \frac{\int_a^b f(x)dx}{\int_a^b dx} = \frac{\int_a^b f(x)dx}{b-a}.$$

In more general terms, if we consider a scalar or vector function $\phi(\mathrm{M})$, defined in every point M of a continuous mechanical set $(\mathcal{D})$, we define the *average of* ϕ *over* $(\mathcal{D})$ *relative to the measure* μ by the relation

$$\underline{\phi}^{(\mathcal{D})/\mu} = \frac{\int_{\mathrm{M}\in(\mathcal{D})} \phi(\mathrm{M})\mathrm{d}\mu(\mathrm{M})}{\int_{\mathrm{M}\in(\mathcal{D})} \mathrm{d}\mu(\mathrm{M})} = \frac{\int_{\mathrm{M}\in(\mathcal{D})} \phi(\mathrm{M})\mathrm{d}\mu(\mathrm{M})}{\mu(\mathcal{D})}.$$

Therefore, the center H of measure, considered from the point of view of its coordinates in $\langle\lambda\rangle$, is none other than the average value of the "vector position" function of the material point M over $(\mathcal{D})$, according to the relation

$$\overrightarrow{\mathrm{O}_\lambda\mathrm{H}} = \frac{\int_{\mathrm{M}\in(\mathcal{D})} \overrightarrow{\mathrm{O}_\lambda\mathrm{M}}\,\mathrm{d}\mu(\mathrm{M})}{\mu(\mathcal{D})}.$$

3.4. Kinetic torsor of a mechanical set $(\mathcal{D})$

3.4.1. *Definition – linear momentum*

Taking into account the field of velocity that characterizes and drives the motion of the mechanical set $(\mathcal{D})$ relative to the system of reference $\langle\lambda\rangle$, and moreover the fact that this mechanical set is continuous and equipped with a measure of the mass, it is tempting to examine the resultant and the moment regarding a point of this field of velocity applied to all the material points M of $(\mathscr{D})$ relative to the measure employed.

Two vector quantities that are a kind of synthesis of kinematic and kinetic characteristics of the moving mechanical set $(\mathscr{D})$ can be defined:

– the *kinetic resultant* of $(\mathscr{D})$ in $\langle\lambda\rangle$ by

$$\vec{s}\{p_{\mathscr{D}}^{\lambda}\} = \int_{M\in(D)} \overrightarrow{v^{(\lambda)}}(M)\,dm(M);$$

– the *kinetic moment* regarding a point Q of $(\mathscr{D})$ in $\langle\lambda\rangle$ by

$$\overrightarrow{\mathscr{M}_Q}\{p_{\mathscr{D}}^{\lambda}\} = \int_{M\in(D)} \overrightarrow{QM}\wedge\overrightarrow{v^{(\lambda)}}(M)\,dm(M).$$

The quantity $\overrightarrow{v^{(\lambda)}}(M)\,dm(M)$ is called the *linear momentum* of the fragment M of matter of $(\mathscr{D})$.

It is important to immediately verify that these are reduction elements of a torsor; therefore, they need to verify a moment relation, which means that in another point P we should have

$$\overrightarrow{\mathscr{M}_P} = \overrightarrow{\mathscr{M}_Q} + \overrightarrow{PQ}\wedge\vec{s}.$$

If we follow the above definition, we then obtain

$$\begin{aligned}
\overrightarrow{\mathscr{M}_P}\{p_{\mathscr{D}}^{\lambda}\} &= \int_{M\in(D)} \overrightarrow{PM}\wedge\overrightarrow{v^{(\lambda)}}(M)\,dm(M)\\
\text{'' ''}\quad &= \int_{M\in(D)} \left(\overrightarrow{PQ}+\overrightarrow{QM}\right)\wedge\overrightarrow{v^{(\lambda)}}(M)\,dm(M)\\
\text{'' ''}\quad &= \overrightarrow{PQ}\wedge\int_{M\in(D)} \overrightarrow{v^{(\lambda)}}(M)\,dm(M) + \int_{M\in(D)} \overrightarrow{QM}\wedge\overrightarrow{v^{(\lambda)}}(M)\,dm(M)
\end{aligned}$$

$$\Rightarrow \quad \overrightarrow{\mathscr{M}_P}\{p_{\mathscr{D}}^{\lambda}\} = \overrightarrow{\mathscr{M}_Q}\{p_{\mathscr{D}}^{\lambda}\} + \overrightarrow{PQ}\wedge\vec{s}\{p_{\mathscr{D}}^{\lambda}\},$$

which is a moment relation. We have, therefore, obtained the reduction elements of a torsor. This torsor $\{p_{\mathscr{D}}^{\lambda}\}$ is called the *kinetic torsor*.

3.4.2. *Kinetic torsor* $\{ p_S^\lambda \}$ *of a rigid solid body*

If the continuous mechanical set is the rigid solid body (S), to which the system of reference $\langle S \rangle = \langle O_S | \vec{x_S}\, \vec{y_S}\, \vec{z_S} \rangle$ is linked, $\forall M \in (S)$ the moment regarding this point of the torsor distributor of velocities is a velocity; we can then write

$$\overrightarrow{v^{(\lambda)}}(M) = \overrightarrow{v^{(\lambda)}}(O_S) + \overrightarrow{\omega_S^\lambda} \wedge \overrightarrow{O_S M},$$

and also:

$$- \ \vec{s}\{ p_S^\lambda \} = \int_{M \in (S)} \left[\overrightarrow{v^{(\lambda)}}(O_S) + \overrightarrow{\omega_S^\lambda} \wedge \overrightarrow{O_S M} \right] \mathrm{d}m(M)$$

$$\vec{s}\{ p_S^\lambda \} = \overrightarrow{v^{(\lambda)}}(O_S) \int_{M \in (S)} \mathrm{d}m(M) + \overrightarrow{\omega_S^\lambda} \wedge \int_{M \in (S)} \overrightarrow{O_S M}\, \mathrm{d}m(M)$$

$$'' \ '' \quad = m(S)\overrightarrow{v^{(\lambda)}}(O_S) + \overrightarrow{\omega_S^\lambda} \wedge m(S)\overrightarrow{O_S G}$$

$$'' \ '' \quad = m(S)\left[\overrightarrow{v^{(\lambda)}}(O_S) + \overrightarrow{\omega_S^\lambda} \wedge \overrightarrow{O_S G} \right]$$

$$\Rightarrow \ \vec{s}\{ p_S^\lambda \} = m(S)\overrightarrow{v^{(\lambda)}}(G);$$

$$- \ \overrightarrow{\mathcal{M}_{O_S}}\{ p_S^\lambda \} = \int_{M \in (S)} \overrightarrow{O_S M} \wedge \left[\overrightarrow{v^{(\lambda)}}(O_S) + \overrightarrow{\omega_S^\lambda} \wedge \overrightarrow{O_S M} \right] dm(M),$$

$$\overrightarrow{\mathcal{M}_{O_S}}\{ p_S^\lambda \} = \int_{M \in (S)} \overrightarrow{O_S M}\, \mathrm{d}m(M) \wedge \overrightarrow{v^{(\lambda)}}(O_S) \ldots$$

$$\ldots + \int_{M \in (S)} \overrightarrow{O_S M} \wedge \left(\overrightarrow{\omega_S^\lambda} \wedge \overrightarrow{O_S M} \right) \mathrm{d}m(M),$$

$$'' \ '' \quad = m(S)\overrightarrow{O_S G} \wedge \overrightarrow{v^{(\lambda)}}(O_S) + I_{O_S}(S|m)\overrightarrow{\omega_S^\lambda}$$

where $I_{O_S}(S|m)$, called *inertia operator* of (S), defined by

$$I_{O_S}(S|m)\overrightarrow{\omega_S^\lambda} = \int_{M \in (S)} \overrightarrow{O_S M} \wedge \left(\overrightarrow{\omega_S^\lambda} \wedge \overrightarrow{O_S M} \right) \mathrm{d}m(M),$$

is a vector operator whose matrix representation in the basis (S) is a (3×3) matrix. This formulation and the existence of this operator are meaningless unless all the material fragments M taken into account in this integral belong to the solid (S).

3.4.3. *Inertia operator* $I_{O_S}(S|m)$

3.4.3.1. *Matrix development of the inertia operator*

Let us develop the double-vector product

$$\overrightarrow{O_SM}\wedge\left(\vec{a}\wedge\overrightarrow{O_SM}\right)=\left(\overrightarrow{O_SM}\cdot\overrightarrow{O_SM}\right)\vec{a}-\left(\overrightarrow{O_SM}\cdot\vec{a}\right)\overrightarrow{O_SM}.$$

Let us consider

$$\overrightarrow{O_SM}=x_S\,\overrightarrow{x_S}+y_S\,\overrightarrow{y_S}+z_S\,\overrightarrow{z_S}\ \text{ and }\ \vec{a}=a_x\overrightarrow{x_S}+a_y\overrightarrow{y_S}+a_z\overrightarrow{z_S}\,.$$

Then, we have

$$\left(\overrightarrow{O_SM}\cdot\overrightarrow{O_SM}\right)\vec{a}=\left(x_S^2+y_S^2+z_S^2\right)\left(a_x\overrightarrow{x_S}+a_y\overrightarrow{y_S}+a_z\overrightarrow{z_S}\right)$$

$$\overrightarrow{O_SM}\cdot\vec{a}=a_xx_S+a_yy_S+a_zz_S\,.$$

The operation $\left(\overrightarrow{O_SM}\cdot\vec{a}\right)\overrightarrow{O_SM}$ can be written in a matrix form in the basis (S)

$$\left[\left(\overrightarrow{O_SM}\cdot\vec{a}\right)\overrightarrow{O_SM}\right]_{/(S)}=\begin{bmatrix}(a_xx_S+a_yy_S+a_zz_S)x_S\\(a_xx_S+a_yy_S+a_zz_S)y_S\\(a_xx_S+a_yy_S+a_zz_S)z_S\end{bmatrix}$$

$$\text{'' ''}\qquad=\begin{bmatrix}x_S^2 & x_Sy_S & x_Sz_S\\x_Sy_S & y_S^2 & y_Sz_S\\x_Sz_S & y_Sz_S & z_S^2\end{bmatrix}\begin{bmatrix}a_x\\a_y\\a_z\end{bmatrix}.$$

We can write this operation as

$$\left(\overrightarrow{\mathrm{O}_S\mathrm{M}}\cdot\vec{a}\right)\overrightarrow{\mathrm{O}_S\mathrm{M}}=\left(\overrightarrow{\mathrm{O}_S\mathrm{M}}\otimes\overrightarrow{\mathrm{O}_S\mathrm{M}}\right)\vec{a},$$

where the matrix representation in the basis (S) of the operator $\overrightarrow{\mathrm{O}_S\mathrm{M}}\otimes\overrightarrow{\mathrm{O}_S\mathrm{M}}$, called the *tensor product*, is

$$\left[\overrightarrow{\mathrm{O}_S\mathrm{M}}\otimes\overrightarrow{\mathrm{O}_S\mathrm{M}}\right]_{/(S)}=\begin{bmatrix} x_S^2 & x_S y_S & x_S z_S \\ x_S y_S & y_S^2 & y_S z_S \\ x_S z_S & y_S z_S & z_S^2 \end{bmatrix}.$$

The double vector product can then be written as

$$\overrightarrow{\mathrm{O}_S\mathrm{M}}\wedge\left(\vec{a}\wedge\overrightarrow{\mathrm{O}_S\mathrm{M}}\right)=\left[\mathrm{I}d\left(\overrightarrow{\mathrm{O}_S\mathrm{M}}^2\right)-\overrightarrow{\mathrm{O}_S\mathrm{M}}\otimes\overrightarrow{\mathrm{O}_S\mathrm{M}}\right]\vec{a},$$

where $\mathrm{I}d$ is the identity vector operator such that $\mathrm{I}d\left(\vec{a}\right)=\vec{a}$, whose matrix representation in the basis (S) is

$$[\mathrm{I}d]_{/(S)}=\begin{bmatrix} 1 & 0 & 0 \\ 0 & 1 & 0 \\ 0 & 0 & 1 \end{bmatrix}.$$

The matrix representation in the basis (S) of the operator $\overrightarrow{\mathrm{O}_S\mathrm{M}}\otimes\overrightarrow{\mathrm{O}_S\mathrm{M}}$ is

$$\begin{bmatrix} y_S^2+z_S^2 & -x_S y_S & -z_S x_S \\ -x_S y_S & z_S^2+x_S^2 & -y_S z_S \\ -z_S x_S & -y_S z_S & x_S^2+y_S^2 \end{bmatrix}.$$

The double vector product $\overrightarrow{\mathrm{O}_S\mathrm{M}}\wedge\left(\overrightarrow{\omega_S^{\lambda}}\wedge\overrightarrow{\mathrm{O}_S\mathrm{M}}\right)$ presents itself as a vector operation applied to the vector $\overrightarrow{\omega_S^{\lambda}}$, with a matrix representation of the operator in the basis (S) as given above.

The operation

$$\int_{M\in(S)} \overrightarrow{O_S M} \wedge \left(\overrightarrow{\omega_S^{\lambda}} \wedge \overrightarrow{O_S M} \right) dm(M)$$

is, therefore, a linear application of the vector $\overrightarrow{\omega_S^{\lambda}}$, whose matrix representation in the basis (S) is

$$\begin{bmatrix} \int_{M\in(S)} (y_S^2 + z_S^2)\,dm(M) & -\int_{M\in(S)} x_S y_S\,dm(M) & -\int_{M\in(S)} z_S x_S\,dm(M) \\ -\int_{M\in(S)} x_S y_S\,dm(M) & \int_{M\in(S)} (z_S^2 + x_S^2)\,dm(M) & -\int_{M\in(S)} y_S z_S\,dm(M) \\ -\int_{M\in(S)} z_S x_S\,dm(M) & -\int_{M\in(S)} y_S z_S\,dm(M) & \int_{M\in(S)} (x_S^2 + y_S^2)\,dm(M) \end{bmatrix}.$$

Called the *inertia matrix*, this matrix is the representation, in the basis (S) linked to the solid (S), of the *inertia operator* of this solid, relative to the measure of the mass m. This operator is denoted as $I_{O_S}(S|m)$.

In the end, the moment regarding O_S of the kinetic torsor of the motion of (S) in the system of reference $\langle \lambda \rangle$ can be written

$$\overrightarrow{\mathcal{M}_{O_S}}\{p_S^{\lambda}\} = m(S)\overrightarrow{O_S G} \wedge \overrightarrow{v^{(\lambda)}}(O_S) + I_{O_S}(S|m)\overrightarrow{\omega_S^{\lambda}}.$$

The inertia matrix of the solid (S) is often represented in a symbolic manner in the form

$$[I_{O_S}(S|m)]_{/(S)} = \begin{bmatrix} A & -F & -E \\ -F & B & -D \\ -E & -D & C \end{bmatrix},$$

a matrix whose various terms are expressed by

– $A = \mathcal{I}_{O_S|\overrightarrow{x_S}}(S) = \int_{M\in(S)} (y_S^2 + z_S^2)\,dm(M)$: *moment of inertia* of (S) relative to the axis $\left(O_S|\overrightarrow{x_S}\right)$;

– $B = \mathcal{I}_{O_S|\overrightarrow{y_S}}(S) = \int_{M\in(S)} (z_S^2 + x_S^2)\,dm(M)$: *moment of inertia* of (S) relative to the axis $\left(O_S|\overrightarrow{y_S}\right)$;

– $C = \mathcal{I}_{O_S|\overrightarrow{z_S}}(S) = \int_{M\in(S)} (x_S^2 + y_S^2)\,dm(M)$: *moment of inertia* of (S) relative to the axis $\left(O_S|\overrightarrow{z_S}\right)$;

– $D = \mathcal{P}_{O_S|\overrightarrow{z_S},\overrightarrow{x_S}}(S) = \int_{M\in(S)} z_S x_S\,dm(M)$: *product of inertia* of (S) relative to the plane $\left(O_S|\overrightarrow{z_S}\,\overrightarrow{x_S}\right)$;

– $E = \mathcal{P}_{O_S|\overrightarrow{y_S},\overrightarrow{z_S}}(S) = \int_{M\in(S)} y_S z_S\,dm(M)$: *product of inertia* of (S) relative to the plane $\left(O_S|\overrightarrow{y_S}\,\overrightarrow{z_S}\right)$;

– $F = \mathcal{P}_{O_S|\overrightarrow{x_S},\overrightarrow{y_S}}(S) = \int_{M\in(S)} x_S y_S\,dm(M)$: *product of inertia* of (S) relative to the plane $\left(O_S|\overrightarrow{x_S}\,\overrightarrow{y_S}\right)$.

3.4.3.2. *Property of the inertia operator – Huygens theorem*

$\overrightarrow{\mathcal{M}_{O_S}}\{p_S^\lambda\}$ is the moment in O_S of a torsor, which means that in any other point P of the system of reference $\langle\lambda\rangle$, the following can be written:

$$\overrightarrow{\mathcal{M}_P}\{p_S^\lambda\} = \overrightarrow{\mathcal{M}_{O_S}}\{p_S^\lambda\} + \overrightarrow{PO_S} \wedge \vec{s}\{p_S^\lambda\}.$$

The center of inertia G of the solid is a particular point, and it is worth examining its importance and the properties that are eventually assigned to it. G belongs to the solid; its velocity can, therefore, be found out starting from the reduction elements in O_S of the torsor distributor of velocities of (S) in $\langle\lambda\rangle$:

$$\overrightarrow{v^{(\lambda)}}(G) = \overrightarrow{v^{(\lambda)}}(O_S) + \overrightarrow{\omega_S^\lambda} \wedge \overrightarrow{O_S G}.$$

Under these conditions, the formula of the kinetic moment established at the particular point O_S of the solid, starting from the relation of velocities, is also applicable to point G ; as we have

$$\overrightarrow{v^{(\lambda)}}(\mathrm{M}) = \overrightarrow{v^{(\lambda)}}(\mathrm{G}) + \overrightarrow{\omega_S^{\lambda}} \wedge \overrightarrow{\mathrm{GM}},$$

we obtain

$$\overrightarrow{\mathcal{M}_{\mathrm{G}}}\{p_S^{\lambda}\} = m(S)\overrightarrow{\mathrm{GG}} \wedge \overrightarrow{v^{(\lambda)}}(\mathrm{G}) + I_{\mathrm{G}}(S|m)\overrightarrow{\omega_S^{\lambda}} = I_{\mathrm{G}}(S|m)\overrightarrow{\omega_S^{\lambda}}.$$

According to the relation of moment applied to the kinetic torsor, as given above, we obtain

$$I_{\mathrm{G}}(S|m)\overrightarrow{\omega_S^{\lambda}} = m(S)\overrightarrow{\mathrm{O}_S\mathrm{G}} \wedge \overrightarrow{v^{(\lambda)}}(\mathrm{O}_S) + I_{\mathrm{O}_S}(S|m)\overrightarrow{\omega_S^{\lambda}} \ldots$$
$$\ldots + \overrightarrow{\mathrm{GO}_S} \wedge m(S)\overrightarrow{v^{(\lambda)}}(\mathrm{G})$$

$$\Rightarrow \quad I_{\mathrm{O}_S}(S|m)\overrightarrow{\omega_S^{\lambda}} = I_{\mathrm{G}}(S|m)\overrightarrow{\omega_S^{\lambda}} + m(S)\overrightarrow{\mathrm{O}_S\mathrm{G}} \wedge \left[\overrightarrow{v^{(\lambda)}}(\mathrm{G}) - \overrightarrow{v^{(\lambda)}}(\mathrm{O}_S)\right].$$
$$'' \; '' \qquad = I_{\mathrm{G}}(S|m)\overrightarrow{\omega_S^{\lambda}} + m(S)\overrightarrow{\mathrm{O}_S\mathrm{G}} \wedge \left(\overrightarrow{\omega_S^{\lambda}} \wedge \overrightarrow{\mathrm{O}_S\mathrm{G}}\right)$$

If in the basis (S), we have $\overrightarrow{\mathrm{O}_S\mathrm{G}} = a\overrightarrow{x_S} + b\overrightarrow{y_S} + c\overrightarrow{z_S}$, the vector

$$\overrightarrow{\mathrm{O}_S\mathrm{G}} \wedge \left(\overrightarrow{\omega_S^{\lambda}} \wedge \overrightarrow{\mathrm{O}_S\mathrm{G}}\right)$$

is the result of a vector operation applied to $\overrightarrow{\omega_S^{\lambda}}$ whose matrix representation in this basis is written

$$\left[\overrightarrow{\mathrm{O}_S\mathrm{G}} \wedge \left(\overrightarrow{\omega_S^{\lambda}} \wedge \overrightarrow{\mathrm{O}_S\mathrm{G}}\right)\right]_{/S} = \begin{bmatrix} b^2 + c^2 & -ab & -ac \\ -ab & c^2 + a^2 & -bc \\ -ac & -bc & a^2 + b^2 \end{bmatrix} \left[\overrightarrow{\omega_S^{\lambda}}\right].$$

This means that, in the matrix form, the representatives in the basis (S) of the inertia operator expressed in O_S and in G are linked by the relation

$$[I_{O_S}(S|m)]_{/S} = [I_G(S|m)]_{/S} + m\begin{bmatrix} b^2+c^2 & -ab & -ac \\ -ab & c^2+a^2 & -bc \\ -ac & -bc & a^2+b^2 \end{bmatrix},$$

which leads, with $\overrightarrow{O_S G} = a\overrightarrow{x_S} + b\overrightarrow{y_S} + c\overrightarrow{z_S}$, to the six relations expressing the *Huygens theorem*,

$$- \mathscr{I}_{O_S|\overrightarrow{x_S}}(S|m) = \mathscr{I}_{G|\overrightarrow{x_S}}(S|m) + m(b^2+c^2),$$

$$- \mathscr{I}_{O_S|\overrightarrow{y_S}}(S|m) = \mathscr{I}_{G|\overrightarrow{y_S}}(S|m) + m(c^2+a^2),$$

$$- \mathscr{I}_{O_S|\overrightarrow{z_S}}(S|m) = \mathscr{I}_{G|\overrightarrow{z_S}}(S|m) + m(a^2+b^2),$$

$$- \mathscr{P}_{O_S|\overrightarrow{x_S}\,\overrightarrow{y_S}}(S|m) = \mathscr{P}_{G|\overrightarrow{x_S}\,\overrightarrow{y_S}}(S|m) + m\,ab,$$

$$- \mathscr{P}_{O_S|\overrightarrow{y_S}\,\overrightarrow{z_S}}(S|m) = \mathscr{P}_{G|\overrightarrow{y_S}\,\overrightarrow{z_S}}(S|m) + m\,bc,$$

$$- \mathscr{P}_{O_S|\overrightarrow{z_S}\,\overrightarrow{x_S}}(S|m) = \mathscr{P}_{G|\overrightarrow{z_S}\,\overrightarrow{x_S}}(S|m) + m\,ca.$$

One can note that, according to these relations,

$$\begin{cases} \mathscr{I}_{O_S|\overrightarrow{x_S}}(S|m) \geq \mathscr{I}_{G|\overrightarrow{x_S}}(S|m) \\ \mathscr{I}_{O_S|\overrightarrow{y_S}}(S|m) \geq \mathscr{I}_{G|\overrightarrow{y_S}}(S|m), \\ \mathscr{I}_{O_S|\overrightarrow{z_S}}(S|m) \geq \mathscr{I}_{G|\overrightarrow{z_S}}(S|m) \end{cases}$$

which means that, if we consider various parallel axes, the one that goes through the center of inertia G is the one relative to which the moment of inertia of the solid is minimal. This shows the particular role played by the center of inertia of a solid.

These six relations can be applied in any point of (S) in place of O_S, provided that the particular role of the center of inertia G is preserved.

In order to express these various terms in an unambiguous form, the first formula of Huygens can, for example, be written in the form

$$\mathscr{I}_{\mathrm{O}_S|\vec{x}_S}(S|m) = \mathscr{I}_{\mathrm{G}|\vec{x}_S}(S|m) + m\left[\left(\overrightarrow{\mathrm{O}_S\mathrm{G}}\cdot\vec{y}_S\right)^2 + \left(\overrightarrow{\mathrm{O}_S\mathrm{G}}\cdot\vec{z}_S\right)^2\right],$$

and also for the example the fifth as

$$\mathscr{P}_{\mathrm{O}_S|\vec{y}_S\,\vec{z}_S}(S|m) = \mathscr{P}_{\mathrm{G}|\vec{y}_S\,\vec{z}_S}(S|m) + m\left(\overrightarrow{\mathrm{O}_S\mathrm{G}}\cdot\vec{y}_S\right)\left(\overrightarrow{\mathrm{O}_S\mathrm{G}}\cdot\vec{z}_S\right).$$

3.4.3.3. *Exercise 24 – Elements of inertia of a solid*

In a direct orthonormal system of reference $\langle \mathrm{O}|\vec{x}\,\vec{y}\,\vec{z}\rangle$ linked to one of its vertices and whose axes are oriented along its edges, according to the diagram below, let us consider a homogeneous rectangular parallelepiped, with density ρ, with sides of lengths ℓ_1, ℓ_2 and ℓ_3, respectively.

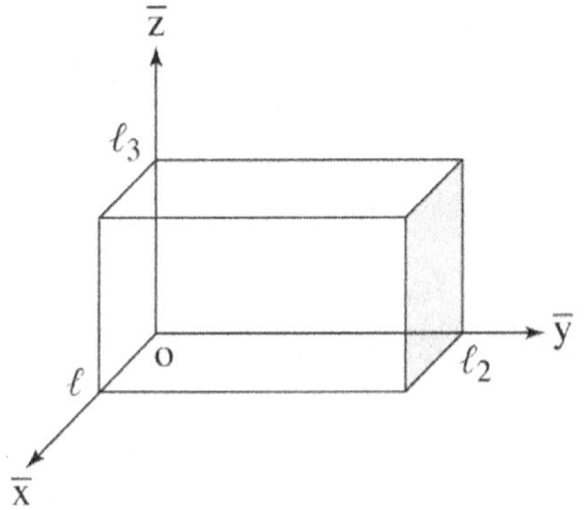

This parallelepiped has the following inertia characteristics

– mass $\mu = \rho\ell_1\ell_2\ell_3$,

– center of inertia G defined by: $\overrightarrow{\mathrm{OG}} = \frac{1}{2}\left(\ell_1\vec{x} + \ell_2\vec{y} + \ell_3\vec{z}\right)$,

– inertia matrix in G in the basis $\left(\vec{x}\,\vec{y}\,\vec{z}\right)$,

$$[I_G] = \begin{bmatrix} \frac{\mu}{12}(\ell_2^2 + \ell_3^2) & 0 & 0 \\ 0 & \frac{\mu}{12}(\ell_3^2 + \ell_1^2) & 0 \\ 0 & 0 & \frac{\mu}{12}(\ell_1^2 + \ell_2^2) \end{bmatrix}.$$

A solid (S) consists of two homogeneous parallelepipeds (S_1) and (S_2), of the same material, whose dimensions are $(2a, 2b, 2c)$ and $(a, b, 2c)$, respectively, as shown by the figure below:

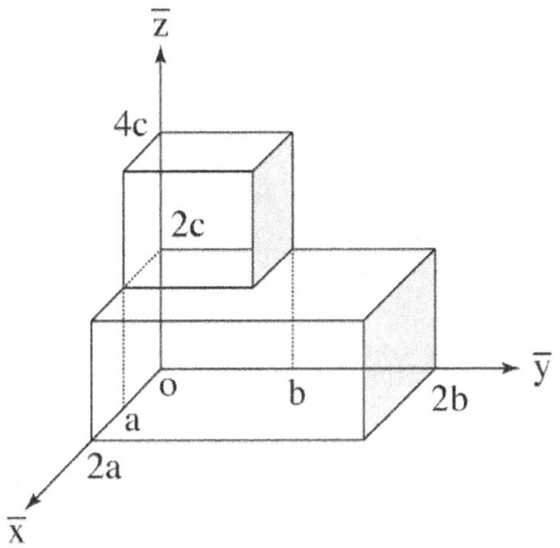

NOTE 3.2.– we shall express the results by means of lengths a, b, c and of the mass m of the solid (S_2).

Question 1: Analyze the masses of the solids (S_1) and (S) depending on m

– mass of (S_2): $m = 2\rho abc$,

– mass of (S_1): $8\rho abc = 4m$,

– mass of $(S) = (S_1) \cup (S_2) = 4m + m = 5m$.

Question 2: Establish the elements of inertia of the solid (S_1): $\overrightarrow{OG_1}$ and $[I_{G_1}(S_1)]$:

$$\overrightarrow{OG_1} = a\vec{x} + b\vec{y} + c\vec{z},$$

$$[I_{G_1}(S_1)] = \frac{4m}{3}\begin{bmatrix} b^2+c^2 & 0 & 0 \\ 0 & c^2+a^2 & 0 \\ 0 & 0 & a^2+b^2 \end{bmatrix}.$$

Question 3: Establish the elements of inertia of the solid (S_2): $\overrightarrow{OG_2}$ and $[I_{G_2}(S_2)]$:

$$\overrightarrow{OG_2} = \overrightarrow{OO'} + \overrightarrow{O'G_2} = 2c\vec{z} + \frac{a}{2}\vec{x} + \frac{b}{2}\vec{y} + c\vec{z} = \frac{a}{2}\vec{x} + \frac{b}{2}\vec{y} + 3c\vec{z},$$

$$[I_{G_2}(S_2)] = \frac{m}{12}\begin{bmatrix} b^2+4c^2 & 0 & 0 \\ 0 & 4c^2+a^2 & 0 \\ 0 & 0 & a^2+b^2 \end{bmatrix}.$$

Question 4: Establish the elements of inertia of the solid (*S*): $\overrightarrow{OG}$ and $[I_G(S)]$:

$$m(S)\overrightarrow{OG} = m(S_1)\overrightarrow{OG_1} + m(S_2)\overrightarrow{OG_2}$$

$$5m\overrightarrow{OG} = 4m\left(a\vec{x} + b\vec{y} + c\vec{z}\right) + m\left(\frac{a}{2}\vec{x} + \frac{b}{2}\vec{y} + 3c\vec{z}\right) = m\left(\frac{9a}{2}\vec{x} + \frac{9b}{2}\vec{y} + 7c\vec{z}\right),$$

$$\overrightarrow{OG} = \frac{9a\vec{x} + 9b\vec{y} + 14c\vec{z}}{10}.$$

According to the formulation of Huygens theorem

$$[I_O(S|m)] = [I_G(S|m)] + m\begin{bmatrix} b^2+c^2 & -ab & -ac \\ -ab & c^2+a^2 & -bc \\ -ac & -bc & a^2+b^2 \end{bmatrix},$$

with $\quad \overrightarrow{OG} = a\vec{x} + b\vec{y} + c\vec{z}$

we can write

$$[I_G(S_1|4m)] = [I_{G_1}(S_1|4m)] + \frac{m}{25}\begin{bmatrix} b^2+16c^2 & -ab & 4ac \\ -ab & 16c^2+a^2 & 4bc \\ 4ac & 4bc & a^2+b^2 \end{bmatrix}$$

with $\quad \overrightarrow{GG_1} = \overrightarrow{OG_1} - \overrightarrow{OG} = \frac{1}{10}\left(a\vec{x} + b\vec{y} - 4c\vec{z}\right)$:

$$[I_G(S_1|4m)] = \frac{4m}{3}\begin{bmatrix} b^2+c^2 & 0 & 0 \\ 0 & c^2+a^2 & 0 \\ 0 & 0 & a^2+b^2 \end{bmatrix}\ldots$$

$$\ldots + \frac{m}{25}\begin{bmatrix} b^2+16c^2 & -ab & 4ac \\ -ab & 16c^2+a^2 & 4bc \\ 4ac & 4bc & a^2+b^2 \end{bmatrix}$$

$$\text{" "} \quad = \frac{m}{75}\begin{bmatrix} 103b^2+148c^2 & -3ab & 12ac \\ -3ab & 148c^2+103a^2 & 12bc \\ 12ac & 12bc & 103(a^2+b^2) \end{bmatrix}.$$

Similarly,

$$[I_G(S_2|m)] = [I_{G_2}(S|m)] + \frac{4m}{25}\begin{bmatrix} b^2+16c^2 & -ab & 4ac \\ -ab & 16c^2+a^2 & 4bc \\ 4ac & 4bc & a^2+b^2 \end{bmatrix}$$

with $\quad \overrightarrow{GG_2} = \overrightarrow{OG_2} - \overrightarrow{OG} = -\frac{2a}{5}\vec{x} - \frac{2b}{5}\vec{y} + \frac{8c}{5}\vec{z}$:

$$[I_G(S_2|m)] = \frac{m}{12}\begin{bmatrix} b^2+4c^2 & 0 & 0 \\ 0 & 4c^2+a^2 & 0 \\ 0 & 0 & a^2+b^2 \end{bmatrix}\ldots$$

$$\ldots + \frac{4m}{25}\begin{bmatrix} b^2+16c^2 & -ab & 4ac \\ -ab & 16c^2+a^2 & 4bc \\ 4ac & 4bc & a^2+b^2 \end{bmatrix},$$

$$\text{" "} \quad = \frac{m}{300}\begin{bmatrix} 73b^2+868c^2 & -48ab & 192ac \\ -48ab & 868c^2+73a^2 & 192bc \\ 192ac & 192bc & 73a^2+73b^2 \end{bmatrix}$$

Taking into account that the two solids (S_1) and (S_2) are disjoint, if the two inertia matrices are calculated at the same point, in this case G, center of inertia of the set $(S)=(S_1)\cup(S_2)$, then the inertia operator of (S) is the sum of the two inertia operators of (S_1) and (S_2).

$$[I_G(S|m)]=[I_G(S_1|m)]+[I_G(S_2|m)]$$

$$\text{" "} \quad =\frac{m}{60}\begin{bmatrix} 97b^2+292c^2 & -12ab & 48ac \\ -12ab & 292c^2+97a^2 & 48bc \\ 48ac & 48bc & 97(a^2+b^2) \end{bmatrix}.$$

3.4.4. *Kinetic torsor and change of basis*

Let us consider the motion of a mechanical set $(\mathscr{D})$ relative to two systems of reference $\langle\lambda\rangle$ and $\langle\mu\rangle$ that are moving relative to one another.

We consider the kinetic torsor according to two points of view, $\{p_{\mathscr{D}}^{\lambda}\}$ and $\{p_{\mathscr{D}}^{\mu}\}$.

In $\langle\lambda\rangle$, we write the following:

– $\vec{s}\{p_{\mathscr{D}}^{\lambda}\}=\int_{M\in\mathscr{D}}\overrightarrow{v^{(\lambda)}}(M)dm(M)$,

– $\overrightarrow{\mathscr{M}_Q}\{p_{\mathscr{D}}^{\lambda}\}=\int_{M\in\mathscr{D}}\overrightarrow{QM}\wedge\overrightarrow{v^{(\lambda)}}(M)dm(M)$.

In $\langle\mu\rangle$, we write the following:

– $\vec{s}\{p_{\mathscr{D}}^{\mu}\}=\int_{M\in\mathscr{D}}\overrightarrow{v^{(\mu)}}(M)dm(M)$,

– $\overrightarrow{\mathscr{M}_Q}\{p_{\mathscr{D}}^{\mu}\}=\int_{M\in\mathscr{D}}\overrightarrow{QM}\wedge\overrightarrow{v^{(\mu)}}(M)dm(M)$.

If we calculate the difference $\{p_{\mathscr{D}}^{\lambda}\}-\{p_{\mathscr{D}}^{\mu}\}$ between the two torsors, we obtain

$$\vec{s}[\{p_{\mathscr{D}}^{\lambda}\}-\{p_{\mathscr{D}}^{\mu}\}]=\vec{s}\{p_{\mathscr{D}}^{\lambda}\}-\vec{s}\{p_{\mathscr{D}}^{\mu}\}$$

$$\text{" "} \quad =\int_{M\in\mathscr{D}}\left[\overrightarrow{v^{(\lambda)}}(M)-\overrightarrow{v^{(\mu)}}(M)\right]dm(M)$$

$$\vec{s}\{p_{\mathscr{D}}^{\lambda}\}-\vec{s}\{p_{\mathscr{D}}^{\mu}\}=\int_{M\in\mathscr{D}}\overrightarrow{\mathscr{M}_M}\{{}^{\lambda}_{\mu}\}dm(M)$$

$$\text{" "} \quad =\int_{M\in\mathscr{D}}\left[\overrightarrow{v^{(\lambda)}}(O_\mu)+\overrightarrow{\omega_\mu^\lambda}\wedge\overrightarrow{O_\mu M}\right]dm(M)$$

$$\text{" "} \quad =m(\mathscr{D})\overrightarrow{v^{(\lambda)}}(O_\mu)+\overrightarrow{\omega_\mu^\lambda}\wedge\int_{M\in\mathscr{D}}\overrightarrow{O_\mu M}\,dm(M)$$

$$\text{" "} \quad =m(\mathscr{D})\left[\overrightarrow{v^{(\lambda)}}(O_\mu)+\overrightarrow{\omega_\mu^\lambda}\wedge\overrightarrow{O_\mu G}\right]=m(\mathscr{D})\overrightarrow{\mathscr{M}_G}\{{}^{\lambda}_{\mu}\}$$

$$\Rightarrow \quad \vec{s}\{p_{\mathscr{D}}^{\lambda}\}-\vec{s}\{p_{\mathscr{D}}^{\mu}\}=m(\mathscr{D})\overrightarrow{\mathscr{M}_G}\{{}^{\lambda}_{\mu}\}.$$

Thus, we obtain the law of combination of velocities established at section 2.6.3.1:

$$\overrightarrow{v^{(\lambda)}}(G)-\overrightarrow{v^{(\mu)}}(G)=\overrightarrow{\mathscr{M}_G}\{{}^{\lambda}_{\mu}\}.$$

$$\overrightarrow{\mathscr{M}_Q}[\{p_{\mathscr{D}}^{\lambda}\}-\{p_{\mathscr{D}}^{\mu}\}]=\overrightarrow{\mathscr{M}_Q}\{p_{\mathscr{D}}^{\lambda}\}-\overrightarrow{\mathscr{M}_Q}\{p_{\mathscr{D}}^{\mu}\}$$

$$\overrightarrow{\mathscr{M}_Q}\{p_{\mathscr{D}}^{\lambda}\}-\overrightarrow{\mathscr{M}_Q}\{p_{\mathscr{D}}^{\mu}\}=\int_{M\in\mathscr{D}}\overrightarrow{QM}\wedge\left[\overrightarrow{v^{(\lambda)}}(M)-\overrightarrow{v^{(\mu)}}(M)\right]dm(M)$$

$$\text{" "} \quad =\int_{M\in\mathscr{D}}\overrightarrow{QM}\wedge\overrightarrow{M_M}\{{}^{\lambda}_{\mu}\}dm(M)$$

$$\text{" "} \quad =\int_{M\in\mathscr{D}}\overrightarrow{QM}\wedge\left[\overrightarrow{v^{(\lambda)}}(O_\mu)+\overrightarrow{\omega_\mu^\lambda}\wedge\overrightarrow{O_\mu M}\right]dm(M)$$

$$\overrightarrow{\mathscr{M}_Q}\{p_{\mathscr{D}}^{\lambda}\}-\overrightarrow{\mathscr{M}_Q}\{p_{\mathscr{D}}^{\mu}\}=\int_{M\in\mathscr{D}}\overrightarrow{QM}\,dm(M)\wedge\overrightarrow{v^{(\lambda)}}(O_\mu)\ldots$$

$$\ldots+\int_{M\in\mathscr{D}}\overrightarrow{QM}\,dm(M)\wedge\left(\overrightarrow{\omega_\mu^\lambda}\wedge\overrightarrow{O_\mu Q}\right)\ldots.$$

$$\ldots+\int_{M\in\mathscr{D}}\overrightarrow{QM}\wedge\left(\overrightarrow{\omega_\mu^\lambda}\wedge\overrightarrow{QM}\right)dm(M)$$

$$\overrightarrow{\mathcal{M}_{\mathrm{Q}}}\{p^{\lambda}_{\mathscr{D}}\}-\overrightarrow{\mathcal{M}_{\mathrm{Q}}}\{p^{\mu}_{\mathscr{D}}\}=m(\mathscr{D})\overrightarrow{\mathrm{QG}}\wedge\overrightarrow{\mathcal{M}_{\mathrm{Q}}}\{{}^{\lambda}_{\mu}\}\ldots$$
$$\ldots+\int_{\mathrm{M}\in\mathscr{D}}\overrightarrow{\mathrm{QM}}\wedge\left(\overrightarrow{\omega^{\lambda}_{\mu}}\wedge\overrightarrow{\mathrm{QM}}\right)\mathrm{d}m(\mathrm{M})\cdot$$

When the mechanical set is the rigid solid body (S), then we have

$$\overrightarrow{\mathcal{M}_{\mathrm{Q}}}\{p^{\lambda}_{S}\}-\overrightarrow{\mathcal{M}_{\mathrm{Q}}}\{p^{\mu}_{S}\}=m(S)\overrightarrow{\mathrm{QG}}\wedge\overrightarrow{\mathcal{M}_{\mathrm{Q}}}\{{}^{\lambda}_{\mu}\}+I_{\mathrm{Q}}(S|m)\overrightarrow{\omega^{\lambda}_{\mu}}$$
$$\text{'' ''}\qquad =m(S)\overrightarrow{\mathrm{QG}}\wedge\left(\overrightarrow{v^{(\lambda)}}(\mathrm{Q})-\overrightarrow{v^{(\mu)}}(\mathrm{Q})\right)+I_{\mathrm{Q}}(S|m)\overrightarrow{\omega^{\lambda}_{\mu}},$$

and, in the center of inertia G

$$\overrightarrow{\mathcal{M}_{\mathrm{G}}}\{p^{\lambda}_{S}\}-\overrightarrow{\mathcal{M}_{\mathrm{G}}}\{p^{\mu}_{S}\}=I_{\mathrm{G}}(S|m)\overrightarrow{\omega^{\lambda}_{\mu}}\,.$$

3.5. Dynamic torsor of a mechanical set ($\mathscr{D}$)

3.5.1. *Definition*

Each material point M of a continuous mechanical set ($\mathscr{D}$) having an elementary mass $\mathrm{d}m(\mathrm{M})$ is submitted, in a system of reference $\langle\lambda\rangle=\langle\mathrm{O}_{\lambda}|\overrightarrow{x_{\lambda}}\,\overrightarrow{y_{\lambda}}\,\overrightarrow{z_{\lambda}}\rangle$, to an acceleration $\overrightarrow{J^{(\lambda)}}(\mathrm{M})$. Its moment relative to an arbitrary point Q of the system of reference can then be obtained. It is then interesting to have a global approach of this field of acceleration by going through the two integrals

$$\int_{\mathrm{M}\in(\mathscr{D})}\overrightarrow{J^{(\lambda)}}(\mathrm{M})\mathrm{d}m(\mathrm{M});\int_{\mathrm{M}\in(\mathscr{D})}\overrightarrow{\mathrm{QM}}\wedge\overrightarrow{J^{(\lambda)}}(\mathrm{M})\mathrm{d}m(\mathrm{M}).$$

They lead to two vectors that, if they are the reduction elements in point Q of a torsor, need to verify the condition for the existence of a torsor

$$\overrightarrow{\mathcal{M}_{\mathrm{P}}}=\overrightarrow{\mathcal{M}_{\mathrm{Q}}}+\overrightarrow{\mathrm{PQ}}\wedge\vec{s}\,.$$

Using the same calculations applied for the field of velocity, we can clearly evidence the existence of this torsor. It is the *dynamic torsor* $\{\mathcal{A}_{\mathcal{D}}^{\lambda}\}$ of the motion of $(\mathcal{D})$ relative to the system of reference $\langle\lambda\rangle$, with the reduction elements

$$\begin{cases} \vec{s}\{A_{\mathcal{D}}^{\lambda}\} = \int_{M\in(\mathcal{D})} \overrightarrow{J^{(\lambda)}}(M)dm(M) \\ \overrightarrow{\mathcal{M}_{Q}}\{A_{\mathcal{D}}^{\lambda}\} = \int_{M\in(\mathcal{D})} \overrightarrow{QM} \wedge \overrightarrow{J^{(\lambda)}}(M)dm(M) \end{cases}.$$

3.5.2. *Dynamic torsor of the rigid solid body (S)*

A system of reference $\langle S\rangle = \langle O_S | \overrightarrow{x_S}\,\overrightarrow{y_S}\,\overrightarrow{z_S}\rangle$ is linked to the solid. Since it is a rigid solid, we have

$$\forall M \in (S) \quad \overrightarrow{J^{(\lambda)}}(M) = \overrightarrow{J^{(\lambda)}}(O_S) + \left(\overrightarrow{\omega_S^{\lambda}}\right)' \wedge \overrightarrow{O_SM} + \overrightarrow{\omega_S^{\lambda}} \wedge \left(\overrightarrow{\omega_S^{\lambda}} \wedge \overrightarrow{O_SM}\right).$$

Under these conditions:

– $\vec{s}\{\mathcal{A}_S^{\lambda}\} = \int_{M\in(S)} \overrightarrow{J^{(\lambda)}}(M)dm(M)$,

$$\vec{s}\{\mathcal{A}_S^{\lambda}\} = \int_{M\in(S)} \left[\overrightarrow{J^{(\lambda)}}(O_S) + \left(\overrightarrow{\omega_S^{\lambda}}\right)' \wedge \overrightarrow{O_SM} + \overrightarrow{\omega_S^{\lambda}} \wedge \left(\overrightarrow{\omega_S^{\lambda}} \wedge \overrightarrow{O_SM}\right)\right] dm(M)$$

$$\text{'' ''} \quad = m(S)\overrightarrow{J^{(\lambda)}}(O_S) + \left(\overrightarrow{\omega_S^{\lambda}}\right)' \wedge \int_{M\in(S)} \overrightarrow{O_SM}dm(M)\ldots$$

$$\ldots + \overrightarrow{\omega_S^{\lambda}} \wedge \left(\overrightarrow{\omega_S^{\lambda}} \wedge \int_{M\in(S)} \overrightarrow{O_SM}dm(M)\right)$$

$$\text{'' ''} \quad = m(S)\left[\overrightarrow{J^{(\lambda)}}(O_S) + \left(\overrightarrow{\omega_S^{\lambda}}\right)' \wedge \overrightarrow{O_SG} + \overrightarrow{\omega_S^{\lambda}} \wedge \left(\overrightarrow{\omega_S^{\lambda}} \wedge \overrightarrow{O_SG}\right)\right]$$

$$\Rightarrow \quad \vec{s}\{\mathcal{A}_S^{\lambda}\} = m(S)\overrightarrow{J^{(\lambda)}}(G);$$

– $\overrightarrow{\mathcal{M}_{O_S}}\{\mathcal{A}_S^{\lambda}\} = \int_{M\in(S)} \overrightarrow{O_SM} \wedge \overrightarrow{J^{(\lambda)}}(M)dm(M)$

$$\overrightarrow{\mathcal{M}_{O_S}}\{\mathcal{A}_S^\lambda\} = \int_{M\in(S)} \overrightarrow{O_SM} \wedge \overrightarrow{J^{(\lambda)}}(O_S)dm(M)\ldots$$

$$\ldots + \int_{M\in(S)} \overrightarrow{O_SM} \wedge \left[\left(\overrightarrow{\omega_S^\lambda}\right)' \wedge \overrightarrow{O_SM} + \overrightarrow{\omega_S^\lambda} \wedge \left(\overrightarrow{\omega_S^\lambda} \wedge \overrightarrow{O_SM}\right)\right]dm(M)$$

$$\text{" "} \quad = m(S)\overrightarrow{O_SG} \wedge \overrightarrow{J^{(\lambda)}}(O_S)\ldots$$

$$\ldots + \int_{M\in(S)} \overrightarrow{O_SM} \wedge \left[\left(\overrightarrow{\omega_S^\lambda}\right)' \wedge \overrightarrow{O_SM}\right]dm(M)$$

$$\ldots + \int_{M\in(S)} \overrightarrow{O_SM} \wedge \left[\overrightarrow{\omega_S^\lambda} \wedge \left(\overrightarrow{\omega_S^\lambda} \wedge \overrightarrow{O_SM}\right)\right]dm(M).$$

NOTE 3.3.– the vector expression $\vec{a} \wedge \left[\vec{b} \wedge \left(\vec{b} \wedge \vec{a}\right)\right]$ can be developed as follows:

$$\vec{a} \wedge \left(\vec{b} \wedge \left(\vec{b} \wedge \vec{a}\right)\right) = \vec{a} \wedge \left(\left(\vec{b}\cdot\vec{a}\right)\vec{b} - \vec{b}^2\vec{a}\right) = \left(\vec{b}\cdot\vec{a}\right)\vec{a} \wedge \vec{b} = \left[\left(\vec{b}\cdot\vec{a}\right)\vec{a}\right] \wedge \vec{b}$$
$$= -\left(\vec{a} \wedge \left(\vec{b} \wedge \vec{a}\right)\right) \wedge \vec{b} = \vec{b} \wedge \left(\vec{a} \wedge \left(\vec{b} \wedge \vec{a}\right)\right).$$

We then have

$$\overrightarrow{\mathcal{M}_{O_S}}\{\mathcal{A}_S^\lambda\} = m(S)\overrightarrow{O_SG} \wedge \overrightarrow{J^{(\lambda)}}(O_S) + I_{O_S}(S|m)\left(\overrightarrow{\omega_S^\lambda}\right)'$$
$$\ldots + \overrightarrow{\omega_S^\lambda} \wedge \int_{M\in(S)} \overrightarrow{O_SM} \wedge \left[\left(\overrightarrow{\omega_S^\lambda} \wedge \overrightarrow{O_SM}\right)\right]dm(M)$$

$$\Rightarrow \quad \overrightarrow{\mathcal{M}_{O_S}}\{\mathcal{A}_S^\lambda\} = m(S)\overrightarrow{O_SG} \wedge \overrightarrow{J^{(\lambda)}}(O_S)\ldots$$
$$\ldots + I_{O_S}(S|m)\overrightarrow{\omega_S^\lambda}' + \overrightarrow{\omega_S^\lambda} \wedge I_{O_S}(S|m)\overrightarrow{\omega_S^\lambda}.$$

3.5.2.1. *Relation between the kinetic torsor and the dynamic torsor*

The kinetic torsor has the following reduction elements in O_S:

$$\begin{cases} \vec{s}\{p_S^\lambda\} = m(S)\overrightarrow{v^{(\lambda)}}(G) \\ \overrightarrow{\mathcal{M}_{O_S}}\{p_S^\lambda\} = m(S)\overrightarrow{O_SG} \wedge \overrightarrow{v^{(\lambda)}}(O_S) + I_{O_S}(S|m)\overrightarrow{\omega_S^\lambda} \end{cases}.$$

3.5.2.1.1. The notion of derivative of a torsor

In the system of reference $\langle \lambda \rangle = \langle O_\lambda | \overrightarrow{x_\lambda}\ \overrightarrow{y_\lambda}\ \overrightarrow{z_\lambda} \rangle$, we consider a torsor $\{\mathscr{T}\}$ with reduction elements in O_λ: $\vec{s}\{\mathscr{T}\}$ and $\overrightarrow{\mathscr{M}_{O_\lambda}}\{\mathscr{T}\}$, functions of the variable q.

We consider the two quantities

$$\frac{d^{(\lambda)}}{dq}\vec{s}\{\mathscr{T}\} \quad \text{and} \quad \frac{d^{(\lambda)}}{dq}\overrightarrow{\mathscr{M}_{O_\lambda}}\{\mathscr{T}\}.$$

In Volume 2 [BOR 16a], in the chapter related to torsors, we show that these quantities are the reduction elements in O_λ of a torsor $\left\{\frac{d^{(\lambda)}}{dq}(\mathscr{T})\right\}$, called the *torsor derived from* $\{\mathscr{T}\}$ *in* $\langle \lambda \rangle$, and that in any arbitrary point Q :

$$\begin{cases} \vec{s}\left\{\dfrac{d^{(\lambda)}}{dq}(\mathscr{T})\right\} = \dfrac{d^{(\lambda)}}{dq}\vec{s}\{\mathscr{T}\} \\ \overrightarrow{\mathscr{M}_Q}\left\{\dfrac{d^{(\lambda)}}{dq}(\mathscr{T})\right\} = \dfrac{d^{(\lambda)}}{dq}\overrightarrow{\mathscr{M}_Q}\{\mathscr{T}\} + \left(\dfrac{d^{(\lambda)}}{dq}\overrightarrow{O_\lambda Q}\right) \wedge \vec{s}\{\mathscr{T}\} \end{cases}.$$

Let us now consider the derivatives with respect to t, in the system of reference $\langle \lambda \rangle$, of the reduction elements of kinetic torsor $\{p_S^\lambda\}$:

– $\dfrac{d^{(\lambda)}}{dt}\vec{s}\{p_S^\lambda\} = m(S)\overrightarrow{J^{(\lambda)}}(G)$

– $\dfrac{d^{(\lambda)}}{dt}\overrightarrow{\mathscr{M}_{O_S}}\{p_S^\lambda\} = m(S)\dfrac{d^{(\lambda)}}{dt}\left[\overrightarrow{O_S G} \wedge \overrightarrow{v^{(\lambda)}}(O_S)\right] + \dfrac{d^{(\lambda)}}{dt}\left[I_{O_S}(S|m)\overrightarrow{\omega_S^\lambda}\right].$

3.5.2.1.2. Derivative of vector $I_{O_S}(S|m)\overrightarrow{\omega_S^\lambda}$

$$\frac{d^{(\lambda)}}{dt}\left[I_{O_S}(S|m)\overrightarrow{\omega_S^\lambda}\right]=\int_{M\in(S)}\frac{d^{(\lambda)}}{dt}\overrightarrow{O_SM}\wedge\left(\overrightarrow{\omega_S^\lambda}\wedge\overrightarrow{O_SM}\right)dm(M)\ldots$$

$$\ldots+\int_{M\in(S)}\overrightarrow{O_SM}\wedge\left(\frac{d^{(\lambda)}}{dt}\overrightarrow{\omega_S^\lambda}\wedge\overrightarrow{O_SM}\right)dm(M)\ldots,$$

$$\ldots+\int_{M\in(S)}\overrightarrow{O_SM}\wedge\left(\overrightarrow{\omega_S^\lambda}\wedge\frac{d^{(\lambda)}}{dt}\overrightarrow{O_SM}\right)dm(M)$$

with $\frac{d^{(\lambda)}}{dt}\overrightarrow{O_SM}=\frac{d^{(S)}}{dt}\overrightarrow{O_SM}+\overrightarrow{\omega_S^\lambda}\wedge\overrightarrow{O_SM}=\overrightarrow{\omega_S^\lambda}\wedge\overrightarrow{O_SM}$

and $\frac{d^{(\lambda)}}{dt}\overrightarrow{\omega_S^\lambda}=\frac{d^{(S)}}{dt}\overrightarrow{\omega_S^\lambda}=\overrightarrow{\omega_S^\lambda}'$.

$$\frac{d^{(\lambda)}}{dt}\left[I_{O_S}(S|m)\overrightarrow{\omega_S^\lambda}\right]=\int_{M\in(S)}\left(\overrightarrow{\omega_S^\lambda}\wedge\overrightarrow{O_SM}\right)\wedge\left(\overrightarrow{\omega_S^\lambda}\wedge\overrightarrow{O_SM}\right)dm(M)\ldots$$

$$\ldots+\int_{M\in(S)}\overrightarrow{O_SM}\wedge\left(\overrightarrow{\omega_S^\lambda}'\wedge\overrightarrow{O_SM}\right)dm(M)\ldots\quad.$$

$$\ldots+\int_{M\in(S)}\overrightarrow{O_SM}\wedge\left[\overrightarrow{\omega_S^\lambda}\wedge\left(\overrightarrow{\omega_S^\lambda}\wedge\overrightarrow{O_SM}\right)\right]dm(M)$$

In the above expression, the first integral is null, the second expresses the application of the inertia operator to the derivative of the rotation vector $\overrightarrow{\omega_S^\lambda}'$. As for the third, it becomes

$$\int_{M\in(S)}\overrightarrow{O_SM}\wedge\left[\overrightarrow{\omega_S^\lambda}\wedge\left(\overrightarrow{\omega_S^\lambda}\wedge\overrightarrow{O_SM}\right)\right]dm(M)\ldots$$

$$\ldots=\overrightarrow{\omega_S^\lambda}\wedge\int_{M\in(S)}\overrightarrow{O_SM}\wedge\left(\overrightarrow{\omega_S^\lambda}\wedge\overrightarrow{O_SM}\right)dm(M),$$

$$\text{"}\quad\text{"}\qquad=\overrightarrow{\omega_S^\lambda}\wedge I_{O_S}(S|m)\overrightarrow{\omega_S^\lambda}$$

$$\Rightarrow\frac{d^{(\lambda)}}{dt}\left[I_{O_S}(S|m)\overrightarrow{\omega_S^\lambda}\right]=I_{O_S}(S|m)\overrightarrow{\omega_S^\lambda}'+\overrightarrow{\omega_S^\lambda}\wedge I_{O_S}(S|m)\overrightarrow{\omega_S^\lambda}.$$

3.5.2.1.3. Derivative of the kinetic torsor $\{p_S^\lambda\}$

We consider the torsor $\left\{\dfrac{d^{(\lambda)}}{dt}(p_S^\lambda)\right\}$, derived from $\{p_S^\lambda\}$, with reduction elements in O_S:

$$-\ \vec{s}\left\{\frac{d^{(\lambda)}}{dt}(p_S^\lambda)\right\}=\frac{d^{(\lambda)}}{dt}\vec{s}\{p_S^\lambda\}=m(S)\overrightarrow{J^{(\lambda)}}(G),$$

$$-\ \overrightarrow{\mathcal{M}_{O_S}}\left\{\frac{d^{(\lambda)}}{dt}(p_S^\lambda)\right\}=\frac{d^{(\lambda)}}{dt}\overrightarrow{\mathcal{M}_{O_S}}\{p_S^\lambda\}+\frac{d^{(\lambda)}}{dt}\left(\overrightarrow{O_\lambda O_S}\right)\wedge\vec{s}\{p_S^\lambda\},$$

$$\Rightarrow\ \overrightarrow{\mathcal{M}_{O_S}}\left\{\frac{d^{(\lambda)}}{dt}(p_S^\lambda)\right\}=\frac{d^{(\lambda)}}{dt}\overrightarrow{\mathcal{M}_{O_S}}\{p_S^\lambda\}+m(S)\overrightarrow{v^{(\lambda)}}(O_S)\wedge\overrightarrow{v^{(\lambda)}}(G),$$

the expression in which

$$\frac{d^{(\lambda)}}{dt}\overrightarrow{\mathcal{M}_{O_S}}\{p_S^\lambda\}=m(S)\frac{d^{(\lambda)}}{dt}\left(\overrightarrow{O_SG}\right)\wedge\overrightarrow{v^{(\lambda)}}(O_S)\ldots$$
$$\ldots+m(S)\overrightarrow{O_SG}\wedge\overrightarrow{J^{(\lambda)}}(O_S)+\frac{d^{(\lambda)}}{dt}\left(I_{O_S}(S|m)\overrightarrow{\omega_S^\lambda}\right)$$

with $\dfrac{d^{(\lambda)}}{dt}\overrightarrow{O_SG}=\dfrac{d^{(S)}}{dt}\overrightarrow{O_SG}+\overrightarrow{\omega_S^\lambda}\wedge\overrightarrow{O_SG}=\overrightarrow{\omega_S^\lambda}\wedge\overrightarrow{O_SG}$.

$$\frac{d^{(\lambda)}}{dt}\overrightarrow{\mathcal{M}_{O_S}}\{p_S^\lambda\}=m(S)\left(\overrightarrow{\omega_S^\lambda}\wedge\overrightarrow{O_SG}\right)\wedge\overrightarrow{v^{(\lambda)}}(O_S)+m(S)\overrightarrow{O_SG}\wedge\overrightarrow{J^{(\lambda)}}(O_S)\ldots$$
$$\ldots+I_{O_S}(S|m)\overrightarrow{\omega_S^\lambda}'+\overrightarrow{\omega_S^\lambda}\wedge I_{O_S}(S|m)\overrightarrow{\omega_S^\lambda}.$$

Consequently,

$$\overrightarrow{\mathcal{M}_{O_S}}\left\{\frac{d^{(\lambda)}}{dt}(p_S^\lambda)\right\}=+m(S)\overrightarrow{O_SG}\wedge\overrightarrow{J^{(\lambda)}}(O_S)+I_{O_S}(S|m)\overrightarrow{\omega_S^\lambda}'\ldots$$
$$\ldots+\overrightarrow{\omega_S^\lambda}\wedge I_{O_S}(S|m)\overrightarrow{\omega_S^\lambda}+m(S)\overrightarrow{v^{(\lambda)}}(O_S)\wedge\left[\overrightarrow{v^{(\lambda)}}(G)-\overrightarrow{\omega_S^\lambda}\wedge\overrightarrow{O_SG}\right],$$

where

$$m(S)\overrightarrow{v^{(\lambda)}}(O_S)\wedge\left[\overrightarrow{v^{(\lambda)}}(G)-\overrightarrow{\omega_S^\lambda}\wedge\overrightarrow{O_SG}\right]=m(S)\overrightarrow{v^{(\lambda)}}(O_S)\wedge\overrightarrow{v^{(\lambda)}}(O_S)=\vec{0}.$$

Therefore, we deduce that

$$\overrightarrow{\mathcal{M}_{O_S}}\left\{\frac{d^{(\lambda)}}{dt}(p_S^\lambda)\right\}=\overrightarrow{\mathcal{M}_{O_S}}\{\mathcal{A}_S^\lambda\}.$$

This equality is verified irrespective of the point in which the moment is calculated. Therefore, we have

$$\{\mathcal{A}_S^\lambda\}=\left\{\frac{d^{(\lambda)}}{dt}(p_S^\lambda)\right\}.$$

3.5.2.2. *Dynamic torsor and change of basis*

Let us consider the motion of a mechanical set $(\mathcal{D})$ relative to two systems of reference $\langle\lambda\rangle$ and $\langle\mu\rangle$ moving relative to one another.

Moreover, let us consider the dynamic torsor according to two points of view, that is $\{\mathcal{A}_{\mathcal{D}}^\lambda\}$ and $\{\mathcal{A}_{\mathcal{D}}^\mu\}$.

In $\langle\lambda\rangle$, we write:

- $\vec{s}\{\mathcal{A}_{\mathcal{D}}^\lambda\}=\int_{M\in\mathcal{D}}\overrightarrow{J^{(\lambda)}}(M)dm(M)$,
- $\overrightarrow{\mathcal{M}_Q}\{\mathcal{A}_{\mathcal{D}}^\lambda\}=\int_{M\in\mathcal{D}}\overrightarrow{QM}\wedge\overrightarrow{J^{(\lambda)}}(M)dm(M)$.

In $\langle\mu\rangle$, we write:

- $\vec{s}\{\mathcal{A}_{\mathcal{D}}^\mu\}=\int_{M\in\mathcal{D}}\overrightarrow{J^{(\mu)}}(M)dm(M)$,
- $\overrightarrow{\mathcal{M}_Q}\{\mathcal{A}_{\mathcal{D}}^\mu\}=\int_{M\in\mathcal{D}}\overrightarrow{QM}\wedge\overrightarrow{J^{(\mu)}}(M)dm(M)$.

If we calculate the difference $\{\mathcal{A}_{\mathcal{D}}^\lambda\}-\{\mathcal{A}_{\mathcal{D}}^\mu\}$ of the two torsors, we obtain:

$$\begin{aligned}\vec{s}[\{\mathcal{A}_{\mathcal{D}}^\lambda\}-\{\mathcal{A}_{\mathcal{D}}^\mu\}]&=\vec{s}\{\mathcal{A}_{\mathcal{D}}^\lambda\}-\vec{s}\{\mathcal{A}_{\mathcal{D}}^\mu\}\\ \text{" "}\qquad&=\int_{M\in\mathcal{D}}\left[\overrightarrow{J^{(\lambda)}}(M)-\overrightarrow{J^{(\mu)}}(M)\right]dm(M)\end{aligned}.$$

According to the formula for the combination of accelerations (see section 2.6.4.1), we can write:

$$\vec{s}\{\mathcal{A}_{\mathcal{D}}^{\lambda}\}-\vec{s}\{\mathcal{A}_{\mathcal{D}}^{\mu}\}=\int_{M\in\mathcal{D}}\left[\overrightarrow{J^{(\lambda)}}(O_{\mu})+\overrightarrow{\omega_{\mu}^{\lambda}}'\wedge\overrightarrow{O_{\mu}M}\right]dm(M)\ldots$$

$$\ldots+\int_{M\in\mathcal{D}}\left[\overrightarrow{\omega_{\mu}^{\lambda}}\wedge\left(\overrightarrow{\omega_{\mu}^{\lambda}}\wedge\overrightarrow{O_{\mu}M}\right)+2\overrightarrow{\omega_{\mu}^{\lambda}}\wedge\overrightarrow{v^{(\mu)}}(M)\right]dm(M)$$

$$\vec{s}\{\mathcal{A}_{\mathcal{D}}^{\lambda}\}-\vec{s}\{\mathcal{A}_{\mathcal{D}}^{\mu}\}=m(\mathcal{D})\left[\overrightarrow{J^{(\lambda)}}(O_{\mu})+\overrightarrow{\omega_{\mu}^{\lambda}}'\wedge\overrightarrow{O_{\mu}G}+\overrightarrow{\omega_{\mu}^{\lambda}}\wedge\left(\overrightarrow{\omega_{\mu}^{\lambda}}\wedge\overrightarrow{O_{\mu}G}\right)\right]\ldots$$

$$\ldots+2\overrightarrow{\omega_{\mu}^{\lambda}}\wedge\int_{M\in\mathcal{D}}\overrightarrow{v^{(\mu)}}(M)\,dm(M).$$

Therefore, we obtain

$$\vec{s}[\{\mathcal{A}_{\mathcal{D}}^{\lambda}\}-\{\mathcal{A}_{\mathcal{D}}^{\mu}\}]=m(\mathcal{D})\left[\overrightarrow{J^{(\lambda)}}(O_{\mu})+\overrightarrow{\omega_{\mu}^{\lambda}}'\wedge\overrightarrow{O_{\mu}G}+\overrightarrow{\omega_{\mu}^{\lambda}}\wedge\left(\overrightarrow{\omega_{\mu}^{\lambda}}\wedge\overrightarrow{O_{\mu}G}\right)\right]\ldots$$

$$\ldots+2m(\mathcal{D})\overrightarrow{\omega_{\mu}^{\lambda}}\wedge\overrightarrow{v^{(\mu)}}(G)$$

$$\overrightarrow{\mathcal{M}_{Q}}[\{\mathcal{A}_{\mathcal{D}}^{\lambda}\}-\{\mathcal{A}_{\mathcal{D}}^{\mu}\}]=\overrightarrow{\mathcal{M}_{Q}}\{\mathcal{A}_{\mathcal{D}}^{\lambda}\}-\overrightarrow{\mathcal{M}_{Q}}\{\mathcal{A}_{\mathcal{D}}^{\mu}\}$$

$$''\quad ''\qquad =\int_{M\in\mathcal{D}}\overrightarrow{QM}\wedge\left[\overrightarrow{J^{(\lambda)}}(M)-\overrightarrow{J^{(\mu)}}(M)\right]dm(M)$$

$$''\quad ''\qquad =\int_{M\in\mathcal{D}}\overrightarrow{QM}\wedge\left[\overrightarrow{J^{(\lambda)}}(O_{\mu})+\overrightarrow{\omega_{\mu}^{\lambda}}'\wedge\overrightarrow{O_{\mu}M}\right]dm(M)\ldots$$

$$\ldots+\int_{M\in\mathcal{D}}\overrightarrow{QM}\wedge\left[\overrightarrow{\omega_{\mu}^{\lambda}}\wedge\left(\overrightarrow{\omega_{\mu}^{\lambda}}\wedge\overrightarrow{O_{\mu}M}\right)+2\overrightarrow{\omega_{\mu}^{\lambda}}\wedge\overrightarrow{v^{(\mu)}}(M)\right]dm(M)$$

$$\overrightarrow{\mathcal{M}_{Q}}\{\mathcal{A}_{\mathcal{D}}^{\lambda}\}-\overrightarrow{\mathcal{M}_{Q}}\{\mathcal{A}_{\mathcal{D}}^{\mu}\}=m(\mathcal{D})\overrightarrow{QG}\wedge\overrightarrow{J^{(\lambda)}}(O_{\mu})\ldots$$

$$\ldots+\int_{M\in\mathcal{D}}\overrightarrow{QM}\wedge\left[\overrightarrow{\omega_{\mu}^{\lambda}}'\wedge\left(\overrightarrow{O_{\mu}Q}+\overrightarrow{QM}\right)\right]dm(M)\ldots$$

$$\ldots+\int_{M\in\mathcal{D}}\overrightarrow{QM}\wedge\left[\overrightarrow{\omega_{\mu}^{\lambda}}\wedge\left(\overrightarrow{\omega_{\mu}^{\lambda}}\wedge\left(\overrightarrow{O_{\mu}Q}+\overrightarrow{QM}\right)\right)\right]dm(M)\ldots$$

$$\ldots+2\int_{M\in\mathcal{D}}\overrightarrow{QM}\wedge\left[\overrightarrow{\omega_{\mu}^{\lambda}}\wedge\overrightarrow{v^{(\mu)}}(M)\right]dm(M)$$

$$\overrightarrow{\mathcal{M}_{\mathrm{Q}}}\{\mathcal{A}_{\mathcal{D}}^{\lambda}\}-\overrightarrow{\mathcal{M}_{\mathrm{Q}}}\{\mathcal{A}_{\mathcal{D}}^{\mu}\}=m(\mathcal{D})\overrightarrow{\mathrm{QG}}\wedge\overrightarrow{J^{(\lambda)}}(\mathrm{O}_{\mu})\ldots$$
$$\ldots+m(\mathcal{D})\overrightarrow{\mathrm{QG}}\wedge\left[\overrightarrow{\omega_{\mu}^{\lambda}}'\wedge\overrightarrow{\mathrm{O}_{\mu}\mathrm{Q}}\right]\ldots$$
$$\ldots+m(\mathcal{D})\overrightarrow{\mathrm{QG}}\wedge\left[\overrightarrow{\omega_{\mu}^{\lambda}}\wedge\left(\overrightarrow{\omega_{\mu}^{\lambda}}\wedge\overrightarrow{\mathrm{O}_{\mu}\mathrm{Q}}\right)\right]\ldots$$
$$\ldots+\int_{\mathrm{M}\in\mathcal{D}}\overrightarrow{\mathrm{QM}}\wedge\left[\overrightarrow{\omega_{\mu}^{\lambda}}'\wedge\overrightarrow{\mathrm{QM}}\right]dm(\mathrm{M})\ldots$$
$$\ldots+\int_{\mathrm{M}\in\mathcal{D}}\overrightarrow{\mathrm{QM}}\wedge\left[\overrightarrow{\omega_{\mu}^{\lambda}}\wedge\left(\overrightarrow{\omega_{\mu}^{\lambda}}\wedge\overrightarrow{\mathrm{QM}}\right)\right]dm(\mathrm{M})\ldots$$
$$\ldots+2\int_{\mathrm{M}\in\mathcal{D}}\overrightarrow{\mathrm{QM}}\wedge\left[\overrightarrow{\omega_{\mu}^{\lambda}}\wedge\overrightarrow{v^{(\mu)}}(\mathrm{M})\right]dm(\mathrm{M}).$$

This leads to:

$$\overrightarrow{\mathcal{M}_{\mathrm{Q}}}[\{\mathcal{A}_{\mathcal{D}}^{\lambda}\}-\{\mathcal{A}_{\mathcal{D}}^{\mu}\}]=m(\mathcal{D})\overrightarrow{\mathrm{QG}}\wedge\overrightarrow{J^{(\lambda)}}(\mathrm{O}_{\mu})\ldots$$
$$\ldots+m(\mathcal{D})\overrightarrow{\mathrm{QG}}\wedge\left[\overrightarrow{\omega_{\mu}^{\lambda}}'\wedge\overrightarrow{\mathrm{O}_{\mu}\mathrm{Q}}+\overrightarrow{\omega_{\mu}^{\lambda}}\wedge\left(\overrightarrow{\omega_{\mu}^{\lambda}}\wedge\overrightarrow{\mathrm{O}_{\mu}\mathrm{Q}}\right)\right]\ldots$$
$$\ldots+I_{\mathrm{Q}}(\mathcal{D}|m)\overrightarrow{\omega_{\mu}^{\lambda}}'+\overrightarrow{\omega_{\mu}^{\lambda}}\wedge I_{\mathrm{Q}}(\mathcal{D}|m)\overrightarrow{\omega_{\mu}^{\lambda}}$$
$$\ldots+2\int_{\mathrm{M}\in\mathcal{D}}\overrightarrow{\mathrm{QM}}\wedge\left[\overrightarrow{\omega_{\mu}^{\lambda}}\wedge\overrightarrow{v^{(\mu)}}(\mathrm{M})\right]dm(\mathrm{M}).$$

It is obvious that the change of basis is not a simple operation at the level of the dynamic torsor and, furthermore, it depends on the possibility to express the motion relative to the system of reference $\langle\mu\rangle$, which means $\overrightarrow{v^{(\mu)}}(\mathrm{M})$, and not only the motion of $\langle\mu\rangle$ relative to $\langle\lambda\rangle$.

3.6. Kinetic energy of a mechanical set ($\mathcal{D}$)

The kinetic energy of a moving body is a key quantity, as it confers to it a new dimension. Up until the present, this motion was perceived only from the standpoint of velocity and its variations. We now approach motion from the perspective of energy and of the exchange of energy between the body and the environment in which it evolves. This is how, at the end of this last stage, the knowledge that we have gained of the kinetics of motion will be completed, and we shall be prepared to approach the study of motion, as we have completed the preliminary stages of its formulation.

3.6.1. *Definition*

The name of kinetic energy of a fragment of matter M, having the elementary mass $dm(\mathrm{M})$ and a velocity $\overrightarrow{v^{(\lambda)}}(\mathrm{M})$ in a system of reference $\langle\lambda\rangle$, is assigned to the quantity

$$\frac{1}{2}\left[\overrightarrow{v^{(\lambda)}}(\mathrm{M})\right]^2 dm(\mathrm{M}).$$

The quantity $T^{(\lambda)}(\mathcal{D})$, called the *kinetic energy* of the mechanical set $(\mathcal{D})$ moving relative to the system of reference $\langle\lambda\rangle$ is defined as the sum of kinetic energies of all the fragments of matter of $(\mathcal{D})$ by the integral relation

$$T^{(\lambda)}(\mathcal{D}) = \frac{1}{2}\int_{\mathrm{M}\in(\mathcal{D})}\left[\overrightarrow{v^{(\lambda)}}(\mathrm{M})\right]^2 \mathrm{d}m(\mathrm{M}).$$

3.6.2. Kinetic energy of the free rigid solid body (*S*)

Let us consider a rigid solid (S), to which a system of reference $\langle S\rangle = \left\langle \mathrm{O}_S \,|\, \overrightarrow{x_S}\,\overrightarrow{y_S}\,\overrightarrow{z_S}\right\rangle$ is linked, moving freely in the system of reference $\langle\lambda\rangle = \left\langle \mathrm{O}_\lambda \,|\, \overrightarrow{x_\lambda}\,\overrightarrow{y_\lambda}\,\overrightarrow{z_\lambda}\right\rangle$. The velocity of any material point M of (S) can be written

$$\overrightarrow{v^{(\lambda)}}(\mathrm{M}) = \overrightarrow{v^{(\lambda)}}(\mathrm{O}_S) + \overrightarrow{\omega_S^{\lambda}} \wedge \overrightarrow{\mathrm{O}_S\mathrm{M}}.$$

3.6.2.1. *Expression of the kinetic energy: Vis Viva*

$$
\begin{aligned}
2T^{(\lambda)}(S) &= \int_{M\in(S)} \left[\overrightarrow{v^{(\lambda)}}(M)\right]^2 dm(M) \\
\text{'' ''} \quad &= \int_{M\in(S)} \left[\overrightarrow{v^{(\lambda)}}(O_S) + \overrightarrow{\omega_S^\lambda} \wedge \overrightarrow{O_S M}\right]^2 dm(M) \\
\text{'' ''} \quad &= m(S)\left[\overrightarrow{v^{(\lambda)}}(O_S)\right]^2 + 2m(S)\left[\overrightarrow{v^{(\lambda)}}(O_S), \overrightarrow{\omega_S^\lambda}, \overrightarrow{O_S G}\right]\ldots \\
&\quad \ldots + \int_{M\in(S)} \left(\overrightarrow{\omega_S^\lambda} \wedge \overrightarrow{O_S M}\right)^2 dm(M).
\end{aligned}
$$

The quantity

$$
\left[\overrightarrow{v^{(\lambda)}}(O_S), \overrightarrow{\omega_S^\lambda}, \overrightarrow{O_S G}\right] = \overrightarrow{v^{(\lambda)}}(O_S) \cdot \left(\overrightarrow{\omega_S^\lambda} \wedge \overrightarrow{O_S G}\right),
$$

is what we call a *mixed product*, which we develop in the chapter related to vectors in Volume 2 [BOR 16a], and which can take various forms, depending on the need, such as

$$
\left[\overrightarrow{v^{(\lambda)}}(O_S), \overrightarrow{\omega_S^\lambda}, \overrightarrow{O_S G}\right] = \left[\overrightarrow{\omega_S^\lambda}, \overrightarrow{O_S G}, \overrightarrow{v^{(\lambda)}}(O_S)\right] = \overrightarrow{\omega_S^\lambda} \cdot \left[\overrightarrow{O_S G} \wedge \overrightarrow{v^{(\lambda)}}(O_S)\right];
$$

The expression $\left(\overrightarrow{\omega_S^\lambda} \wedge \overrightarrow{O_S M}\right)^2$ can be considered a mixed product, in the form

$$
\left(\overrightarrow{\omega_S^\lambda} \wedge \overrightarrow{O_S M}\right)^2 = \left(\overrightarrow{\omega_S^\lambda} \wedge \overrightarrow{O_S M}\right) \cdot \left(\overrightarrow{\omega_S^\lambda} \wedge \overrightarrow{O_S M}\right) = \overrightarrow{\omega_S^\lambda} \cdot \left[\overrightarrow{O_S M} \wedge \left(\overrightarrow{\omega_S^\lambda} \wedge \overrightarrow{O_S M}\right)\right],
$$

and, consequently

$$
\begin{aligned}
\int_{M\in(S)} \left(\overrightarrow{\omega_S^\lambda} \wedge \overrightarrow{O_S M}\right)^2 dm(M) &= \overrightarrow{\omega_S^\lambda} \cdot \int_{M\in(S)} \left[\overrightarrow{O_S M} \wedge \left(\overrightarrow{\omega_S^\lambda} \wedge \overrightarrow{O_S M}\right)\right] dm(M) \\
\text{'' ''} \quad &= \overrightarrow{\omega_S^\lambda} \cdot I_{O_S}(S|m)\overrightarrow{\omega_S^\lambda}.
\end{aligned}
$$

From this we obtain the expression of the quantity $2T^{(\lambda)}(S)$ called *vis viva* of (S) in its motion relative to $\langle\lambda\rangle$

$$2T^{(\lambda)}(S) = m(S)\left[\overrightarrow{v^{(\lambda)}}(\mathrm{O}_S)\right]^2 \ldots$$

$$\ldots + 2m(S)\left[\overrightarrow{\omega_S^\lambda}, \overrightarrow{\mathrm{O}_S\mathrm{G}}, \overrightarrow{v^{(\lambda)}}(\mathrm{O}_S)\right] + \overrightarrow{\omega_S^\lambda} \cdot I_{\mathrm{O}_S}(S|m)\overrightarrow{\omega_S^\lambda}$$

In case O_S, is a fixed point of $\langle \lambda \rangle \Rightarrow \overrightarrow{v^{(\lambda)}}(\mathrm{O}_S) = \vec{0}$:

$$2T^{(\lambda)}(S) = \overrightarrow{\omega_S^\lambda} \cdot I_{\mathrm{O}_S}(S|m)\overrightarrow{\omega_S^\lambda} \, .$$

If $\mathrm{O}_S \equiv \mathrm{G} \Rightarrow \overrightarrow{\mathrm{O}_S\mathrm{G}} = \vec{0}$:

$$2T^{(\lambda)}(S) = m(S)\left[\overrightarrow{v^{(\lambda)}}(\mathrm{G})\right]^2 + \overrightarrow{\omega_S^\lambda} \cdot I_{\mathrm{G}}(S|m)\overrightarrow{\omega_S^\lambda} \, .$$

3.6.2.2. *Torsor expression of the kinetic energy.*

$$2T^{(\lambda)}(S) = \int_{\mathrm{M}\in(S)} \left[\overrightarrow{v^{(\lambda)}}(\mathrm{M})\right]^2 \mathrm{d}m(\mathrm{M})$$

$$'' \; '' \quad = \int_{\mathrm{M}\in(S)} \left[\overrightarrow{v^{(\lambda)}}(\mathrm{O}_S) + \overrightarrow{\omega_S^\lambda} \wedge \overrightarrow{\mathrm{O}_S\mathrm{M}}\right] \cdot \overrightarrow{v^{(\lambda)}}(\mathrm{M})\mathrm{d}m(\mathrm{M})$$

$$'' \; '' \quad = \overrightarrow{v^{(\lambda)}}(\mathrm{O}_S) \cdot \int_{\mathrm{M}\in(S)} \overrightarrow{v^{(\lambda)}}(\mathrm{M})\mathrm{d}m(\mathrm{M}) \ldots$$

$$\ldots + \overrightarrow{\omega_S^\lambda} \cdot \int_{\mathrm{M}\in(S)} \overrightarrow{\mathrm{O}_S\mathrm{M}} \wedge \overrightarrow{v^{(\lambda)}}(\mathrm{M})\mathrm{d}m(\mathrm{M})$$

$$\Rightarrow \; 2T^{(\lambda)}(S) = \overrightarrow{\mathcal{M}_{\mathrm{O}_S}}\{\,{}^{\lambda}_{S}\,\} \cdot \vec{s}\{p_S^\lambda\} + \vec{s}\{\,{}^{\lambda}_{S}\,\} \cdot \overrightarrow{\mathcal{M}_{\mathrm{O}_S}}\{p_S^\lambda\} \, .$$

According to the calculation rules for torsors[1]:

$$2T^{(\lambda)}(S) = \{\,{}^{\lambda}_{S}\,\} \otimes \{p_S^\lambda\} \, .$$

3.6.2.3. *Decomposition of the expression of kinetic energy*

3.6.2.3.1. Partial torsors

If the motion relative to the system of reference $\langle \lambda \rangle$, of origin O_λ, of the solid (S) to which the system of reference $\langle S \rangle$ is linked, with

---

1 See the corresponding chapter of Volume 2 [BOR 16a].

origin O_S, is governed by six parameters q_α, the corresponding torsor distributor of velocities can be decomposed, according to Einstein notation, in the form:

$$\left\{ {}^{\lambda}_{S} \right\} = \left\{ {}^{\lambda}_{S,\alpha} \right\} q'_\alpha,$$

where $\left\{ {}^{\lambda}_{S,\alpha} \right\}$ is the *partial torsor distributor* relative to the variable q_α. The reduction elements of this torsor in a point Q are the respective factors of q'_α in the reduction elements in this same point of the torsor distributor of velocities $\left\{ {}^{\lambda}_{S} \right\}$.

NOTE 3.4.– the notation employed above in the expression of the torsor distributor, in which, in the right-hand side of the equation, the index α appears twice, indicates a summation of the expression over the six values of α, which means

$$\left\{ {}^{\lambda}_{S,\alpha} \right\} q'_\alpha \approx \sum_{\alpha=1}^{6} \left\{ {}^{\lambda}_{S,\alpha} \right\} q'_\alpha .$$

When applicable, this simplified notation (Einstein notation) will be commonly used in the following.

The kinetic torsor can thus be expressed in a condensed form as a function of these six parameters. In effect

$$\vec{s}\left\{ p^{\lambda}_{S} \right\} = m(S)\overrightarrow{v^{(\lambda)}}(G) = m(S)\left[\overrightarrow{v^{(\lambda)}}(O_S) + \overrightarrow{\omega^{\lambda}_{S}} \wedge \overrightarrow{O_S G} \right],$$

with

$$- \overrightarrow{O_\lambda O_S} = \vec{f}(q_\alpha) \quad \Rightarrow \quad \overrightarrow{v^{(\lambda)}}(O_S) = \frac{\partial^{(\lambda)} \vec{f}}{\partial q_\alpha} q'_\alpha = \overrightarrow{{}^{\alpha}d_{O_S}} q'_\alpha$$

$$- \overrightarrow{\omega^{\lambda}_{S}} = {}^{\alpha}\vec{\delta}(\lambda, S) q'_\alpha$$

$$\Rightarrow \quad \vec{s}\left\{ p^{\lambda}_{S} \right\} = m(S)\left[\overrightarrow{{}^{\alpha}d_{O_S}} + {}^{\alpha}\vec{\delta}(\lambda, S) \wedge \overrightarrow{O_S G} \right] q'_\alpha = \vec{s}\left\{ p^{\lambda}_{S,\beta} \right\} q'_\beta .$$

And, similarly

$$\overrightarrow{\mathcal{M}_{O_S}}\{p_S^{\lambda}\} = m(S)\overrightarrow{O_SG}\wedge\overrightarrow{v^{(\lambda)}}(O_S) + I_{O_S}(S|m)\overrightarrow{\omega_S^{\lambda}}$$

or

$$\overrightarrow{\mathcal{M}_{O_S}}\{p_S^{\lambda}\} = m(S)\overrightarrow{O_SG}\wedge\left[\overrightarrow{{}^{\alpha}d_{O_S}}q'_{\alpha}\right] + I_{O_S}(S|m)\left[\overrightarrow{{}^{\alpha}\delta}(\lambda,S)q'_{\alpha}\right],$$

with

$$I_{O_S}(S|m)\left[\overrightarrow{{}^{\alpha}\delta}(\lambda,S)q'_{\alpha}\right] = \int_{M\in(S)}\overline{O_SM}\wedge\left(\overrightarrow{{}^{\alpha}\delta}(\lambda,S)q'_{\alpha}\wedge\overline{O_SM}\right)dm(M)$$

$$\text{" "} \qquad = \left[\int_{M\in(S)}\overline{O_SM}\wedge\left(\overrightarrow{{}^{\alpha}\delta}(\lambda,S)\wedge\overline{O_SM}\right)\right]q'_{\alpha}\,dm(M)$$

$$\Rightarrow \quad \overrightarrow{\mathcal{M}_{O_S}}\{p_S^{\lambda}\} = \left[m(S)\overrightarrow{O_SG}\wedge\overrightarrow{{}^{\alpha}d_{O_S}} + I_{O_S}(S|m)\overrightarrow{{}^{\alpha}\delta}(\lambda,S)\right]q'_{\alpha}$$

$$\text{" "} \qquad = \overrightarrow{\mathcal{M}_{O_S}}\{p_{S,\beta}^{\lambda}\}q'_{\beta}.$$

Given the form of the reduction elements of the kinetic torsor, this can also be written in a condensed form:

$$\{p_S^{\lambda}\} = \{p_{S,\beta}^{\lambda}\}q'_{\beta}.$$

3.6.2.3.2. Kinetic energy

The expression of the kinetic energy of the motion in $\langle\lambda\rangle$ of the free rigid solid (S) can thus be written in the generic condensed form

$$2T^{(\lambda)}(S) = \{{}^{\lambda}_{S}\}\otimes\{p_S^{\lambda}\} = [\{{}^{\lambda}_{S,\alpha}\}\otimes\{p_{S,\beta}^{\lambda}\}]q'_{\alpha}q'_{\beta} = A_{\alpha\beta}q'_{\alpha}q'_{\beta},$$

with

– $A_{\alpha\beta} = \{{}^{\lambda}_{S,\alpha}\}\otimes\{p_{S,\beta}^{\lambda}\}$

– $\{{}^{\lambda}_{S,\alpha}\}_{O_S} = \left[\overrightarrow{{}^{\alpha}\delta}(\lambda,S)\middle|\overrightarrow{{}^{\alpha}d_{O_S}}\right]$

– $\{p_{S,\beta}^{\lambda}\}_{O_S} = \left[m(S)\left(\overrightarrow{{}^{\beta}d_{O_S}} + \overrightarrow{{}^{\beta}\delta}(\lambda,S)\wedge\overline{O_SG}\right)\middle|\ldots\right.$

$$\left.\ldots m(S)\overline{O_SG}\wedge\overrightarrow{{}^{\beta}d_{O_S}} + I_{O_S}(S|m)\overrightarrow{{}^{\beta}\delta}(\lambda,S)\right].$$

NOTE 3.5.– the product of two torsors written both at the same point of application Q can be written as

$$\{\mathscr{T}_1\}\otimes\{\mathscr{T}_2\}=\vec{s}\{\mathscr{T}_1\}\cdot\overrightarrow{\mathscr{M}_{\mathrm{Q}}}\{\mathscr{T}_2\}+\vec{s}\{\mathscr{T}_2\}\cdot\overrightarrow{\mathscr{M}_{\mathrm{Q}}}\{\mathscr{T}_1\}.$$

Under these conditions,

$$A_{\alpha\beta}=m(S)\left(\overrightarrow{{}^{\alpha}\delta},\overrightarrow{\mathrm{O}_S\mathrm{G}},\overrightarrow{{}^{\beta}d_{\mathrm{O}_S}}\right)+\overrightarrow{{}^{\alpha}\delta}\cdot I_{\mathrm{O}_S}(S|m)\overrightarrow{{}^{\beta}\delta}\ldots$$
$$\ldots+m(S)\left(\overrightarrow{{}^{\alpha}d_{\mathrm{O}_S}},\overrightarrow{{}^{\beta}\delta},\overrightarrow{\mathrm{O}_S\mathrm{G}}\right)+m(S)\overrightarrow{{}^{\alpha}d_{\mathrm{O}_S}}\cdot\overrightarrow{{}^{\beta}d_{\mathrm{O}_S}}$$

$$\Rightarrow\ A_{\alpha\beta}=m(S)\left(\overrightarrow{\mathrm{O}_S\mathrm{G}},\overrightarrow{{}^{\alpha}d_{\mathrm{O}_S}},\overrightarrow{{}^{\beta}\delta}\right)+m(S)\left(\overrightarrow{\mathrm{O}_S\mathrm{G}},\overrightarrow{{}^{\beta}d_{\mathrm{O}_S}},\overrightarrow{{}^{\alpha}\delta}\right)\ldots$$
$$\ldots+m(S)\overrightarrow{{}^{\alpha}d_{\mathrm{O}_S}}\cdot\overrightarrow{{}^{\beta}d_{\mathrm{O}_S}}+\overrightarrow{{}^{\alpha}\delta}\cdot I_{\mathrm{O}_S}(S|m)\overrightarrow{{}^{\beta}\delta}.$$

The first three terms of the above expression present a symmetry between α and β. We shall see if the fourth term also verifies this property.

$$\overrightarrow{{}^{\alpha}\delta}\cdot I_{\mathrm{O}_S}(S|m)\overrightarrow{{}^{\beta}\delta}=\overrightarrow{{}^{\alpha}\delta}\cdot\int_{M\in(S)}\overrightarrow{\mathrm{O}_S\mathrm{M}}\wedge\left(\overrightarrow{{}^{\beta}\delta}\wedge\overrightarrow{\mathrm{O}_S\mathrm{M}}\right)\mathrm{d}m(\mathrm{M})$$
$$\text{'' ''}\qquad=\int_{M\in(S)}\left(\overrightarrow{{}^{\alpha}\delta}\wedge\overrightarrow{\mathrm{O}_S\mathrm{M}}\right)\cdot\left(\overrightarrow{{}^{\beta}\delta}\wedge\overrightarrow{\mathrm{O}_S\mathrm{M}}\right)\mathrm{d}m(\mathrm{M})$$
$$\text{'' ''}\qquad=\overrightarrow{{}^{\beta}\delta}\cdot\int_{M\in(S)}\overrightarrow{\mathrm{O}_S\mathrm{M}}\wedge\left(\overrightarrow{{}^{\alpha}\delta}\wedge\overrightarrow{\mathrm{O}_S\mathrm{M}}\right)\mathrm{d}m(\mathrm{M})$$
$$\text{'' ''}\qquad=\overrightarrow{{}^{\beta}\delta}\cdot I_{\mathrm{O}_S}(S|m)\overrightarrow{{}^{\alpha}\delta}.$$

From this, we deduce that

$$A_{\alpha\beta}-A_{\beta\alpha},$$

with $A_{\alpha\beta}=\{{}^{\lambda}_{S,\alpha}\}\otimes\{p^{\lambda}_{S,\beta}\}$, $A_{\beta\alpha}=\{{}^{\lambda}_{S,\beta}\}\otimes\{p^{\lambda}_{S,\alpha}\}$,

and consequently $\{{}^{\lambda}_{S,\alpha}\}\otimes\{p^{\lambda}_{S,\beta}\}=\{{}^{\lambda}_{S,\beta}\}\otimes\{p^{\lambda}_{S,\alpha}\}$.

Given this result, by developing the generic form of vis viva, we obtain

$$2T^{(\lambda)}(S) = A_{11}q_1'^2 + A_{22}q_2'^2 + \ldots + A_{66}q_6'^2 \ldots$$
$$\ldots + 2A_{12}q_1'q_2' + 2A_{13}q_1'q_3' + \ldots + 2A_{56}q_5'q_6'.$$

Such an expression is well suited for matrix formulation:

$$2T^{(\lambda)}(S) = [q_1' \quad q_2' \quad q_3' \quad q_4' \quad q_5' \quad q_6'] \begin{bmatrix} A_{11} & A_{12} & A_{13} & A_{14} & A_{15} & A_{16} \\ A_{12} & A_{22} & A_{23} & A_{24} & A_{25} & A_{26} \\ A_{13} & A_{23} & A_{33} & A_{34} & A_{35} & A_{36} \\ A_{14} & A_{24} & A_{34} & A_{44} & A_{45} & A_{46} \\ A_{15} & A_{25} & A_{35} & A_{45} & A_{55} & A_{56} \\ A_{16} & A_{26} & A_{36} & A_{46} & A_{56} & A_{66} \end{bmatrix} \begin{bmatrix} q_1' \\ q_2' \\ q_3' \\ q_4' \\ q_5' \\ q_6' \end{bmatrix}$$

or, still:

$$2T^{(\lambda)}(S) = \overline{[q']}[M][q'].$$

The matrix $[M]$ (6,6) is called the *mass* matrix and the column $[q']$ (6,1) is called the *velocity column* or *velocity vector*.

3.6.3. *Derivatives of the kinetic energy – Lagrangian*

In addition to the tools that will be needed for expressing the dynamics of solids, we shall now present the derivative with respect to time of the kinetic energy. Since this energy term is a scalar, its derivative is mathematically speaking a simple derivative, which does not depend on the system of reference. Let us, therefore, examine the expression

$$\frac{dT^{(\lambda)}(S)}{dt}.$$

3.6.3.1. *Derivative of a scalar product*

Without getting here into a detailed presentation of the proof made in Volume 2 [BOR 16a], in the chapter corresponding to vectors and vector calculus, we shall summarize the rules.

A scalar product is effectively a scalar, whose derivative does not depend on the system of reference, but is the product of two vectors whose derivatives depend on the choice of a basis.

Let us consider the scalar product $f = \vec{a} \cdot \vec{b}$, in which the components of the two vectors $\vec{a}$ and $\vec{b}$, both functions of the variable q, are known in two distinct bases (e) and (E). By differentiating f with respect to q, the following two configurations can be obtained, in which the two vectors are differentiated in the same basis:

$$\begin{cases} \dfrac{df}{dq} = \dfrac{d^{(e)}}{dq}\vec{a} \cdot \vec{b} + \vec{a} \cdot \dfrac{d^{(e)}}{dq}\vec{b} \\ \dfrac{df}{dq} = \dfrac{d^{(E)}}{dq}\vec{a} \cdot \vec{b} + \vec{a} \cdot \dfrac{d^{(E)}}{dq}\vec{b} \end{cases} \quad \text{with} \quad \begin{cases} \dfrac{d^{(E)}}{dq}\vec{a} = \dfrac{d^{(e)}}{dq}\vec{a} + \overrightarrow{{}^{q}\delta}(E,e) \wedge \vec{a} \\ \dfrac{d^{(E)}}{dq}\vec{b} = \dfrac{d^{(e)}}{dq}\vec{b} + \overrightarrow{{}^{q}\delta}(E,e) \wedge \vec{b}. \end{cases}$$

This type of configuration will be used to define the derivative of a scalar product.

We note that

$$\frac{df}{dq} = \frac{d^{(E)}}{dq}\vec{a} \cdot \vec{b} + \vec{a} \cdot \frac{d^{(E)}}{dq}\vec{b} = \left(\frac{d^{(e)}}{dq}\vec{a} + \overrightarrow{{}^{q}\delta}(E,e) \wedge \vec{a} \right) \cdot \vec{b} + \vec{a} \cdot \left(\frac{d^{(e)}}{dq}\vec{b} + \overrightarrow{{}^{q}\delta}(E,e) \wedge \vec{b} \right)$$

$$'' \; '' = \frac{d^{(e)}}{dq}\vec{a} \cdot \vec{b} + \vec{a} \cdot \frac{d^{(e)}}{dq}\vec{b} + \left[\left(\overrightarrow{{}^{q}\delta}(E,e), \vec{a}, \vec{b} \right) + \left(\vec{a}, \overrightarrow{{}^{q}\delta}(E,e), \vec{b} \right) \right]$$

$$'' \; '' = \frac{d^{(e)}}{dq}\vec{a} \cdot \vec{b} + \vec{a} \cdot \frac{d^{(e)}}{dq}\vec{b}$$

which means that the derivative of a scalar product, as it is defined, does not depend on the basis in which we differentiate the two vectors that compose it, provided that they are both in the same basis.

3.6.3.2. *Derivative of a product of torsors*

Let us consider the two torsors $\{\mathcal{T}_1\}$ and $\{\mathcal{T}_2\}$, their product $\mathrm{T}=\{\mathcal{T}_1\}\otimes\{\mathcal{T}_2\}$ which is a scalar, and the derivative of this product with respect to a variable q:

$$\mathrm{T}=\{\mathcal{T}_1\}\otimes\{\mathcal{T}_2\}=\vec{s}\{\mathcal{T}_1\}\cdot\overrightarrow{\mathcal{M}_{\mathrm{Q}}}\{\mathcal{T}_2\}+\vec{s}\{\mathcal{T}_2\}\cdot\overrightarrow{\mathcal{M}_{\mathrm{Q}}}\{\mathcal{T}_1\},$$

in any point Q of the system of reference $\langle\lambda\rangle$. We shall consider the derivative in $\langle\lambda\rangle$ of the two torsors of this product:

$$\frac{d\mathrm{T}}{dq}=\frac{d^{(\lambda)}}{dq}\vec{s}\{\mathcal{T}_1\}\cdot\overrightarrow{\mathcal{M}_{\mathrm{Q}}}\{\mathcal{T}_2\}+\vec{s}\{\mathcal{T}_1\}\cdot\frac{d^{(\lambda)}}{dq}\overrightarrow{\mathcal{M}_{\mathrm{Q}}}\{\mathcal{T}_2\}\ldots$$
$$\ldots+\frac{d^{(\lambda)}}{dq}\vec{s}\{\mathcal{T}_2\}\cdot\overrightarrow{\mathcal{M}_{\mathrm{Q}}}\{\mathcal{T}_1\}+\vec{s}\{\mathcal{T}_2\}\cdot\frac{d^{(\lambda)}}{dq}\overrightarrow{\mathcal{M}_{\mathrm{Q}}}\{\mathcal{T}_1\}.$$

According to the definition of the derivative of a torsor[2]

$$\begin{cases}\dfrac{d^{(\lambda)}}{dq}\vec{s}\{\mathcal{T}\}=\vec{s}\left\{\dfrac{d^{(\lambda)}}{dq}(\mathcal{T})\right\}\\[2ex] \overrightarrow{\mathcal{M}_{\mathrm{Q}}}\left\{\dfrac{d^{(\lambda)}}{dq}(\mathcal{T})\right\}=\dfrac{d^{(\lambda)}}{dq}\overrightarrow{\mathcal{M}_{\mathrm{Q}}}\{\mathcal{T}\}+\dfrac{d^{(\lambda)}}{dq}\overrightarrow{\mathrm{O}_\lambda\mathrm{Q}}\wedge\vec{s}\{\mathcal{T}\}.\end{cases}$$

we obtain

$$\frac{d\mathrm{T}}{dq}=\vec{s}\left\{\frac{d^{(\lambda)}}{dq}(\mathcal{T}_1)\right\}\cdot\overrightarrow{\mathcal{M}_{\mathrm{Q}}}\{\mathcal{T}_2\}+\vec{s}\left\{\frac{d^{(\lambda)}}{dq}(\mathcal{T}_2)\right\}\cdot\overrightarrow{\mathcal{M}_{\mathrm{Q}}}\{\mathcal{T}_1\}\ldots$$
$$\ldots+\vec{s}\{\mathcal{T}_1\}\cdot\left[\overrightarrow{\mathcal{M}_{\mathrm{Q}}}\left\{\frac{d^{(\lambda)}}{dq}(\mathcal{T}_2)\right\}-\frac{d^{(\lambda)}}{dq}\overrightarrow{\mathrm{O}_\lambda\mathrm{Q}}\wedge\vec{s}\{\mathcal{T}_2\}\right]\ldots$$
$$\ldots+\vec{s}\{\mathcal{T}_2\}\cdot\left[\overrightarrow{\mathcal{M}_{\mathrm{Q}}}\left\{\frac{d^{(\lambda)}}{dq}(\mathcal{T}_1)\right\}-\frac{d^{(\lambda)}}{dq}\overrightarrow{\mathrm{O}_\lambda\mathrm{Q}}\wedge\vec{s}\{\mathcal{T}_1\}\right]$$

2 See the corresponding chapter of Volume 2 [BOR 16a].

or

$$\frac{d\mathrm{T}}{dq}=\vec{s}\left\{\frac{d^{(\lambda)}}{dq}(\mathscr{T}_1)\right\}\cdot\overrightarrow{\mathscr{M}_{\mathrm{Q}}}\{\mathscr{T}_2\}+\vec{s}\{\mathscr{T}_1\}\cdot\overrightarrow{\mathscr{M}_{\mathrm{Q}}}\left\{\frac{d^{(\lambda)}}{dq}(\mathscr{T}_2)\right\}\ldots$$

$$\ldots+\vec{s}\left\{\frac{d^{(\lambda)}}{dq}(\mathscr{T}_2)\right\}\cdot\overrightarrow{\mathscr{M}_{\mathrm{Q}}}\{\mathscr{T}_1\}+\vec{s}\{\mathscr{T}_2\}\cdot\overrightarrow{\mathscr{M}_{\mathrm{Q}}}\left\{\frac{d^{(\lambda)}}{dq}(\mathscr{T}_1)\right\}\ldots$$

$$\ldots-\left(\vec{s}\{\mathscr{T}_2\},\frac{d^{(\lambda)}}{dq}\overrightarrow{\mathrm{O}_\lambda\mathrm{Q}},\vec{s}\{\mathscr{T}_1\}\right)-\left(\vec{s}\{\mathscr{T}_1\},\frac{d^{(\lambda)}}{dq}\overrightarrow{\mathrm{O}_\lambda\mathrm{Q}},\vec{s}\{\mathscr{T}_2\}\right).$$

The two mixed products of the above expression being opposite, we finally obtain the formulation of the derivative of a product of torsors, which have each been differentiated in the system of reference $\langle\lambda\rangle$. We shall consider this expression as a basis of the definition of the derivative of a product of torsors,

$$\frac{d}{dq}[\{\mathscr{T}_1\}\otimes\{\mathscr{T}_2\}]=\left\{\frac{d^{(\lambda)}}{dq}(\mathscr{T}_1)\right\}\otimes\{\mathscr{T}_2\}+\{\mathscr{T}_1\}\otimes\left\{\frac{d^{(\lambda)}}{dq}(\mathscr{T}_2)\right\}.$$

3.6.3.3. *Derivative of $T^{(\lambda)}(S)$ with respect to time*

For a rigid solid whose mass is conserved

$$\frac{dT}{dt}=\frac{dT^{(\lambda)}(S)}{dt}=\frac{1}{2}\frac{d}{dt}\int_{\mathrm{M}\in(S)}\left[\overrightarrow{v^{(\lambda)}}(\mathrm{M})\right]^2\mathrm{d}m(\mathrm{M})$$

$$''\ ''\quad=\int_{\mathrm{M}\in(S)}\overrightarrow{v^{(\lambda)}}(\mathrm{M})\cdot\frac{d^{(\lambda)}}{dt}\overrightarrow{v^{(\lambda)}}(\mathrm{M})\mathrm{d}m(\mathrm{M})$$

$$''\ ''\quad=\int_{\mathrm{M}\in(S)}\overrightarrow{v^{(\lambda)}}(\mathrm{M})\cdot\overrightarrow{J^{(\lambda)}}(\mathrm{M})\mathrm{d}m(\mathrm{M})$$

$$''\ ''\quad=\int_{\mathrm{M}\in(S)}\left[\overrightarrow{v^{(\lambda)}}(\mathrm{O}_S)+\overrightarrow{\omega_S^{\lambda}}\wedge\overrightarrow{\mathrm{O}_S\mathrm{M}}\right]\cdot\overrightarrow{J^{(\lambda)}}(\mathrm{M})\mathrm{d}m(\mathrm{M})$$

$$\frac{dT}{dt}=\overrightarrow{v^{(\lambda)}}(\mathrm{O}_S)\cdot\int_{\mathrm{M}\in(S)}\overrightarrow{J^{(\lambda)}}(\mathrm{M})\mathrm{d}m(\mathrm{M})\ldots$$

$$\ldots+\overrightarrow{\omega_S^{\lambda}}\cdot\int_{\mathrm{M}\in(S)}\overrightarrow{\mathrm{O}_S\mathrm{M}}\wedge\overrightarrow{J^{(\lambda)}}(\mathrm{M})\mathrm{d}m(\mathrm{M})$$

$$\Rightarrow\quad\frac{dT}{dt}=\{\mathscr{V}_S^{\lambda}\}\otimes\{\mathscr{A}_S^{\lambda}\}.$$

Furthermore,

$$\frac{dT}{dt}=\frac{1}{2}\frac{d}{dt}[\{ {}^{\lambda}_{S} \}\otimes\{ p^{\lambda}_{S} \}]=\frac{1}{2}\left[\left\{\frac{d^{(\lambda)}}{dt}({}^{\lambda}_{S})\right\}\otimes\{ p^{\lambda}_{S} \}+\{ {}^{\lambda}_{S} \}\otimes\left\{\frac{d^{(\lambda)}}{dt}(p^{\lambda}_{S})\right\}\right]$$

$$\text{" "}=\frac{1}{2}\left\{\frac{d^{(\lambda)}}{dt}({}^{\lambda}_{S})\right\}\otimes\{ p^{\lambda}_{S} \}+\frac{1}{2}\{ {}^{\lambda}_{S} \}\otimes\{ \mathscr{A}^{\lambda}_{S} \}$$

and hence, we deduce that

$$\left\{\frac{d^{(\lambda)}}{dt}({}^{\lambda}_{S})\right\}\otimes\{ p^{\lambda}_{S} \}=\{ {}^{\lambda}_{S} \}\otimes\{ \mathscr{A}^{\lambda}_{S} \}.$$

3.6.3.4. *Derivative with respect to the time derivative* q'_α

$$\frac{\partial T}{\partial q'_\alpha}=\frac{\partial T^{(\lambda)}(S)}{\partial q'_\alpha}=\frac{1}{2}\frac{\partial}{\partial q'_\alpha}\int_{\mathrm{M}\in(S)}\left[\overrightarrow{v^{(\lambda)}}(\mathrm{M})\right]^2 \mathrm{d}m(M)$$

$$\frac{\partial T}{\partial q'_\alpha}=\int_{\mathrm{M}\in(S)}\frac{\partial^{(\lambda)}}{\partial q'_\alpha}\left[\overrightarrow{v^{(\lambda)}}(\mathrm{M})\right]\cdot\overrightarrow{v^{(\lambda)}}(\mathrm{M})\mathrm{d}m(M)$$

$$\text{"}\quad=\int_{\mathrm{M}\in(S)}\frac{\partial^{(\lambda)}}{\partial q'_\alpha}\left[\overrightarrow{v^{(\lambda)}}(\mathrm{O}_S)+\overrightarrow{\omega^{\lambda}_{S}}\wedge\overrightarrow{\mathrm{O}_S\mathrm{M}}\right]\cdot\overrightarrow{v^{(\lambda)}}(\mathrm{M})\mathrm{d}m(M)$$

$$\text{"}\quad=\int_{\mathrm{M}\in(S)}\frac{\partial^{(\lambda)}}{\partial q'_\alpha}\left[\overrightarrow{{}^{\alpha}d_{\mathrm{O}_S}}q'_\alpha+\overrightarrow{{}^{\alpha}\delta}q'_\alpha\wedge\overrightarrow{\mathrm{O}_S\mathrm{M}}\right]\cdot\overrightarrow{v^{(\lambda)}}(\mathrm{M})\mathrm{d}m(M)$$

where $\overrightarrow{{}^{\alpha}d_{\mathrm{O}_S}}$ and $\overrightarrow{{}^{\alpha}\delta}$ are independent of q'_α.

$$\frac{\partial T}{\partial q'_\alpha}=\int_{\mathrm{M}\in(S)}\left(\overrightarrow{{}^{\alpha}d_{\mathrm{O}_S}}+\overrightarrow{{}^{\alpha}\delta}\wedge\overrightarrow{\mathrm{O}_S\mathrm{M}}\right)\cdot\overrightarrow{v^{(\lambda)}}(\mathrm{M})\mathrm{d}m(M)$$

$$\text{"}\quad=\overrightarrow{{}^{\alpha}d_{\mathrm{O}_S}}\cdot\int_{\mathrm{M}\in(S)}\overrightarrow{v^{(\lambda)}}(\mathrm{M})\mathrm{d}m(M)+\overrightarrow{{}^{\alpha}\delta}\cdot\int_{\mathrm{M}\in(S)}\overrightarrow{\mathrm{O}_S\mathrm{M}}\wedge\overrightarrow{v^{(\lambda)}}(\mathrm{M})\mathrm{d}m(M)$$

$$\Rightarrow\quad\frac{\partial T}{\partial q'_\alpha}=\{ {}^{\lambda}_{S,\alpha} \}\otimes\{ p^{\lambda}_{S} \}.$$

Moreover,

$$\frac{\partial}{\partial q'_\alpha}\left[\frac{1}{2}\{ {}^{\lambda}_{S} \}\otimes\{ p^{\lambda}_{S} \}\right]=\frac{1}{2}\left[\left\{\frac{\partial^{(\lambda)}}{\partial q'_\alpha}({}^{\lambda}_{S})\right\}\otimes\{ p^{\lambda}_{S} \}+\{ {}^{\lambda}_{S} \}\otimes\left\{\frac{\partial^{(\lambda)}}{\partial q'_\alpha}(p^{\lambda}_{S})\right\}\right]$$

$$\text{" "}\quad=\frac{1}{2}\{ {}^{\lambda}_{S,\alpha} \}\otimes\{ p^{\lambda}_{S} \}+\frac{1}{2}\{ {}^{\lambda}_{S} \}\otimes\{ p^{\lambda}_{S,\alpha} \}=\{ {}^{\lambda}_{S,\alpha} \}\otimes\{ p^{\lambda}_{S} \}$$

$$\Rightarrow \left\{ {}^{\lambda}_{S,\alpha} \right\} \otimes \left\{ p^{\lambda}_{S} \right\} = \left\{ {}^{\lambda}_{S} \right\} \otimes \left\{ p^{\lambda}_{S,\alpha} \right\}.$$

3.6.3.5. *Derivative with respect to* q_α

We have to develop the following expression

$$\frac{\partial T^{(\lambda)}(S)}{\partial q_\alpha} = \int_{M \in (S)} \frac{\partial^{(S)}}{\partial q_\alpha} \left[\overrightarrow{v^{(\lambda)}}(M) \right] \cdot \overrightarrow{v^{(\lambda)}}(M) dm(M).$$

The choice of the basis for differentiation having no influence on the differentiation of vectors of the scalar product, the choice of the basis (S), linked to the solid, allows for a simplification of the calculations, since, for example, $\frac{\partial^{(S)}}{\partial q_\alpha} \overrightarrow{O_S M} = \vec{0}$.

$$\frac{\partial T^{(\lambda)}(S)}{\partial q_\alpha} = \int_{M \in (S)} \frac{\partial^{(S)}}{\partial q_\alpha} \left[\overrightarrow{v^{(\lambda)}}(M) \right] \cdot \overrightarrow{v^{(\lambda)}}(M) dm(M).$$

$$\begin{aligned} \frac{\partial T^{(\lambda)}(S)}{\partial q_\alpha} &= \int_{M \in (S)} \frac{\partial^{(S)}}{\partial q_\alpha} \left[\overrightarrow{v^{(\lambda)}}(O_S) + \overrightarrow{\omega^{\lambda}_{S}} \wedge \overrightarrow{O_S M} \right] \cdot \overrightarrow{v^{(\lambda)}}(M) dm(M) \\ \text{" "} \quad &= \frac{\partial^{(S)}}{\partial q_\alpha} \overrightarrow{v^{(\lambda)}}(O_S) \cdot \int_{M \in (S)} \overrightarrow{v^{(\lambda)}}(M) dm(M) \ldots \\ &\ldots + \frac{\partial^{(S)}}{\partial q_\alpha} \overrightarrow{\omega^{\lambda}_{S}} \cdot \int_{M \in (S)} \overrightarrow{O_S M} \wedge \overrightarrow{v^{(\lambda)}}(M) dm(M) \end{aligned}.$$

Considering the composition of this last expression, which contains reduction elements in O_S of the kinetic torsor, it is useful to identify now the two terms $\frac{\partial^{(S)}}{\partial q_\alpha} \overrightarrow{v^{(\lambda)}}(O_S)$ and $\frac{\partial^{(S)}}{\partial q_\alpha} \overrightarrow{\omega^{\lambda}_{S}}$, with a particular focus on the reduction elements of a torsor.

3.6.3.5.1. Identification of $\frac{\partial^{(S)}}{\partial q_\alpha} \overrightarrow{v^{(\lambda)}}(O_S)$ and $\frac{\partial^{(S)}}{\partial q_\alpha} \overrightarrow{\omega^{\lambda}_{S}}$

In an arbitrary point P of (S)

$$\overrightarrow{v^{(\lambda)}}(P) = \overrightarrow{v^{(\lambda)}}(O_S) + \overrightarrow{\omega^{\lambda}_{S}} \wedge \overrightarrow{O_S P},$$

$$\frac{\partial^{(S)}}{\partial q_\alpha}\overrightarrow{v^{(\lambda)}}(\mathrm{P})=\frac{\partial^{(S)}}{\partial q_\alpha}\overrightarrow{v^{(\lambda)}}(\mathrm{O}_S)+\frac{\partial^{(S)}}{\partial q_\alpha}\left[\overrightarrow{\omega_S^\lambda}\wedge\overrightarrow{\mathrm{O}_S\mathrm{P}}\right]\text{ with }\frac{\partial^{(S)}}{\partial q_\alpha}\overrightarrow{\mathrm{O}_S\mathrm{P}}=\vec{0},$$

$$\Rightarrow\quad\frac{\partial^{(S)}}{\partial q_\alpha}\overrightarrow{v^{(\lambda)}}(\mathrm{P})=\frac{\partial^{(S)}}{\partial q_\alpha}\overrightarrow{v^{(\lambda)}}(\mathrm{O}_S)+\frac{\partial^{(S)}}{\partial q_\alpha}\overrightarrow{\omega_S^\lambda}\wedge\overrightarrow{\mathrm{O}_S\mathrm{P}}.$$

Given the form of this relation, we can suppose that, if $\frac{\partial^{(S)}}{\partial q_\alpha}\overrightarrow{\omega_S^\lambda}$ is the sum and $\frac{\partial^{(S)}}{\partial q_\alpha}\overrightarrow{v^{(\lambda)}}(\mathrm{O}_S)$ the moment in O_S of a torsor $\{\mathscr{T}\}$, the quantity $\frac{\partial^{(S)}}{\partial q_\alpha}\overrightarrow{v^{(\lambda)}}(\mathrm{P})$ will be its moment in P.

Under these conditions, we shall have

$$\frac{\partial T^{(\lambda)}(S)}{\partial q_\alpha}=\overrightarrow{\mathscr{M}_{\mathrm{O}_S}}\{\mathscr{T}\}\cdot\vec{s}\{p_S^\lambda\}+\vec{s}\{\mathscr{T}\}\cdot\overrightarrow{\mathscr{M}_{\mathrm{O}_S}}\{p_S^\lambda\}=\{\mathscr{T}\}\otimes\{p_S^\lambda\}.$$

Now, this torsor $\{\mathscr{T}\}$ should be identified.

3.6.3.5.2. Development of $\frac{\partial^{(S)}}{\partial q_\alpha}\overrightarrow{\omega_S^\lambda}$

$$\frac{\partial^{(S)}}{\partial q_\alpha}\overrightarrow{\omega_S^\lambda}=\frac{\partial^{(S)}}{\partial q_\alpha}\left(\overrightarrow{{}^\beta\delta}\,q'_\beta\right)=\left(\frac{\partial^{(S)}}{\partial q_\alpha}\overrightarrow{{}^\beta\delta}\right)q'_\beta.$$

Let us consider a fixed vector $\vec{w}$ in (S) $\Rightarrow$ $\frac{\partial^{(S)}}{\partial q_\alpha}\vec{w}=\vec{0}$, $\forall\alpha$

$$\begin{cases}\frac{\partial^{(\lambda)}}{\partial q_\alpha}\vec{w}=\overrightarrow{{}^\alpha\delta}(\lambda,S)\wedge\vec{w}\;\Rightarrow\;\frac{\partial^{(\lambda)2}}{\partial q_\alpha\partial q_\beta}\vec{w}=\frac{\partial^{(\lambda)}}{\partial q_\beta}\left(\frac{\partial^{(\lambda)}}{\partial q_\alpha}\vec{w}\right)=\frac{\partial^{(\lambda)}}{\partial q_\beta}\left(\overrightarrow{{}^\alpha\delta}\wedge\vec{w}\right)\\ \frac{\partial^{(\lambda)}}{\partial q_\beta}\vec{w}=\overrightarrow{{}^\beta\delta}(\lambda,S)\wedge\vec{w}\;\Rightarrow\;\frac{\partial^{(\lambda)2}}{\partial q_\alpha\partial q_\beta}\vec{w}=\frac{\partial^{(\lambda)}}{\partial q_\alpha}\left(\frac{\partial^{(\lambda)}}{\partial q_\beta}\vec{w}\right)=\frac{\partial^{(\lambda)}}{\partial q_\alpha}\left(\overrightarrow{{}^\beta\delta}\wedge\vec{w}\right)\end{cases}.$$

We obtain a first intermediary result,

$$\frac{\partial^{(\lambda)}}{\partial q_\beta}\left(\overrightarrow{{}^\alpha\delta}\wedge\vec{w}\right)=\frac{\partial^{(\lambda)}}{\partial q_\alpha}\left(\overrightarrow{{}^\beta\delta}\wedge\vec{w}\right).$$

Then,

$$\frac{\partial^{(\lambda)}}{\partial q_\beta}\left({}^\alpha\vec{\delta}\wedge\vec{w}\right)=\frac{\partial^{(\lambda)}}{\partial q_\beta}\left({}^\alpha\vec{\delta}\right)\wedge\vec{w}+{}^\alpha\vec{\delta}\wedge\frac{\partial^{(\lambda)}}{\partial q_\beta}\left(\vec{w}\right)$$

$$\text{'' ''}\qquad=\frac{\partial^{(\lambda)}}{\partial q_\beta}\left({}^\alpha\vec{\delta}\right)\wedge\vec{w}+{}^\alpha\vec{\delta}\wedge\left({}^\beta\vec{\delta}\wedge\vec{w}\right)$$

and similarly

$$\frac{\partial^{(\lambda)}}{\partial q_\alpha}\left({}^\beta\vec{\delta}\wedge\vec{w}\right)=\frac{\partial^{(\lambda)}}{\partial q_\alpha}\left({}^\beta\vec{\delta}\right)\wedge\vec{w}+{}^\beta\vec{\delta}\wedge\left({}^\alpha\vec{\delta}\wedge\vec{w}\right).$$

This leads to the equality

$$\frac{\partial^{(\lambda)}}{\partial q_\beta}\left({}^\alpha\vec{\delta}\right)\wedge\vec{w}+{}^\alpha\vec{\delta}\wedge\left({}^\beta\vec{\delta}\wedge\vec{w}\right)=\frac{\partial^{(\lambda)}}{\partial q_\alpha}\left({}^\beta\vec{\delta}\right)\wedge\vec{w}+{}^\beta\vec{\delta}\wedge\left({}^\alpha\vec{\delta}\wedge\vec{w}\right)$$

and consequently

$$\frac{\partial^{(\lambda)}}{\partial q_\beta}\left({}^\alpha\vec{\delta}\right)\wedge\vec{w}+\left({}^\alpha\vec{\delta}\cdot\vec{w}\right){}^\beta\vec{\delta}-\left({}^\alpha\vec{\delta}\cdot{}^\beta\vec{\delta}\right)\vec{w}\ldots$$

$$\ldots=\frac{\partial^{(\lambda)}}{\partial q_\alpha}\left({}^\beta\vec{\delta}\right)\wedge\vec{w}+\left({}^\beta\vec{\delta}\cdot\vec{w}\right){}^\alpha\vec{\delta}-\left({}^\beta\vec{\delta}\cdot{}^\alpha\vec{\delta}\right)\vec{w}.$$

However, the term $\left({}^\beta\vec{\delta}\cdot\vec{w}\right){}^\alpha\vec{\delta}-\left({}^\alpha\vec{\delta}\cdot\vec{w}\right){}^\beta\vec{\delta}$ is the development of the double vector product $\vec{w}\wedge\left({}^\alpha\vec{\delta}\wedge{}^\beta\vec{\delta}\right)=-\left({}^\alpha\vec{\delta}\wedge{}^\beta\vec{\delta}\right)\wedge\vec{w}$.

The previous expression can, therefore, be written

$$\left[\frac{\partial^{(\lambda)}}{\partial q_\beta}\left({}^\alpha\vec{\delta}\right)-\frac{\partial^{(\lambda)}}{\partial q_\alpha}\left({}^\beta\vec{\delta}\right)+\left({}^\alpha\vec{\delta}\wedge{}^\beta\vec{\delta}\right)\right]\wedge\vec{w}=\vec{0},$$

and is verified for any vector $\vec{w}$. Given that $\vec{w}$ is arbitrary, the condition for the vector product to be null is that the first term is null. It can be deduced that

$$\frac{\partial^{(\lambda)}}{\partial q_\beta}\left({}^\alpha\vec{\delta}\right)=\frac{\partial^{(\lambda)}}{\partial q_\alpha}\left({}^\beta\vec{\delta}\right)-\left({}^\alpha\vec{\delta}\wedge{}^\beta\vec{\delta}\right),$$

with $\dfrac{\partial^{(\lambda)}}{\partial q_\alpha}\left(\overrightarrow{{}^{\beta}\delta}\right) = \dfrac{\partial^{(S)}}{\partial q_\alpha}\left(\overrightarrow{{}^{\beta}\delta}\right) + \overrightarrow{{}^{\alpha}\delta}(\lambda, S) \wedge \overrightarrow{{}^{\beta}\delta}$.

The following remarkable result is obtained

$$\frac{\partial^{(\lambda)}}{\partial q_\beta}\left(\overrightarrow{{}^{\alpha}\delta}\right) = \frac{\partial^{(S)}}{\partial q_\alpha}\left(\overrightarrow{{}^{\beta}\delta}\right),$$

and the searched identification of the term $\dfrac{\partial^{(S)}}{\partial q_\alpha}\overrightarrow{\omega_S^\lambda}$

$$\frac{\partial^{(S)}}{\partial q_\alpha}\overrightarrow{\omega_S^\lambda} = \left(\frac{\partial^{(\lambda)}}{\partial q_\beta}\overrightarrow{{}^{\alpha}\delta}\right) q'_\beta = \frac{d^{(\lambda)}}{dt}\overrightarrow{{}^{\alpha}\delta} = \vec{s}\left\{\frac{d^{(\lambda)}}{dt}\left(\begin{smallmatrix}\lambda\\ S,\alpha\end{smallmatrix}\right)\right\}.$$

3.6.3.5.3. Development of the term $\dfrac{\partial^{(S)}}{\partial q_\alpha}\overrightarrow{v^{(\lambda)}}(\mathrm{O}_S)$

$$\frac{\partial^{(S)}}{\partial q_\alpha}\overrightarrow{v^{(\lambda)}}(\mathrm{O}_S) = \frac{\partial^{(\lambda)}}{\partial q_\alpha}\overrightarrow{v^{(\lambda)}}(\mathrm{O}_S) - \overrightarrow{{}^{\alpha}\delta}(\lambda, S) \wedge \overrightarrow{v^{(\lambda)}}(\mathrm{O}_S).$$

$$\frac{\partial^{(S)}}{\partial q_\alpha}\overrightarrow{v^{(\lambda)}}(\mathrm{O}_S) = \frac{\partial^{(\lambda)}}{\partial q_\alpha}\left[\left(\frac{\partial^{(\lambda)}}{\partial q_\beta}\overrightarrow{\mathrm{O}_\lambda \mathrm{O}_S}\right) q'_\beta\right] - \overrightarrow{{}^{\alpha}\delta} \wedge \overrightarrow{v^{(\lambda)}}(\mathrm{O}_S)$$

$$\text{'' ''} \qquad = \frac{\partial^{(\lambda)}}{\partial q_\beta}\left(\frac{\partial^{(\lambda)}}{\partial q_\alpha}\overrightarrow{\mathrm{O}_\lambda \mathrm{O}_S}\right) q'_\beta - \overrightarrow{{}^{\alpha}\delta} \wedge \overrightarrow{v^{(\lambda)}}(\mathrm{O}_S)$$

$$\text{'' ''} \qquad = \frac{d^{(\lambda)}}{dt}\left(\frac{\partial^{(\lambda)}}{\partial q_\alpha}\overrightarrow{\mathrm{O}_\lambda \mathrm{O}_S}\right) + \overrightarrow{v^{(\lambda)}}(\mathrm{O}_S) \wedge \overrightarrow{{}^{\alpha}\delta}$$

with $\overrightarrow{v^{(\lambda)}}(\mathrm{O}_S) = \dfrac{d^{(\lambda)}}{dt}\overrightarrow{\mathrm{O}_\lambda \mathrm{O}_S} = \left(\dfrac{\partial^{(\lambda)}}{\partial q_\alpha}\overrightarrow{\mathrm{O}_\lambda \mathrm{O}_S}\right) q'_\alpha = \overrightarrow{{}^{\alpha}d_{\mathrm{O}_S}}\, q'_\alpha$.

Therefore we have

$$\frac{\partial^{(S)}}{\partial q_\alpha}\overrightarrow{v^{(\lambda)}}(\mathrm{O}_S) = \frac{d^{(\lambda)}}{dt}\left(\overrightarrow{{}^{\alpha}d_{\mathrm{O}_S}}\right) + \overrightarrow{v^{(\lambda)}}(\mathrm{O}_S) \wedge \overrightarrow{{}^{\alpha}\delta}$$

$$\text{'' ''} \qquad = \frac{d^{(\lambda)}}{dt}\overrightarrow{\mathcal{M}_{\mathrm{O}_S}}\left\{\begin{smallmatrix}\lambda\\ S,\alpha\end{smallmatrix}\right\} + \overrightarrow{v^{(\lambda)}}(\mathrm{O}_S) \wedge \vec{s}\left\{\begin{smallmatrix}\lambda\\ S,\alpha\end{smallmatrix}\right\} = \overrightarrow{\mathcal{M}_{\mathrm{O}_S}}\left\{\frac{d^{(\lambda)}}{dt}\left(\begin{smallmatrix}\lambda\\ S,\alpha\end{smallmatrix}\right)\right\}.$$

3.6.3.5.4. Identification of torsor $\{\mathscr{T}\}$

The torsor $\{\mathscr{T}\}$ to be found is thus the torsor $\left\{\frac{d^{(\lambda)}}{dt}\left(\overset{\lambda}{S,\alpha}\right)\right\}$ and the term $\frac{\partial T^{(\lambda)}(S)}{\partial q_\alpha}$ can be written:

$$\frac{\partial T^{(\lambda)}(S)}{\partial q_\alpha}=\overrightarrow{\mathscr{M}_{O_S}}\left\{\frac{d^{(\lambda)}}{dt}\left(\overset{\lambda}{S,\alpha}\right)\right\}\cdot\vec{s}\{p_S^\lambda\}+\vec{s}\left\{\frac{d^{(\lambda)}}{dt}\left(\overset{\lambda}{S,\alpha}\right)\right\}\cdot\overrightarrow{\mathscr{M}_{O_S}}\{p_S^\lambda\}$$

$$\text{'' ''}\quad=\left\{\frac{d^{(\lambda)}}{dt}\left(\overset{\lambda}{S,\alpha}\right)\right\}\otimes\{p_S^\lambda\}.$$

3.6.6.4. *Derivative with respect to time of the kinetic energy – Lagrangian*

$$\frac{d}{dt}\left[\{\overset{\lambda}{S,\alpha}\}\otimes\{p_S^\lambda\}\right]=\left\{\frac{d^{(\lambda)}}{dt}\left(\overset{\lambda}{S,\alpha}\right)\right\}\otimes\{p_S^\lambda\}+\{\overset{\lambda}{S,\alpha}\}\otimes\left\{\frac{d^{(\lambda)}}{dt}(p_S^\lambda)\right\},$$

or, according to the previous results

$$\frac{d}{dt}\left(\frac{\partial T^{(\lambda)}(S)}{\partial q'_\alpha}\right)=\frac{\partial T^{(\lambda)}(S)}{\partial q_\alpha}+\{\overset{\lambda}{S,\alpha}\}\otimes\{\mathscr{A}_S^\lambda\}$$

or $$\frac{dT}{dt}=\frac{dT^{(\lambda)}(S)}{dt}=\{\overset{\lambda}{S}\}\otimes\{\mathscr{A}_S^\lambda\}=\{\overset{\lambda}{S,\alpha}\}\otimes\{\mathscr{A}_S^\lambda\}q'_\alpha,$$

which leads to the searched expression

$$\frac{dT^{(\lambda)}(S)}{dt}=\left[\frac{d}{dt}\left(\frac{\partial T^{(\lambda)}(S)}{\partial q'_\alpha}\right)-\frac{\partial T^{(\lambda)}(S)}{\partial q_\alpha}\right]q'_\alpha.$$

The term

$$(L_\alpha)=\frac{d}{dt}\left(\frac{\partial T^{(\lambda)}(S)}{\partial q'_\alpha}\right)-\frac{\partial T^{(\lambda)}(S)}{\partial q_\alpha}=\{\overset{\lambda}{S,\alpha}\}\otimes\{\mathscr{A}_S^\lambda\}$$

is the *Lagrangian* (L_α) *relative to* q_α of the motion of (S) relative to $\langle\lambda\rangle$.

3.6.4. *Exercises*

3.6.4.1. *Exercise 25 – Kinematics and kinetics of a solid*

Let us consider, in the system of reference $\langle\lambda\rangle \equiv \langle \mathrm{O}_\lambda | \vec{x_\lambda}\, \vec{y_\lambda}\, \vec{z_\lambda} \rangle$, the motion of a solid (S) to which the system of reference $\langle \mathrm{O}_S | \vec{x_S}\, \vec{y_S}\, \vec{z_S} \rangle$ is linked, so that

– the point O_S describes in the plane $\Pi(\mathrm{O}_\lambda | \vec{x_\lambda}\, \vec{y_\lambda})$ the trajectory defined by

$$\overrightarrow{\mathrm{O}_\lambda \mathrm{O}_S} = a\vec{u}(\alpha) = a\left(\cos\alpha\,\vec{x_\lambda} + \sin\alpha\,\vec{y_\lambda}\right),$$

where a is a given positive constant.

– the orientation of the basis $(\vec{x_S}\, \vec{y_S}\, \vec{z_S})$ relative to the basis $(\vec{x_\lambda}\, \vec{y_\lambda}\, \vec{z_\lambda})$ is defined by

$$\begin{aligned}
\vec{n} &= \cos\alpha\,\vec{x_\lambda} + \sin\alpha\,\vec{y_\lambda} = \vec{u}(\alpha) \\
\vec{z_S} &= \vec{z_\lambda} \wedge \vec{n} \\
\vec{x_S} &= \cos\beta\,\vec{z_\lambda} + \sin\beta\,\vec{n} \\
\vec{y_S} &= \vec{z_S} \wedge \vec{x_S}.
\end{aligned}$$

where α and β are two functions that are twice-derivable with respect to time.

– m is the mass of solid (S), G is its center of the mass defined by $\overrightarrow{\mathrm{O}_S\mathrm{G}} = b\,\vec{y_S}$. The matrix representative of its inertia operator in O_S in the basis $(\vec{x_S}\, \vec{y_S}\, \vec{z_S})$ is

$$[I_{\mathrm{O}_S}(S)] = \begin{bmatrix} A & 0 & 0 \\ 0 & B & -D \\ 0 & -D & C \end{bmatrix}.$$

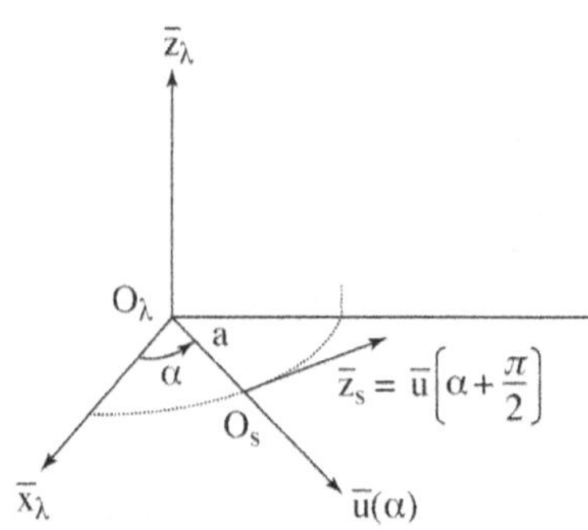

Question 1: Characterize the trajectory of O_S in $\langle \lambda \rangle$

The trajectory of O_S is defined in $\langle \lambda \rangle$ by

$$\overrightarrow{O_\lambda O_S} = a\vec{u}(\alpha) = a\left(\cos\alpha\,\overrightarrow{x_\lambda} + \sin\alpha\,\overrightarrow{y_\lambda}\right) = x\overrightarrow{x_\lambda} + y\overrightarrow{y_\lambda}\,.$$

Consequently, the coordinates of this point verify the equation

$$x^2 + y^2 = a^2,$$

which is the equation of a circle with center O_λ and radius a contained in the plane $\Pi\left(O_\lambda | \overrightarrow{x_\lambda}\ \overrightarrow{y_\lambda}\right)$.

Question 2: Using α and β, determine the three Euler angles ψ, θ, φ that help locate the bases $\left(\overrightarrow{x_\lambda}\ \overrightarrow{y_\lambda}\ \overrightarrow{z_\lambda}\right)$ and $\left(\overrightarrow{x_S}\ \overrightarrow{y_S}\ \overrightarrow{z_S}\right)$ relative to each other:

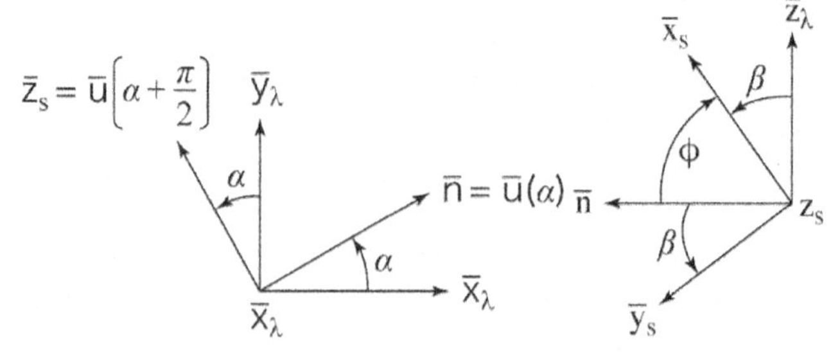

$\vec{n} = \vec{u}(\alpha)$ being in fact the nodal vector of the Eulerian representation, we have:

– $\psi \equiv \alpha = \widehat{\vec{x_\lambda}, \vec{n}}$, measured along $\vec{z_\lambda}$,

– $\vec{z_S} = \vec{z_\lambda} \wedge \vec{n} \quad \Rightarrow \quad \theta = \widehat{\vec{z_\lambda}, \vec{z_S}} = \frac{\pi}{2}$, measured along $\vec{n}$,

– $\vec{x_S} = \cos\beta\,\vec{z_\lambda} + \sin\beta\,\vec{n}$ and $\varphi = \widehat{\vec{n}, \vec{x_S}} = \widehat{\vec{n}, \vec{z_\lambda}} + \widehat{\vec{z_\lambda}, \vec{x_S}} = -\frac{\pi}{2} + \beta$, measured along $\vec{z_S}$.

Question 3: Determine, by means of their components in the basis $(e) \equiv \left(\vec{n}\,\vec{z_S}\,\vec{z_\lambda}\right)$, the reduction elements in O_S of the torsor distributor of velocities $\left\{ {}^{\lambda}_{S} \right\}$ of the solid (S) relative to $\langle \lambda \rangle$. Deduce from them the partial torsors distributors $\left\{ {}^{\lambda}_{S,\alpha} \right\}$ and $\left\{ {}^{\lambda}_{S,\beta} \right\}$ and characterize these torsors.

$$\left\{ {}^{\lambda}_{S} \right\}_{\mathrm{O}_S} = \left[\overrightarrow{\omega_S^{\lambda}} \,\middle|\, \overrightarrow{v^{(\lambda)}}(\mathrm{O}_S) \right].$$

In the Eulerian representation:

$$\overrightarrow{\omega_S^{\lambda}} = \psi'\vec{z_\lambda} + \theta'\vec{n} + \varphi'\vec{z_S} = \alpha'\vec{z_\lambda} + \beta'\vec{z_S}$$

$$\overrightarrow{v^{(\lambda)}}(O_S) = \frac{d^{(\lambda)}}{dt}\overrightarrow{\mathrm{O}_\lambda \mathrm{O}_S} = a\alpha'\vec{u}\left(\alpha + \frac{\pi}{2}\right) = a\alpha'\vec{z_S},$$

$$\Rightarrow \quad \left\{ {}^{\lambda}_{S} \right\}_{\mathrm{O}_S} = \left[\alpha'\vec{z_\lambda} + \beta'\vec{z_S} \,\middle|\, a\alpha'\vec{z_S} \right].$$

As the torsor distributor of velocities uses first derivatives of the two parameters α and β, we can express it by means of partial torsors:

$$\left\{ {}^{\lambda}_{S} \right\} = \left\{ {}^{\lambda}_{S,\alpha} \right\}\alpha' + \left\{ {}^{\lambda}_{S,\beta} \right\}\beta',$$

in which the two partial torsors distributors are expressed as

$$\left\{ {}_{S,\alpha}^{\lambda} \right\}_{O_S} = \left[\overrightarrow{z_\lambda} \middle| a\overrightarrow{z_S} \right] \quad \text{and} \quad \left\{ {}_{S,\beta}^{\lambda} \right\}_{O_S} = \left[\overrightarrow{z_S} \middle| \vec{0} \right].$$

The scalar invariant of the first of these torsors $I\left\{ {}_{S,\alpha}^{\lambda} \right\} = \overrightarrow{z_\lambda} \cdot a\overrightarrow{z_S}$ being null, as the two components of this torsor are orthogonal, this is a sliding torsor.

Its instantaneous center of rotation Q_0 is defined by

$$\overrightarrow{O_S Q_0} = \frac{\vec{s}\left\{ {}_{S,\alpha}^{\lambda} \right\} \wedge \overrightarrow{\mathcal{M}_{O_S}\left\{ {}_{S,\alpha}^{\lambda} \right\}}}{\vec{s}\left\{ {}_{S,\alpha}^{\lambda} \right\}^2} = \overrightarrow{z_\lambda} \wedge a\overrightarrow{z_S} = -a\vec{n} = \overrightarrow{O_s O_\lambda}.$$

It is, therefore, a sliding torsor of axis $\left(O_\lambda \middle| \overrightarrow{z_\lambda} \right)$.

The second torsor is also a sliding torsor of axis $\left(O_S \middle| \overrightarrow{z_S} \right)$.

The motion of (S) relative to $\langle \lambda \rangle$ is the combination of two rotations, one of angle α regarding the axis $\left(O_\lambda \middle| \overrightarrow{z_\lambda} \right)$, the other of angle β regarding the axis $\left(O_S \middle| \overrightarrow{z_S} \right)$.

Question 4: Determine the reduction elements in O_S of the kinetic torsor $\left\{ p_S^\lambda \right\}$ of the motion of (S) relative to $\langle \lambda \rangle$, by their components in the basis $(S) \equiv \left(\overrightarrow{x_S}\ \overrightarrow{y_S}\ \overrightarrow{z_S} \right)$. Subsequently, deduce the partial kinetic torsors $\left\{ p_{S,\alpha}^{\lambda} \right\}$ and $\left\{ p_{S,\beta}^{\lambda} \right\}$:

$$\vec{S}\left\{ p_S^\lambda \right\} = m\overrightarrow{v^{(\lambda)}}(G) = m\left[\overrightarrow{v^{(\lambda)}}(O_S) + \overrightarrow{\omega_S^\lambda} \wedge \overrightarrow{O_S G} \right],$$

as O_S and G both belong to the solid (S):

$$\vec{S}\left\{ p_S^\lambda \right\} = m\left[a\alpha'\overrightarrow{z_S} + \left(\alpha'\cos\beta\,\overrightarrow{x_S} - \alpha'\sin\beta\,\overrightarrow{y_S} + \beta'\overrightarrow{z_S} \right) \wedge b\overrightarrow{y_S} \right]$$

$$\text{" "} \quad = m\left[-b\beta'\overrightarrow{x_S} + (a + b\cos\beta)\alpha'\overrightarrow{z_S} \right]$$

$$\overrightarrow{\mathcal{M}_{O_S}}\{p_S^\lambda\} = m\overrightarrow{O_S G} \wedge \overrightarrow{v^{(\lambda)}}(O_S) + \mathcal{I}_{O_S}(S)\overrightarrow{\omega_S^\lambda}$$

$$\left[\overrightarrow{\mathcal{M}_{O_S}}\{p_S^\lambda\}\right] = \left[mb\overrightarrow{y_S} \wedge a\alpha'\overrightarrow{z_S}\right] + \begin{bmatrix} A & 0 & 0 \\ 0 & B & -D \\ 0 & -D & C \end{bmatrix}\begin{bmatrix} \alpha'\cos\beta \\ -\alpha'\sin\beta \\ \beta' \end{bmatrix}$$

$$\left[\overrightarrow{\mathcal{M}_{O_S}}\{p_S^\lambda\}\right] = mab\alpha'\begin{bmatrix} 1 \\ 0 \\ 0 \end{bmatrix} + \begin{bmatrix} A\alpha'\cos\beta \\ -B\alpha'\sin\beta - D\beta' \\ D\alpha'\sin\beta + C\beta' \end{bmatrix}$$

$$\overrightarrow{\mathcal{M}_{O_S}}\{p_S^\lambda\} = (mab + A\cos\beta)\alpha'\overrightarrow{x_S} - (B\alpha'\sin\beta + D\beta')\overrightarrow{y_S}\ldots$$
$$\ldots + (D\alpha'\sin\beta + C\beta')\overrightarrow{z_S}$$

$$\Rightarrow \{p_S^\lambda\}_{O_S} = \left[m\left(-b\beta'\overrightarrow{x_S} + (a + b\cos\beta)\alpha'\overrightarrow{z_S}\right)\right|\ldots$$
$$\ldots(mab + A\cos\beta)\alpha'\overrightarrow{x_S} - (B\alpha'\sin\beta + D\beta')\overrightarrow{y_S} + (D\alpha'\sin\beta + C\beta')\overrightarrow{z_S}\Big]^{\cdot}$$

Since $\{p_S^\lambda\}$ is a linear function of α' and β', it can be written by means of partial torsors

$$\{p_S^\lambda\} = \{p_{S,\alpha}^\lambda\}\alpha' + \{p_{S,\beta}^\lambda\}\beta',$$

where the partial kinetic torsors $\{p_{S,\alpha}^\lambda\}$ and $\{p_{S,\beta}^\lambda\}$ have the expressions

$$\{p_{S,\alpha}^\lambda\}_{O_S} = \left[m(a + b\cos\beta)\overrightarrow{z_S}\ \middle|\ (mab + A\cos\beta)\overrightarrow{x_S} - B\sin\beta\,\overrightarrow{y_S} + D\sin\beta\,\overrightarrow{z_S}\right],$$

$$\{p_{S,\beta}^\lambda\}_{O_S} = \left[-mb\overrightarrow{x_S}\ \middle|\ -D\overrightarrow{y_S} + C\overrightarrow{z_S}\right].$$

Question 5: Calculate the vis viva $2T^{(\lambda)}(S)$ of the motion of (S) relative to $\langle\lambda\rangle$ and write the result in the form

$$2T^{(\lambda)}(S) = [\alpha' \quad \beta']\begin{bmatrix} M_{11} & M_{12} \\ M_{12} & M_{22} \end{bmatrix}\begin{bmatrix} \alpha' \\ \beta' \end{bmatrix},$$

where the symmetrical matrix $[M]$, with M_{ij} as the current element, is to be analyzed.

$$2T^{(\lambda)}(S) = m\left[\overrightarrow{v^{(\lambda)}}(O_S)\right]^2 + 2m\overrightarrow{v^{(\lambda)}}(O_S)\cdot\left(\overrightarrow{\omega_S^\lambda}\wedge\overrightarrow{O_S G}\right) + \overrightarrow{\omega_S^\lambda}\cdot\mathcal{I}_{O_S}(S)\overrightarrow{\omega_S^\lambda}$$

$$m\left[\overrightarrow{v^{(\lambda)}}(O_S)\right]^2 = m\left(a\alpha'\overrightarrow{z_S}\right)^2 = ma^2\alpha'^2$$

$$2m\overrightarrow{v^{(\lambda)}}(O_S)\cdot\left(\overrightarrow{\omega_S^\lambda}\wedge\overrightarrow{O_S G}\right) = 2ma\alpha'\overrightarrow{z_S}\cdot\left[\left(\alpha'\cos\beta\overrightarrow{x_S} - \alpha'\sin\beta\overrightarrow{y_S} + \beta'\overrightarrow{z_S}\right)\wedge b\overrightarrow{y_S}\right]$$

$$\text{'' ''} \quad = 2ma\alpha'\overrightarrow{z_S}\cdot\left(-b\beta'\overrightarrow{x_S} + b\alpha'\cos\beta\overrightarrow{z_S}\right)$$

$$\text{'' ''} \quad = 2mab\alpha'^2\cos\beta$$

$$\overrightarrow{\omega_S^\lambda}\cdot I_{O_S}(S)\overrightarrow{\omega_S^\lambda} = \left(\alpha'\cos\beta\overrightarrow{x_S} - \alpha'\sin\beta\overrightarrow{y_S} + \beta'\overrightarrow{z_S}\right)\ldots$$
$$\ldots\cdot\left[A\cos\beta\alpha'\overrightarrow{x_S} - (B\alpha'\sin\beta + D\beta')\overrightarrow{y_S} + (D\alpha'\sin\beta + C\beta')\overrightarrow{z_S}\right],$$

$$\Rightarrow\ 2T^{(\lambda)}(S) = \left(ma^2 + 2mab\cos\beta + A\cos^2\beta + B\sin^2\beta\right)\alpha'^2\ldots$$
$$\ldots + C\beta'^2 + 2D\alpha'\beta'\sin\beta,$$

a quadratic expression that can be written in matrix form:

$$2T^{(\lambda)}(S) = [\alpha' \quad \beta'][M]\begin{bmatrix}\alpha'\\ \beta'\end{bmatrix}$$

$$\text{'' ''} \quad = [\alpha' \quad \beta']\begin{bmatrix} ma^2 + 2mab\cos\beta + A\cos^2\beta + B\sin^2\beta & D\sin\beta \\ D\sin\beta & C\end{bmatrix}\begin{bmatrix}\alpha'\\ \beta'\end{bmatrix}.$$

The terms M_{ij} of this matrix $[M]$ can also be obtained by the following relations:

$$-\ M_{11} = \left\{\begin{smallmatrix}\lambda\\ S,\alpha\end{smallmatrix}\right\}\otimes\left\{p\begin{smallmatrix}\lambda\\ S,\alpha\end{smallmatrix}\right\} = \vec{s}\left\{\begin{smallmatrix}\lambda\\ S,\alpha\end{smallmatrix}\right\}\cdot\overrightarrow{\mathcal{M}_{O_S}}\left\{p\begin{smallmatrix}\lambda\\ S,\alpha\end{smallmatrix}\right\} + \vec{s}\left\{p\begin{smallmatrix}\lambda\\ S,\alpha\end{smallmatrix}\right\}\cdot\overrightarrow{\mathcal{M}_{O_S}}\left\{\begin{smallmatrix}\lambda\\ S,\alpha\end{smallmatrix}\right\},$$

$$\Rightarrow M_{11} = \left(\cos\beta\,\overrightarrow{x_S} - \sin\beta\,\overrightarrow{y_S}\right)\cdot\left[(mab + A\cos\beta)\overrightarrow{x_S} - B\sin\beta\,\overrightarrow{y_S} + D\sin\beta\,\overrightarrow{z_S}\right]\ldots$$

$$\ldots + m(a + b\cos\beta)\overrightarrow{z_S}\cdot a\,\overrightarrow{z_S} \qquad ,$$

$$'' \; '' \quad = ma^2 + 2mab\cos\beta + A\cos^2\beta + B\sin^2\beta$$

$$- \; M_{22} = \left\{\mathcal{}_{S,\beta}^{\lambda}\right\}\otimes\left\{p_{S,\beta}^{\lambda}\right\} = C\,,$$

$$- \; M_{12} = \left\{\mathcal{}_{S,\alpha}^{\lambda}\right\}\otimes\left\{p_{S,\beta}^{\lambda}\right\} = M_{21} = \left\{\mathcal{}_{S,\beta}^{\lambda}\right\}\otimes\left\{p_{S,\alpha}^{\lambda}\right\} = D\sin\beta\,.$$

Question 6: Determine the reduction elements in O_S of the dynamic torsor $\{\mathcal{A}_S^{\lambda}\}$ of the motion of (S) relative to $\langle\lambda\rangle$, by their components in the basis (S):

$$- \; \vec{S}\{\mathcal{A}_S^{\lambda}\} = m\overrightarrow{J^{(\lambda)}}(G) = m\left[\overrightarrow{J^{(\lambda)}}(O_S) + \overrightarrow{\omega_S^{\lambda}}' \wedge \overrightarrow{O_S G} + \overrightarrow{\omega_S^{\lambda}} \wedge \left(\overrightarrow{\omega_S^{\lambda}} \wedge \overrightarrow{O_S G}\right)\right]$$

$$\overrightarrow{J^{(\lambda)}}(O_S) = \frac{d^{(\lambda)}}{dt}\overrightarrow{v^{(\lambda)}}(O_S) = \frac{d^{(\lambda)}}{dt}\left(a\alpha'\overrightarrow{z_S}\right) = a\alpha''\overrightarrow{z_S} + a\alpha'\frac{d^{(\lambda)}}{dt}\overrightarrow{z_S}$$

$$'' \; '' \quad = a\alpha''\overrightarrow{z_S} + a\alpha'\overrightarrow{\omega_S^{\lambda}} \wedge \overrightarrow{z_S}$$

$$'' \; '' \quad = a\alpha''\overrightarrow{z_S} + a\alpha'\left(\alpha'\cos\beta\,\overrightarrow{x_S} - \alpha'\sin\beta\,\overrightarrow{y_S} + \beta'\overrightarrow{z_S}\right)\wedge\overrightarrow{z_S}$$

$$'' \; '' \quad = a\left(-\alpha'^2\sin\beta\,\overrightarrow{x_S} - \alpha'^2\cos\beta\,\overrightarrow{y_S} + \alpha''\overrightarrow{z_S}\right)$$

$$\overrightarrow{\omega_S^{\lambda}}' \wedge \overrightarrow{O_S G} = \frac{d^{(\lambda)}}{dt}\overrightarrow{\omega_S^{\lambda}} \wedge \overrightarrow{O_S G} = \frac{d^{(S)}}{dt}\overrightarrow{\omega_S^{\lambda}} \wedge \overrightarrow{O_S G}$$

$$'' \; '' \quad = \frac{d^{(S)}}{dt}\left(\alpha'\cos\beta\,\overrightarrow{x_S} - \alpha'\sin\beta\,\overrightarrow{y_S} + \beta'\overrightarrow{z_S}\right)\wedge b\,\overrightarrow{y_S}$$

$$'' \; '' \quad = -b\beta''\overrightarrow{x_S} + b(\alpha''\cos\beta - \alpha'\beta'\sin\beta)\overrightarrow{z_S}$$

$$\overrightarrow{\omega_S^{\lambda}} \wedge \left(\overrightarrow{\omega_S^{\lambda}} \wedge \overrightarrow{O_S G}\right) = \left(\alpha'\cos\beta\,\overrightarrow{x_S} - \alpha'\sin\beta\,\overrightarrow{y_S} + \beta'\overrightarrow{z_S}\right)\ldots$$

$$\ldots \wedge \left(-b\beta'\overrightarrow{x_S} + b\alpha'\cos\beta\,\overrightarrow{z_S}\right)$$

$$'' \; '' \quad = -b\alpha'^2\sin\beta\cos\beta\,\overrightarrow{x_S} - b(\beta'^2 + \alpha'^2\cos^2\beta)\overrightarrow{y_S}\ldots$$

$$\ldots - b\alpha'\beta'\sin\beta\,\overrightarrow{z_S}$$

$$\Rightarrow \quad \vec{S}\{\mathscr{A}_S^\lambda\} = -m[b\beta'' + \alpha'^2(a + b\cos\beta)\sin\beta]\overrightarrow{x_S}\ldots$$
$$\ldots - m[b\beta'^2 + \alpha'^2(a + b\cos\beta)\cos\beta]\overrightarrow{y_S}\ldots.$$
$$\ldots + m[\alpha''(a + b\cos\beta) - 2\alpha'\beta'\sin\beta]\overrightarrow{z_S};$$

$$-\overrightarrow{\mathscr{M}_{O_S}}\{\mathscr{A}_S^\lambda\} = m\overrightarrow{O_S G} \wedge \overrightarrow{J^{(\lambda)}}(O_S) + I_{O_S}\left(\overrightarrow{\omega_S^\lambda}'\right) + \overrightarrow{\omega_S^\lambda} \wedge I_{O_S}\left(\overrightarrow{\omega_S^\lambda}\right)$$

$$m\overrightarrow{O_S G} \wedge \overrightarrow{J^{(\lambda)}}(O_S) = mb\overrightarrow{y_S} \wedge a\left(-\alpha'^2\sin\beta\overrightarrow{x_S} - \alpha'^2\cos\beta\overrightarrow{y_S} + \alpha''\overrightarrow{z_S}\right)$$
$$\text{'' ''} \qquad = mab\left(\alpha''\overrightarrow{x_S} + \alpha'^2\sin\beta\overrightarrow{z_S}\right)$$

$$\left[I_{O_S}\left(\overrightarrow{\omega_S^\lambda}'\right)\right] = \begin{bmatrix} A & 0 & 0 \\ 0 & B & -D \\ 0 & -D & C \end{bmatrix}\begin{bmatrix} \alpha''\cos\beta - \alpha'\beta'\sin\beta \\ -\alpha''\sin\beta - \alpha'\beta'\cos\beta \\ \beta'' \end{bmatrix}$$

$$\Rightarrow \quad I_{O_S}\left(\overrightarrow{\omega_S^\lambda}'\right) = A(\alpha''\cos\beta - \alpha'\beta'\sin\beta)\overrightarrow{x_S}\ldots$$
$$\ldots + [-B(\alpha''\sin\beta + \alpha'\beta'\cos\beta) - D\beta'']\overrightarrow{y_S}$$
$$\ldots + [D(\alpha''\sin\beta + \alpha'\beta'\cos\beta) + C\beta'']\overrightarrow{z_S}$$

$$\overrightarrow{\omega_S^\lambda} \wedge I_{O_S}\left(\overrightarrow{\omega_S^\lambda}\right) = \left(\alpha'\cos\beta\overrightarrow{x_S} - \alpha'\sin\beta\overrightarrow{y_S} + \beta'\overrightarrow{z_S}\right)\ldots$$
$$\ldots \wedge \left[A\cos\beta\alpha'\overrightarrow{x_S} - (B\alpha'\sin\beta + D\beta')\overrightarrow{y_S} + (D\alpha'\sin\beta + C\beta')\overrightarrow{z_S}\right],$$

$$\overrightarrow{\omega_S^\lambda} \wedge I_{O_S}\left(\overrightarrow{\omega_S^\lambda}\right) = [-\alpha'\sin\beta(D\alpha'\sin\beta + C\beta') + \beta'(B\alpha'\sin\beta + D\beta')]\overrightarrow{x_S}\ldots$$
$$\ldots + [A\alpha'\beta'\cos\beta - \alpha'\cos\beta(D\alpha'\sin\beta + C\beta')]\overrightarrow{y_S}\ldots$$
$$\ldots + [-\alpha'\cos\beta(B\alpha'\sin\beta + D\beta') + A\alpha'^2\sin\beta\cos\beta]\overrightarrow{z_S}$$

$$\overrightarrow{\omega_S^\lambda} \wedge I_{O_S}\left(\overrightarrow{\omega_S^\lambda}\right) = [(B - C)\alpha'\beta'\sin\beta + D(\beta'^2 - \alpha'^2\sin^2\beta)]\overrightarrow{x_S}\ldots$$
$$\ldots + [(A - C)\alpha'\beta'\cos\beta - D\alpha'^2\sin\beta\cos\beta]\overrightarrow{y_S}\ldots.$$
$$\ldots + [(A - B)\alpha'^2\sin\beta\cos\beta - D\alpha'\beta'\cos\beta]\overrightarrow{z_S}.$$

In this way, we obtain the expression in O_S of the moment of the dynamic torsor of the motion of (S) relative to $\langle \lambda \rangle$:

$$\overrightarrow{\mathcal{M}_{O_s}}\{\mathcal{A}_s^{\lambda}\} = [(mab + A\cos\beta)\alpha'' + (B - A - C)\alpha'\beta'\sin\beta + D(\beta'^2 - \alpha'^2\sin^2\beta)]\vec{x_s}\dots$$
$$\dots + [-B\alpha''\sin\beta - D\beta'' + (A - B - C)\alpha'\beta'\cos\beta - D\alpha'^2\sin\beta\cos\beta]\vec{y_s}\dots \quad .$$
$$\dots + [C\beta'' + D\alpha''\sin\beta + (mab + (A - B)\cos\beta)\alpha'^2\sin\beta]\vec{z_s}$$

The reduction elements of the dynamic torsor could also have been established starting from the derivatives of the reduction elements of the kinetic torsor.

$$\begin{cases} \vec{s}\{\mathcal{A}_S^{\lambda}\} = \dfrac{d^{(\lambda)}}{dt}\vec{s}\{p_S^{\lambda}\} = m(S)\overrightarrow{J^{(\lambda)}}(G) \\ \overrightarrow{\mathcal{M}_{O_S}}\{\mathcal{A}_S^{\lambda}\} = \dfrac{d^{(\lambda)}}{dt}\overrightarrow{\mathcal{M}_{O_S}}\{p_S^{\lambda}\} + m(S)\overrightarrow{v^{(\lambda)}}(O_S) \wedge \overrightarrow{v^{(\lambda)}}(G) \end{cases}$$

Question 7: Calculate the expressions of Lagrangians relative to the parameters α and β

The Lagrangian (L_α) can be calculated starting from one of these two relations:

$$-(L_\alpha) = \{\mathcal{A}_S^{\lambda}\} \otimes \{{}^{\lambda}_{S,\alpha}\}$$

$$-(L_\alpha) = \frac{d}{dt}\frac{\partial T^{(\lambda)}(S)}{\partial \alpha'} - \frac{\partial T^{(\lambda)}(S)}{\partial \alpha}.$$

We shall rather use the second formula, particularly in this case, where in one of the Lagrangians the term $\dfrac{\partial T^{(\lambda)}(S)}{\partial \alpha} = 0$.

$$(L_\alpha) = (ma^2 + 2mab\cos\beta + A\cos^2\beta + B\sin^2\beta)\alpha''\dots$$
$$\dots - 2[mab + (A - B)\cos\beta]\alpha'\beta'\sin\beta + D\beta''\sin\beta + D\beta'^2\cos\beta$$
$$(L_\beta) = C\beta'' + D\alpha''\sin\beta + [mab + (A - B)\cos\beta]\alpha'^2\sin\beta$$

3.6.4.2. *Exercise 26 – Elements of dynamics*

In the Galilean system of reference $\langle g \rangle \equiv \langle \mathrm{O}_g \,|\, \overrightarrow{x_g}\ \overrightarrow{y_g}\ \overrightarrow{z_g} \rangle$, we consider the motion of a solid (S) to which the system of reference $\langle \mathrm{O}_S \,|\, \overrightarrow{x_S}\ \overrightarrow{y_S}\ \overrightarrow{z_S} \rangle$ is linked. This solid is composed of:

– a sphere of center C and radius *R*;

– a rod of length R with one end in O_S and the other end welded on the sphere, and perpendicular to the sphere surface. We consider that $\overrightarrow{\mathrm{CO}_S} = 2R\overrightarrow{z_S}$.

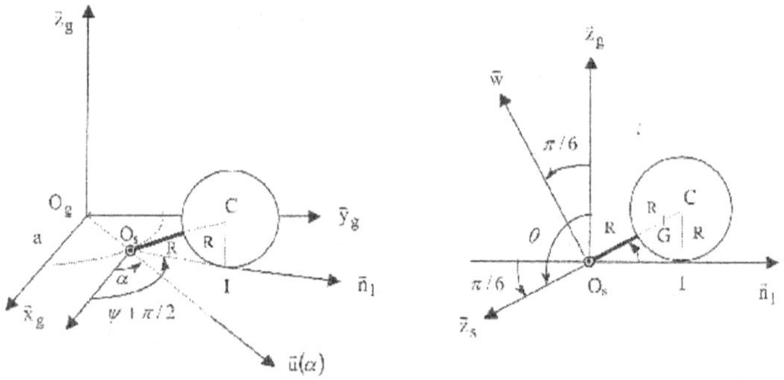

The motion of (S) is such that:

– O_S describes a circle in the plane $\Pi\left(\mathrm{O}_g \,|\, \overrightarrow{x_g}, \overrightarrow{y_g}\right)$, of center O_g and given radius *a*,

– the sphere remains in contact with the plane $\Pi\left(\mathrm{O}_g \,|\, \overrightarrow{x_g}, \overrightarrow{y_g}\right)$ in a variable point I.

We consider;

– $\overrightarrow{\mathrm{O}_g\mathrm{O}_S} = a\vec{u}(\alpha)$

– $\dfrac{\overrightarrow{\mathrm{O}_S\mathrm{I}}}{\left\|\overrightarrow{\mathrm{O}_S\mathrm{I}}\right\|} = \overrightarrow{n_1} = \vec{u}\left(\psi + \dfrac{\pi}{2}\right)$

– $\vec{n} = \vec{n_1} \wedge \vec{z_g} = \vec{u}(\psi)$, nodal vector of the Eulerian representation

– $\vec{z_S} \wedge \vec{n} = \vec{w}$

– $\varphi = \widehat{\vec{n}, \vec{x_S}} = \widehat{\vec{w}, \vec{y_S}}$, spin of (S).

The location of (S) is thus established by means of parameters α, ψ and φ, which are twice-derivable functions of time.

Question 1: What is the value of the Euler nutation angle θ?

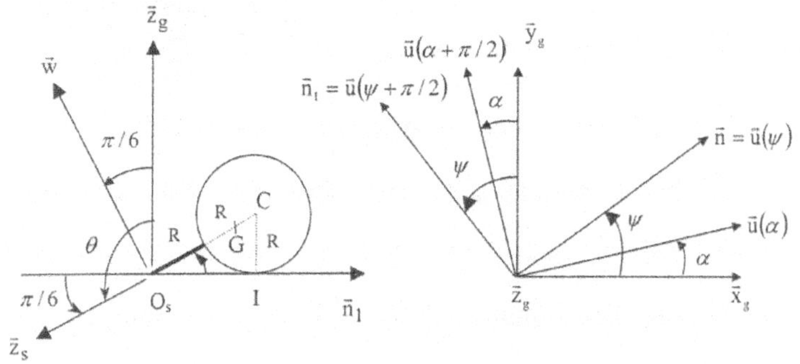

If we consider the triangle $O_S CI$, with a right angle in I, we have

$$\sin \widehat{O_S} = \frac{IC}{O_S C} = \frac{R}{2R} = \frac{1}{2} = \sin \frac{\pi}{6}$$

$$\theta = \widehat{\vec{z_g}, \vec{z_S}} = \widehat{\vec{z_g}, -\vec{n_1}} + \widehat{-\vec{n_1}, \vec{z_S}} = \frac{\pi}{2} + \frac{\pi}{6} = \frac{2\pi}{3}.$$

Question 2: Determine the kinematic torsor $\{ {}^g_S \}$; its sum will be expressed in the basis $\left(\vec{n}\,\vec{w}\,\vec{z_S}\right)$, its moment in O_S in

$$\left(\vec{u}(\alpha)\,\vec{u}\left(\alpha + \frac{\pi}{2}\right)\vec{z_g}\right)$$

$$\{ {}^g_S \}_{O_S} = \left[\vec{\omega_S^g} \,\middle|\, \vec{v^{(g)}}(O_S) \right]$$

$$\overrightarrow{\omega_S^g} = \psi' \overrightarrow{z_g} + \varphi' \overrightarrow{z_S} = \psi' \left(\cos\frac{\pi}{6}\vec{w} - \sin\frac{\pi}{6}\overrightarrow{z_S} \right) + \varphi' \overrightarrow{z_S} \text{ because } \theta' = 0,$$

$$\overrightarrow{v^{(g)}}(\mathrm{O}_S) = \frac{d^{(g)}}{dt}\overrightarrow{\mathrm{O}_g\mathrm{O}_S} = a\alpha' \vec{u}\left(\alpha + \frac{\pi}{2} \right),$$

$$\Rightarrow \left\{ {}^g_S \right\}_{\mathrm{O}_S} = \left[\frac{\sqrt{3}}{2}\psi'\vec{w} + \left(\varphi' - \frac{\psi'}{2} \right)\overrightarrow{z_S} \middle| a\alpha'\vec{u}\left(\alpha + \frac{\pi}{2} \right) \right].$$

Question 3: Let m be the mass of (S), G the center of mass of (S), defined by $\overrightarrow{\mathrm{O}_S\mathrm{G}} = -l\,\overrightarrow{z_S}$ with $\ell > 0$ the inertia operator in O_S of (S) represented by the matrix

$$[I_{\mathrm{O}_S}(S)]_{(\overrightarrow{x_S}\,\overrightarrow{y_S}\,\overrightarrow{z_S})} = \begin{bmatrix} A & 0 & 0 \\ 0 & A & 0 \\ 0 & 0 & C \end{bmatrix}.$$

Determine the dynamic torsor $\{\mathcal{A}_S^g\}$, its sum on the basis $\left(\vec{u}(\alpha)\ \vec{u}\left(\alpha + \frac{\pi}{2} \right)\ \overrightarrow{z_g} \right)$, its moment in O_S in the basis $\left(\vec{n}\ \overrightarrow{n_1}\ \overrightarrow{z_g} \right)$:

$$\overrightarrow{v^{(g)}}(\mathrm{G}) = \overrightarrow{v^{(g)}}(\mathrm{O}_S) + \overrightarrow{\omega_S^g} \wedge \overrightarrow{\mathrm{O}_S\mathrm{G}}$$

$$'' \quad = a\alpha'\vec{u}\left(\alpha + \frac{\pi}{2} \right) + \left[\frac{\sqrt{3}}{2}\psi'\vec{w} + \left(\varphi' - \frac{\psi'}{2} \right)\overrightarrow{z_S} \right] \wedge -\ell\overrightarrow{z_S}$$

$$'' \quad = a\alpha'\vec{u}\left(\alpha + \frac{\pi}{2} \right) - \ell\frac{\sqrt{3}}{2}\psi'\vec{n}$$

– $\vec{S}\{\mathcal{A}_S^g\} = m\overrightarrow{J^{(g)}}(\mathrm{G}) = m\dfrac{d^{(g)}}{dt}\overrightarrow{v^{(g)}}(\mathrm{G})$

– $\overrightarrow{\mathcal{M}_{\mathrm{O}_S}}\{\mathcal{A}_S^g\} = m\overrightarrow{\mathrm{O}_S\mathrm{G}} \wedge \overrightarrow{J^{(g)}}(\mathrm{O}_S) + I_{\mathrm{O}_S}\left(\overrightarrow{\omega_S^g}' \right) + \overrightarrow{\omega_S^g} \wedge I_{\mathrm{O}_S}\left(\overrightarrow{\omega_S^g} \right)$

$$\overrightarrow{J^{(g)}}(\mathrm{G}) = \frac{d^{(g)}}{dt}\left[a\alpha'\vec{u}\left(\alpha + \frac{\pi}{2}\right) - \ell\frac{\sqrt{3}}{2}\psi'\vec{n} \right]$$

$$'' \ '' \quad = a\alpha''\vec{u}\left(\alpha + \frac{\pi}{2}\right) + a\alpha'\frac{d^{(\lambda)}}{dt}\vec{u}\left(\alpha + \frac{\pi}{2}\right)\ldots$$

$$\ldots - \ell\frac{\sqrt{3}}{2}\psi''\vec{n} - \ell\frac{\sqrt{3}}{2}\psi'\frac{d^{(\lambda)}}{dt}\vec{u}(\psi)$$

$$'' \ '' \quad = a\alpha''\vec{u}\left(\alpha + \frac{\pi}{2}\right) - a\alpha'^2\vec{u}(\alpha) - \ell\frac{\sqrt{3}}{2}\psi''\vec{n} - \ell\frac{\sqrt{3}}{2}\psi'^2\vec{u}\left(\psi + \frac{\pi}{2}\right)$$

$$\Rightarrow \quad \vec{S}\{\mathscr{A}_S^g\} = m\left[-a\alpha'^2\vec{u}(\alpha) + a\alpha''\vec{u}\left(\alpha + \frac{\pi}{2}\right) - \ell\frac{\sqrt{3}}{2}\psi''\vec{n} - \ell\frac{\sqrt{3}}{2}\psi'^2 n_1 \right].$$

$$m\overrightarrow{\mathrm{O}_S\mathrm{G}} \wedge \overrightarrow{J^{(g)}}(\mathrm{O}_S) = -m\ell\vec{z_S} \wedge \frac{d^{(g)}}{dt}\overrightarrow{v^{(g)}}(\mathrm{O}_S)$$

$$'' \ '' \quad = -m\ell\left(-\frac{\sqrt{3}}{2}\vec{n_1} - \frac{1}{2}\vec{z_g} \right)\ldots$$

$$\ldots \wedge a\left([\alpha''\sin(\psi - \alpha) - \alpha'^2\cos(\psi - \alpha)]\vec{n}\ldots\right.$$

$$\left.\ldots + [\alpha''\cos(\psi - \alpha) + \alpha'^2\sin(\psi - \alpha)]\overrightarrow{n_1} \right)$$

$$m\overrightarrow{\mathrm{O}_S\mathrm{G}} \wedge \overrightarrow{J^{(g)}}(\mathrm{O}_S) = -\frac{ma\ell}{2}[\alpha''\cos(\psi - \alpha) + \alpha'^2\sin(\psi - \alpha)]\vec{n}\ldots$$

$$\ldots + \frac{ma\ell}{2}[\alpha''\sin(\psi - \alpha) - \alpha'^2\cos(\psi - \alpha)]\overrightarrow{n_1}\ldots$$

$$\ldots - \frac{ma\ell\sqrt{3}}{2}[\alpha''\sin(\psi - \alpha) - \alpha'^2\cos(\psi - \alpha)]\overrightarrow{z_g}$$

$$\overrightarrow{\omega_S^g}' = \frac{d^{(g)}}{dt}\left[\overrightarrow{\omega_S^g}\right] = \frac{d^{(S)}}{dt}\left[\overrightarrow{\omega_S^g}\right]$$

$$'' \ '' = \frac{d^{(S)}}{dt}\left[\frac{\sqrt{3}}{2}\psi'\sin\varphi\overrightarrow{x_S} + \frac{\sqrt{3}}{2}\psi'\cos\varphi\overrightarrow{y_S} + \left(\varphi' - \frac{\psi'}{2}\right)\overrightarrow{z_S} \right]$$

$$'' \ '' = \frac{\sqrt{3}}{2}(\psi''\sin\varphi + \psi'\varphi'\cos\varphi)\overrightarrow{x_S}\ldots$$

$$\ldots + \frac{\sqrt{3}}{2}(\psi''\cos\varphi - \psi'\varphi'\sin\varphi)\overrightarrow{y_S} + \left(\varphi'' - \frac{\psi''}{2}\right)\overrightarrow{z_S}$$

$$\left[I_{O_S}\left(\overrightarrow{\omega_S^g}'\right)\right] = \begin{bmatrix} A & 0 & 0 \\ 0 & A & 0 \\ 0 & 0 & C \end{bmatrix} \begin{bmatrix} \frac{\sqrt{3}}{2}(\psi''\sin\varphi + \psi'\varphi'\cos\varphi) \\ \frac{\sqrt{3}}{2}(\psi''\cos\varphi - \psi'\varphi'\sin\varphi) \\ \varphi'' - \frac{\psi''}{2} \end{bmatrix}$$

$$I_{O_S}\left(\overrightarrow{\omega_S^g}'\right) = A\frac{\sqrt{3}}{2}(\psi''\sin\varphi + \psi'\varphi'\cos\varphi)\overrightarrow{x_S}\ldots$$

$$\ldots + A\frac{\sqrt{3}}{2}(\psi''\cos\varphi - \psi'\varphi'\sin\varphi)\overrightarrow{y_S} + C\left(\varphi'' - \frac{\psi''}{2}\right)\overrightarrow{z_S}$$

$$I_{O_S}\left(\overrightarrow{\omega_S^g}'\right) = A\frac{\sqrt{3}}{2}(\psi''\sin\varphi + \psi'\varphi'\cos\varphi)\left(\cos\varphi\vec{n} + \frac{\sqrt{3}}{2}\sin\varphi\vec{n_1} + \frac{1}{2}\sin\varphi\vec{z_g}\right)\ldots$$

$$\ldots + A\frac{\sqrt{3}}{2}(\psi''\cos\varphi - \psi'\varphi'\sin\varphi)\left(-\sin\varphi\vec{n} + \frac{\sqrt{3}}{2}\cos\varphi\vec{n_1} + \frac{1}{2}\cos\varphi\vec{z_g}\right)\ldots$$

$$\ldots - C\left(\varphi'' - \frac{\psi''}{2}\right)\left(\frac{\sqrt{3}}{2}\vec{n_1} + \frac{1}{2}\vec{z_g}\right)$$

$$\Rightarrow \quad I_{O_S}\left(\overrightarrow{\omega_S^g}'\right) = \frac{A\sqrt{3}}{2}\psi'\varphi'\vec{n} + \left[\frac{3A}{4}\psi'' - \frac{C\sqrt{3}}{2}\left(\varphi'' - \frac{\psi''}{2}\right)\right]\vec{n_1}\ldots$$

$$\ldots + \left[\frac{A\sqrt{3}}{4}\psi'' - \frac{C}{2}\left(\varphi'' - \frac{\psi''}{2}\right)\right]\vec{z_g}$$

NOTE 3.6.– We could, and usefully so, do without a part of the changes of basis effected in the previous calculation by noting that:

– (S) is a solid of revolution, of axis $\left(\mathrm{C}|\overrightarrow{z_S}\right)$;

– the moment of inertia by any axis that is orthogonal on the axis of revolution has the same value A;

– and, consequently, the basis $\left(\vec{n}\,\vec{w}\,\overrightarrow{z_S}\right)$ is also the main inertia basis of (S) with

$$[I_{O_S}(S)]_{(\vec{n}\,\vec{w}\,\vec{z_S})} = \begin{bmatrix} A & 0 & 0 \\ 0 & A & 0 \\ 0 & 0 & C \end{bmatrix}.$$

We can, therefore, use this basis $\left(\vec{n}\,\vec{w}\,\vec{z_S}\right)$ to calculate the expression $\overrightarrow{\omega_S^g} \wedge I_{O_S}\left(\overrightarrow{\omega_S^g}\right)$:

$$\overrightarrow{\omega_S^g} = \psi'\,\vec{z_g} + \varphi'\,\vec{z_S} = \frac{\sqrt{3}}{2}\psi'\,\vec{w} + \left(\varphi' - \frac{\psi'}{2}\right)\vec{z_S}$$

$$\left[I_{O_S}\left(\overrightarrow{\omega_S^g}\right)\right] = \begin{bmatrix} A & 0 & 0 \\ 0 & A & 0 \\ 0 & 0 & C \end{bmatrix} \begin{bmatrix} 0 \\ \frac{\sqrt{3}}{2}\psi' \\ \varphi' - \frac{\psi'}{2} \end{bmatrix}$$

$$I_{O_S}\left(\overrightarrow{\omega_S^g}\right) = A\frac{\sqrt{3}}{2}\psi'\,\vec{w} + C\left(\varphi' - \frac{\psi'}{2}\right)\vec{z_S}$$

$$\overrightarrow{\omega_S^g} \wedge I_{O_S}\left(\overrightarrow{\omega_S^g}\right) = \left[\frac{\sqrt{3}}{2}\psi'\,\vec{w} + \left(\varphi' - \frac{\psi'}{2}\right)\vec{z_S}\right] \wedge \left[A\frac{\sqrt{3}}{2}\psi'\,\vec{w} + C\left(\varphi' - \frac{\psi'}{2}\right)\vec{z_S}\right]$$

$$\Rightarrow \quad \overrightarrow{\omega_S^g} \wedge I_{O_S}\left(\overrightarrow{\omega_S^g}\right) = (C - A)\frac{\sqrt{3}}{2}\psi'\left(\varphi' - \frac{\psi'}{2}\right)\vec{n}.$$

and in the end we obtain

$$\overrightarrow{\mathcal{M}_{O_s}}\{\mathcal{A}_s^g\} = \ldots$$

$$\ldots\left(-\frac{mal}{2}[\alpha''\cos(\psi-\alpha) + \alpha'^2\sin(\psi-\alpha)] + \frac{A\sqrt{3}}{4}\psi'^2 + \frac{C\sqrt{3}}{2}\psi'\left(\varphi' - \frac{\psi'}{2}\right)\right)\vec{n}\ldots$$

$$\ldots + \left(\frac{mal}{2}[\alpha''\sin(\psi-\alpha) - \alpha'^2\cos(\psi-\alpha)] + \frac{3A}{4}\psi'' - \frac{C\sqrt{3}}{2}\left(\varphi'' - \frac{\psi''}{2}\right)\right)\vec{n_1}\ldots$$

$$\ldots + \left(\frac{mal\sqrt{3}}{2}[-\alpha''\sin(\psi-\alpha) + \alpha'^2\cos(\psi-\alpha)] + \frac{A\sqrt{3}}{4}\psi'' - \frac{C}{2}\left(\varphi'' - \frac{\psi''}{2}\right)\right)\vec{z_g}$$

3.6.5. *Kinetic energy of a constrained solid*

A solid is said to be constrained when its motion is subjected to external stresses or *constraints*, which determine or limit its evolution. These constraints will be the subject of a more in-depth study when we shall develop the dynamics of solids (in Volume 3 of this set of books [BOR 16b]).

A constraint can be *parameterized* when the motion of the solid (S) relative to the system of reference $\langle \lambda \rangle$ can be described by means of parameters whose number $n < 6$ defines the number of *degrees of freedom* of this solid. When $n = 6$, we say that the solid is free; in this case, its motion depends uniquely on the forces exerted on it by the external environment.

We say that a constraint is time-dependent when time is explicitly involved in how the constraint is exerted on the moving solid; it is part of the descriptive parameters of this motion and is consequently involved in its expression. In the case of the free solid, time is not among these parameters; only the evolution of the latter depends on time, since it is in time that it unfolds.

Here, we shall only present the case in which the constraints can be parameterized and are time-dependent. The torsor distributor of velocities of the motion of the rigid solid (S) in $\langle \lambda \rangle$ can be written

$$\left\{ {}^{\lambda}_{S} \right\} = \left\{ {}^{\lambda}_{S,\alpha} \right\} q'_{\alpha} + \left\{ {}^{\lambda}_{S,t} \right\}.$$

Consequently, the kinetic torsor can also be explicitly expressed as a function of time:

$$\left\{ p^{\lambda}_{S} \right\} = \left\{ p^{\lambda}_{S,\beta} \right\} q'_{\beta} + \left\{ p^{\lambda}_{S,t} \right\}.$$

This leads to the expression of the kinetic energy $2T^{(\lambda)}(S) = \left\{ {}^{\lambda}_{S} \right\} \otimes \left\{ p^{\lambda}_{S} \right\}$ in which time is explicitly involved:

$$2T^{(\lambda)}(S) = \left[\left\{ {}^{\lambda}_{S,\alpha} \right\} q'_{\alpha} + \left\{ {}^{\lambda}_{S,t} \right\} \right] \otimes \left[\left\{ p^{\lambda}_{S,\beta} \right\} q'_{\beta} + \left\{ p^{\lambda}_{S,t} \right\} \right],$$

$$\Rightarrow \; 2T^{(\lambda)}(S) = \left\{ {}^{\lambda}_{S,\alpha} \right\} \otimes \left\{ p^{\lambda}_{S,\beta} \right\} q'_{\alpha} q'_{\beta} \cdots$$
$$\cdots + \left[\left\{ {}^{\lambda}_{S,\alpha} \right\} \otimes \left\{ p^{\lambda}_{S,t} \right\} + \left\{ p^{\lambda}_{S,\alpha} \right\} \otimes \left\{ {}^{\lambda}_{S,t} \right\} \right] q'_{\alpha} + \left\{ {}^{\lambda}_{S,t} \right\} \otimes \left\{ p^{\lambda}_{S,t} \right\}$$

This expression takes the form $2T^{(\lambda)}(S) = A_{\alpha\beta} q'_{\alpha} q'_{\beta} + B_{\alpha} q'_{\alpha} + 2T_0$, which is the sum:

– of a term quadratic in $q'_{\alpha} q'_{\beta}$

$$2T_2 = A_{\alpha\beta} q'_{\alpha} q'_{\beta}$$

with symmetric coefficients $A_{\alpha\beta} = A_{\beta\alpha}$;

– of a term linear in q'_{α}

$$2T_1 = B_{\alpha} q'_{\alpha}$$

with coefficients $B_{\alpha} = \left\{ {}^{\lambda}_{S,\alpha} \right\} \otimes \left\{ p^{\lambda}_{S,t} \right\} + \left\{ p^{\lambda}_{S,\alpha} \right\} \otimes \left\{ {}^{\lambda}_{S,t} \right\}$;

– of a term which is explicitly time-dependent

$$2T_0 = \left\{ {}^{\lambda}_{S,t} \right\} \otimes \left\{ p^{\lambda}_{S,t} \right\}.$$

The kinetic energy can thus be more synthetically expressed in the form

$$2T^{(\lambda)}(S) = 2T_2 + 2T_1 + 2T_0.$$

3.7. Partition of a continuous mechanical set ($\mathscr{D}$)

Let us consider the partition of a continuous mechanical set ($\mathscr{D}$) into two complementary and disjoint sets ($\mathscr{D}_1$) and ($\mathscr{D}_2$) so that

$$(\mathscr{D}) = (\mathscr{D}_1) \cup (\mathscr{D}_2) \quad \text{and} \quad (\mathscr{D}_1) \cap (\mathscr{D}_2) = \varnothing.$$

According to the additivity of the integral and given the disjunction of the sets:

$$-\int_{M\in(\mathscr{D})}\overrightarrow{v^{(\lambda)}}(M)dm(M)=\int_{M\in(\mathscr{D}_1)}\overrightarrow{v^{(\lambda)}}(M)dm(M)\ldots$$
$$\ldots+\int_{M\in(\mathscr{D}_2)}\overrightarrow{v^{(\lambda)}}(M)dm(M)$$

$$-\int_{M\in(\mathscr{D})}\overrightarrow{QM}\wedge\overrightarrow{v^{(\lambda)}}(M)dm(M)=\int_{M\in(\mathscr{D}_1)}\overrightarrow{QM}\wedge\overrightarrow{v^{(\lambda)}}(M)dm(M)\ldots$$
$$\ldots+\int_{M\in(\mathscr{D}_2)}\overrightarrow{QM}\wedge\overrightarrow{v^{(\lambda)}}(M)dm(M).$$

This leads to

$$\begin{cases}\vec{s}\{p^{\lambda}_{\mathscr{D}}\}=\vec{s}\{p^{\lambda}_{\mathscr{D}_1}\}+\vec{s}\{p^{\lambda}_{\mathscr{D}_2}\}\\ \overrightarrow{\mathcal{M}_Q}\{p^{\lambda}_{\mathscr{D}}\}=\overrightarrow{\mathcal{M}_Q}\{p^{\lambda}_{\mathscr{D}_1}\}+\overrightarrow{\mathcal{M}_Q}\{p^{\lambda}_{\mathscr{D}_2}\}\end{cases}$$
$$\Rightarrow\quad\{p^{\lambda}_{\mathscr{D}}\}=\{p^{\lambda}_{\mathscr{D}_1}\}+\{p^{\lambda}_{\mathscr{D}_2}\}.$$

Similarly, we have

$$\begin{cases}\vec{s}\{\mathcal{A}^{\lambda}_{\mathscr{D}}\}=\vec{s}\{\mathcal{A}^{\lambda}_{\mathscr{D}_1}\}+\vec{s}\{\mathcal{A}^{\lambda}_{\mathscr{D}_2}\}\\ \overrightarrow{\mathcal{M}_Q}\{\mathcal{A}^{\lambda}_{\mathscr{D}}\}=\overrightarrow{\mathcal{M}_Q}\{\mathcal{A}^{\lambda}_{\mathscr{D}_1}\}+\overrightarrow{\mathcal{M}_Q}\{\mathcal{A}^{\lambda}_{\mathscr{D}_2}\}\end{cases}$$
$$\Rightarrow\quad\{\mathcal{A}^{\lambda}_{\mathscr{D}}\}=\{\mathcal{A}^{\lambda}_{\mathscr{D}_1}\}+\{\mathcal{A}^{\lambda}_{\mathscr{D}_2}\}.$$

and for the kinetic energy

$$\int_{M\in(\mathscr{D})}\left[\overrightarrow{v^{(\lambda)}}(M)\right]^2dm(M)=\int_{M\in(\mathscr{D}_1)}\left[\overrightarrow{v^{(\lambda)}}(M)\right]^2dm(M)\ldots$$
$$\ldots+\int_{M\in(\mathscr{D}_2)}\left[\overrightarrow{v^{(\lambda)}}(M)\right]^2dm(M)$$

$$\Rightarrow\ T^{(\lambda)}(\mathscr{D})=T^{(\lambda)}(\mathscr{D}_1)+T^{(\lambda)}(\mathscr{D}_2).$$

Bibliography

[AGA 96] AGATI P., BREMONT Y., *Mécanique des solides – Applications industrielles*, Dunod, 1996.

[ANG 07] ANGELES J., *Fundamentals of Rigid Body Mechanics*, Springer, 2007.

[BOR 16a] BOREL M., VÉNIZÉLOS G., *Movement Equations 2*, ISTE, London and John Wiley & Sons, New York, 2016.

[BOR 16b] BOREL M., VÉNIZÉLOS G., *Movement Equations 3*, ISTE, London and John Wiley & Sons, New York, 2016.

[BRE 96] BREMONT Y., REOCREUX P., *Mécanique 2 : mécanique des solides indéformables. Statique. Cours et exercices résolus*, Editions Ellipses, 1996.

[DEL 12] DEL PEDRO M., GMÜR T., *Introduction à la mécanique des solides et des structures*, Presses polytechniques et universitaires romandes, 2012.

[GOU 11] GOURINAT Y., *Exercices et problèmes de mécanique des solides et des structures. Applications à l'aéronautique et à l'aérospatiale*, Dunod, 2011.

[HEA 05] HEARD W.B., *Rigid Body Mechanics: Mathematics, Physics and Applications*, Wiley, 2005.

[REO 98] REOCREUX P., BREMONT Y., *Mécanique 3: mécanique du solide indéformable. Cinématique-dynamique. Cours et exercices résolus*, Editions Ellipses, 1998.

Index

A, B, C

L, M

N, O, P

R, S

T, V

Other titles from

iSTE

in

Mechanical Engineering and Solid Mechanics

2016

BOYARD Nicolas
Heat Transfer in Polymer Composite Materials

CARDON Alain, ITMI Mhamed
New Autonomous Systems
(Reliability of Multiphysical Systems Set - Volume 1)

DAHOO Pierre Richard, POUGNET Philippe, EL HAMI Abdelkhalak
Nanometer-scale Defect Detection Using Polarized Light
(Reliability of Multiphysical Systems Set – Volume 2)

DE SAXCÉ Géry
Galilean Mechanics and Thermodynamics of Continua

DORMIEUX Luc, KONDO Djimédo
Micromechanics of Fracture and Damage (Micromechanics Set - Volume 1)

MOLIMARD Jérôme
Experimental Mechanics of Solids and Structures

2015

KARLIČIĆ Danilo, MURMU Tony, ADHIKARI Sondipon, MCCARTHY Michael
Non-local Structural Mechanics

SAB Karam, LEBÉE Arthur
Homogenization of Heterogeneous Thin and Thick Plates

2014

ATANACKOVIC M. Teodor, PILIPOVIC Stevan, STANKOVIC Bogoljub, ZORICA Dusan
Fractional Calculus with Applications in Mechanics: Vibrations and Diffusion Processes

ATANACKOVIC M. Teodor, PILIPOVIC Stevan, STANKOVIC Bogoljub, ZORICA Dusan
Fractional Calculus with Applications in Mechanics: Wave Propagation, Impact and Variational Principles

CIBLAC Thierry, MOREL Jean-Claude
Sustainable Masonry: Stability and Behavior of Structures

ILANKO Sinniah, MONTERRUBIO Luis E., MOCHIDA Yusuke
The Rayleigh–Ritz Method for Structural Analysis

LALANNE Christian
Mechanical Vibration and Shock Analysis – 5-volume series – 3rd edition
Sinusoidal Vibration – volume 1
Mechanical Shock – volume 2
Random Vibration – volume 3
Fatigue Damage – volume 4
Specification Development – volume 5

LEMAIRE Maurice
Uncertainty and Mechanics

2013

ADHIKARI Sondipon
Structural Dynamic Analysis with Generalized Damping Models: Analysis

ADHIKARI Sondipon
Structural Dynamic Analysis with Generalized Damping Models: Identification

BAILLY Patrice
Materials and Structures under Shock and Impact

BASTIEN Jérôme, BERNARDIN Frédéric, LAMARQUE Claude-Henri
Non-smooth Deterministic or Stochastic Discrete Dynamical Systems: Applications to Models with Friction or Impact

EL HAMI Abdelkhalak, BOUCHAIB Radi
Uncertainty and Optimization in Structural Mechanics

KIRILLOV Oleg N., PELINOVSKY Dmitry E.
Nonlinear Physical Systems: Spectral Analysis, Stability and Bifurcations

LUONGO Angelo, ZULLI Daniele
Mathematical Models of Beams and Cables

SALENÇON Jean
Yield Design

2012

DAVIM J. Paulo
Mechanical Engineering Education

DUPEUX Michel, BRACCINI Muriel
Mechanics of Solid Interfaces

ELISHAKOFF Isaac *et al.*
Carbon Nanotubes and Nanosensors: Vibration, Buckling and Ballistic Impact

GRÉDIAC Michel, HILD François
Full-Field Measurements and Identification in Solid Mechanics

GROUS Ammar
Fracture Mechanics – 3-volume series
Analysis of Reliability and Quality Control – volume 1
Applied Reliability – volume 2
Applied Quality Control – volume 3

RECHO Naman
Fracture Mechanics and Crack Growth

2011

KRYSINSKI Tomasz, MALBURET François
Mechanical Instability

SOUSTELLE Michel
An Introduction to Chemical Kinetics

2010

BREITKOPF Piotr, FILOMENO COELHO Rajan
Multidisciplinary Design Optimization in Computational Mechanics

DAVIM J. Paulo
Biotribolgy

PAULTRE Patrick
Dynamics of Structures

SOUSTELLE Michel
Handbook of Heterogenous Kinetics

2009

BERLIOZ Alain, TROMPETTE Philippe
Solid Mechanics using the Finite Element Method

LEMAIRE Maurice
Structural Reliability

2007

GIRARD Alain, ROY Nicolas
Structural Dynamics in Industry

GUINEBRETIÈRE René
X-ray Diffraction by Polycrystalline Materials

KRYSINSKI Tomasz, MALBURET François
Mechanical Vibrations

KUNDU Tribikram
Advanced Ultrasonic Methods for Material and Structure Inspection

SIH George C. *et al.*
Particle and Continuum Aspects of Mesomechanics

CPSIA information can be obtained
at www.ICGtesting.com
Printed in the USA
BVOW06*1948020217
475195BV00007B/24/P